CORAL REEF ANIMALS OF THE INDO-PACIFIC

Animal life from Africa to Hawai'i exclusive of the vertebrates

by

Terrence M. Gosliner
David W. Behrens
and
Gary C. Williams

Sea Challengers   •   Monterey, California

1996

A SEA CHALLENGERS PUBLICATION

Copy Editors - Hans Bertsch and Katie Martin

Front Cover

Petrosid sponge by Terry Schuller
Heteractis magnifica by Nick Galluzzi   *Calpurnus verrucosus* by Roy Eisenhardt
Odontodactylus scyllarus by Mike Miller
Three species of didemnid tunicates by Marc Chamberlain   *Pseudobiceros gratus* by Leslie Newman & Andrew Flowers
Iconaster longimanus by Mike Severns
Cover Background - Pig Island, Papua New Guinea by Leslie Newman & Andrew Flowers

Cover design by Gary C. Williams and Dave Behrens

Back Cover

Underwater photographer in the Solomon Islands by Gary C. Williams

Library of Congress Cataloging-in-Publication Data

Gosliner, Terrence M.
    Coral reef animals of the Indo-Pacific: animal life from Africa to Hawai'i exclusive of the vertebrates/by Terrence M. Gosliner, David W. Behrens, and Gary C. Williams.
    p. 26 cm.
Includes bibliographical references (p. 303-305) and index.
ISBN 0-930118-21-9
1. Marine invertebrates--Indo-Pacific Region. 2. Coral reef animals--Indo-Pacific Region. 3. Marine invertebrates--Indo-Pacific Region--Pictorial works. 4. Coral reef animals--Indo-Pacific Region--Pictorial works. I. Behrens, David W. II. Williams, Gary C. III. Title.
QL 137.5.G67 1996
592-092'59–dc20   96-1822
                                                                              CIP

SEA CHALLENGERS
4 Sommerset Rise, Monterey, CA 93940
Printed in Hong Kong through Global InterPrint, Petaluma, CA, U.S.A.
Kodak Photo CD Imaging by Faulkner Color Lab, San Francisco, CA, U.S.A.
Typography and prepress production by Colorgraphics, Monterey, CA, U.S.A.

FOREWORD

The tropical coral reefs ranging from the Indian Ocean coast of Africa to the islands of the central Pacific (Hawai'i in the Northern Hemisphere and French Polynesia and Pitcairn Group in the Southern Hemisphere) support the richest marine life found on our planet. This vast area, termed the Indo-Pacific region, is estimated to have five to ten times the species of marine animals of the Caribbean Sea. Studies of many groups of marine animals in the region are far from complete. In some, such as the flatworms, more than 90% are believed to be undescribed. The many common species in this volume identified only to genus (indicated by "sp." behind the name) are a testimonial to the taxonomic research that is still needed on most groups of Indo-Pacific animals.

The marine vertebrate animals of the Indo-Pacific, i.e., the fishes, reptiles, birds, and mammals, are much better known, in general, than the remaining animals (those without backbones, hence the invertebrates). Numerous guide books dealing with the fishes and other vertebrates of the Indo-Pacific or its subregions have been published in recent years. The present volume is restricted to the invertebrate animals.

Obviously it is far too soon for a comprehensive guide to be attempted on the invertebrates of the Indo-Pacific. If it could be done, it would take a series of volumes, many authors, and many years to complete. A critical need, however, has arisen for an authoritative text covering the most common animals that inhabit the coral reefs of the tropical Indian Ocean and the central and western Pacific. It is very important for marine biologists studying Indo-Pacific marine life and pharmaceutical researchers searching for chemicals of medical value to obtain the correct scientific names of the organisms. The ever-growing army of divers and snorkelers and underwater photographers who are enjoying the immense diversity of marine life of Indo-Pacific coral reefs also want to know the names of the animals they see, the basis for their identification, and obtain information on their habits and ecology. This volume provides accurate taxonomic, natural historical, and distributional information on eleven hundred of the common species of the Indo-Pacific.

Terry Gosliner, Dave Behrens and Gary Williams have compiled a collection of superb photographs of each of the species treated in this book. They have selected ones that show the characteristics important for the identification of these animals, as well as depict them in their natural habitat. The photographs herein are not just the best of those taken by the authors but include ones of more than 52 divers, photographers and biologists (I am pleased that 27 of mine were chosen).

The group effort that was necessary to present these photographs was also needed to compile information on the taxonomy and biology of the animals. Not only have the authors relied on their own expertise, but they have consulted authorities on the difficult groups of marine animals from around the world. The quality of this volume therefore reflects the contribution of many individuals (see Acknowledgments).

Of particular value in this volume is the information on the distribution of the species. Many have proven to be far more wide-ranging than previously realized. The authors have extended the range of some species known before from a single locality. Several species described a century or more ago but not reported since have been rediscovered while preparing this book.

Also special compliments are due the authors for their coverage of such subjects as interspecific associations, commensalism, mimicry, feeding preferences, and reproductive behavior.

The reader is encouraged to take the time to go through the Introduction of this book with care because it will give him a broad understanding of the basis for the classification of animals, the evolution of marine life, how coral reefs are formed, and advice on how to use the book. There is a wealth of information in the Introduction that provides for a better appreciation of the rest of the volume.

Terry, Dave and Gary have provided students of coral reefs, whether they be amateur divers or photographers or PhD. scientists, with a very valuable and attractive resource to promote their interest and provide for a better understanding of the animals in the incredibly rich but still fragile coral reef ecosystems. I am confident that this volume will long serve as the foundation for much-needed future taxonomic research and studies on the biology of Indo-Pacific invertebrate animals. It should also create a better appreciation of coral reefs and the need to establish more marine preserves to insure their protection for future generations.

John E. Randall
Bishop Museum and Hawai'i Institute of Marine Biology
University of Hawai'i, Honolulu

DEDICATION

To Dustin Dale Chivers ("Dusty"): late Senior Curatorial Assistant in the Department of
Invertebrate Zoology and Geology, California Academy of Sciences:
a friend, a colleague, and an inspiration.

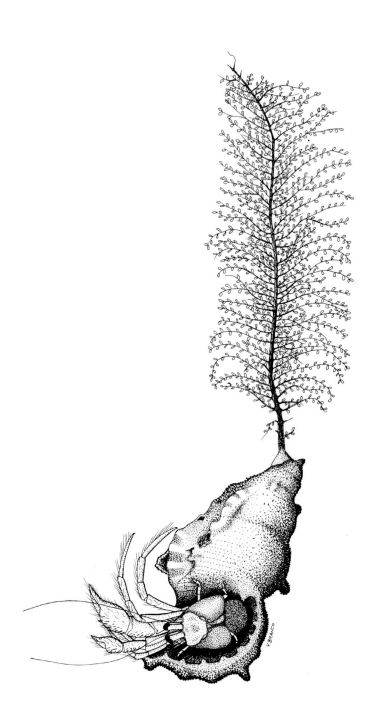

Table of Contents

INTRODUCTION

"The first examining of volcanic rocks, must to a geologist be a memorable epoch, and little less so to the naturalist is the first burst of admiration at seeing corals growing on their native rock."

Charles Darwin

Anyone who has visited a marine aquarium, donned a mask and snorkel or waded in the shallow tide pools along the edges of tropical reefs has been struck by the beauty and seemingly endless diversity of marine life inhabiting tropical waters.

The tropical Indo-Pacific (Figure 1) spans two entire ocean basins from the east coast of Africa to the margins of the Americas. The region is inhabited by the greatest diversity of species known in the marine realm. The waters of the Caribbean appear rich with life. In contrast to temperate regions, the reefs of the western Atlantic support abundant life. The Indo-Pacific is far richer, with about five to ten times the number of species as the Caribbean. For example, there are about 60 species of reef-building corals in the Caribbean and perhaps 600 species in the Indo-Pacific tropics.

Many field guides have been written about the fish and other vertebrate life of the Indo-Pacific tropics. This book focuses on the other bottom-dwelling animals that inhabit these tropical reefs, animals that have been traditionally called invertebrates. Invertebrates are not a natural group of organisms, as they are only united by something they lack, a vertebral column. Instead, natural groups such as coelenterates, mollusks and echinoderms are considered. This book is designed to allow the user to readily identify the more common species that inhabit tropical habitats in the Indo-Pacific. The geographical focus of the book is the tropical western Pacific, from the Philippines, Indonesia, and Papua New Guinea to Australia. This area supports the richest biota of all of the Indo-Pacific. Many of the common species present elsewhere in the Indo-Pacific are also found in this region. Thus, this guide permits the user to identify most of the common animals that would be found from such remote localities as Kenya, the Seychelles, the Maldives, Fiji, Belau, Okinawa, Guam, Tahiti and Hawai'i. However, there are many more species present in the western Pacific than in any other portion of the Indo-Pacific. No book smaller than the New York City phone directories can possibly include all of the species from the region. For example, we have found almost 600 species of nudibranchs from a single bay in Papua New Guinea. The emphasis in our guide is to identify the most common species that a diver or snorkeler would find in the upper 50 meters of the western Pacific. Detailed references allow the reader to explore the scientific literature to find more in depth information about other species not included in this book, and to find additional information about the biology of included species.

Classification of Animals

Classification and evolution of marine animals: What's in a name?
The classification of living organisms is called systematics. It is one of the most fundamental components of biology. One of the first things anyone needs to know about an organism is its name. Systematic biologists, scientists who study classification, believe that the classification of organisms should reflect their evolutionary history. In other words, things that are more closely related to each other than to other organisms would be included in the same group. This kind of classification is often referred to as natural classification, since it reflects nature. Other classifications are considered as "artificial." Artificial classifications are simply collections of objects that have some superficial feature uniting them. For example, all yellow objects in your house could be kept together. However, this does not suggest that they have more in common than their color. All animals with wings could be included in the same group, despite the fact we know that insect, bird and bat wings have evolved independently in these different groups. Artificial classifications may be convenient ways of storing things, but they tell us nothing about who is related to whom.

Determining evolutionary relationships: Are you out of your tree?
The first step in classifying organisms is to look at their similarities and try to construct a phylogeny, or evolutionary tree. This is similar to putting together a family tree of a person's relatives and ancestors. The major difference is that people generally know who their parents, grandparents and cousins are, and how they are related. In putting together an evolutionary tree of organisms, one is inferring relationships, based on features that organisms share. Members of any group of organisms should be characterized by newly evolved features that are unique to that group. For example, all mammals have mammary glands and hair. These are new innovations that are found in all mammals but not in other vertebrates. Such groups that include all of the descendants of a common ancestor are called **monophyletic.** Students of phylogenetic systematics only consider monophyletic groups in their classifications. Groups that are not monophyletic include corals, fishes and invertebrates. Many of the groups that we have traditionally learned are not natural groups since they do not include all of the descendants of the common ancestor. For example, some of the descendants of fishes include members of some monophyletic groups such as

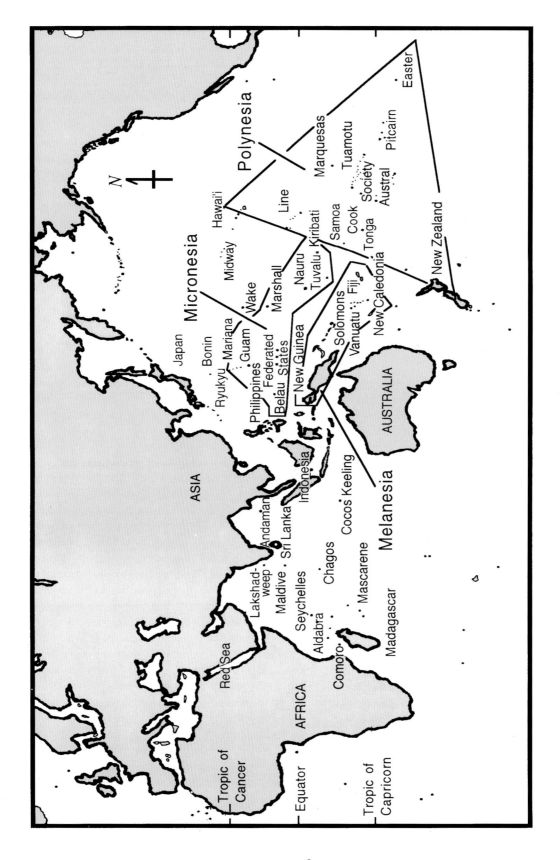

Figure 1. Islands and Archipelagos of the Indo-Pacific region.

mammals and birds. Fishes have no unique features that they share, that are absent in mammals and birds. On the other hand, birds have many unique modifications that are shared by all birds, but are absent in all other vertebrates.

Traditionally, animals have been placed into two distinct groups, invertebrates and vertebrates. The vertebrates are well understood by most people. Vertebrates all have a uniquely evolved structure, the vertebral column, that unites all members of this group. On the other hand, the only thing which unites invertebrates is their lack of a vertebral column. This would be like calling all things which lack a shell "non-mollusks" or all mammals that are not primates "as non-primates." Since the remainder of animals which lack a backbone do not share any feature that is not also found in vertebrates, we must conclude that they do not form a natural group. Rather than perpetuating the word "inverte-brates" we have chosen to focus on the natural groups that are contained within the animal kingdom, groups like sponges and mollusks.

Animal phylogeny: How are all these bizarre creatures related to each other?

The evolutionary relationships of the major groups of animals still remain largely unresolved. Many different hypotheses have been put forth, based on traditional morphological evidence as well as studies that have incorpo-rated molecular evidence from **DNA** and **RNA** sequences. Molecular evidence has provided a new source of data for determining relationships, but to date has provided as many questions as it has answers. One problem in determining the evolutionary relationships of animal groups, is that most major groups diverged from each other in the Early Cam-brian, approximately 600 million years ago. Reconstructing the events that took place so far back in some of the earliest portions of the fossil record is especially difficult. Virtually every textbook dealing with animals presents a somewhat different view of how animals are related to each other. Most phylogenetic studies indicate that there are many groups of animals such as sponges and coelenterates that are probably not closely related to any other group of animals, although some workers have considered the ctenophores (comb jellies) to be on the same evolutionary line as the coelenter-ates. Most zoologists agree that there is a major division between two groups of higher animals. These two groups are called **protostomes** and **deuterostomes**. The names are derived from the development of the mouth in the transition from larval to adult forms. In protostomes, the adult mouth develops directly from the larval mouth. In deuterostomes, the adult mouth develops independently from the larval one. The protostomes include organisms that exhibit some evi-dence of segmentation, such as the annelids, mollusks and

arthropods. In mollusks, segmentation is evident only in the more primitive representatives such as chitons, monoplacophorans and *Nautilus*. In other groups, such as the sipunculans and echiurans, there is no evidence of segmenta-tion. It is unclear whether these animals were primitively unsegmented or whether they have lost any trace of segmen-tation. The deuterostomes include the echinoderms and chordates. In most of these organisms the body cavity, or **coelom,** is divided into two or three portions. They have a fundamentally different larval stage than protostomes and have **radial** rather than **spiral cleavage.** Figure 2 shows our hypothesis of how the major animal groups treated in this book may be related to each other.

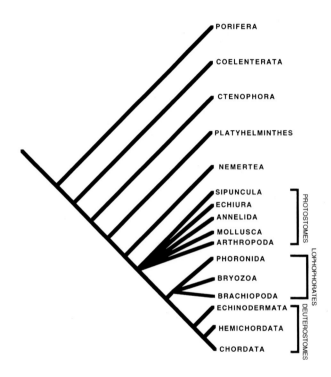

Figure 2. Phylogenetic relationships of the major animal groups.

Unequal classification: How rank can you get?

Classifications have traditionally used a series of ranks to depict natural groups of organisms. Phylum, class, order, family, genus and species are the major ranks that zoologists have used to order animals into a system of classification. Modern phylogenetic methods not only restrict one to recog-nition of monophyletic groups, but tell us that not all ranks are equal. For example, a family of tunicates may not be the same as a family of bryozoans. Phylogenetic studies indicate that not even within any group are families of consistent rank, based on different number of branches in their phylogenetic trees. This is not only true for families, but all ranks above the species level, from genus to phylum. The other problem is that systematic biologists since Linnaeus have named species by a two part scientific name - the genus followed by the species name. If genera, the plural of genus, are not equal, that can have an impact on the scientific names we give to animals. This is further complicated by the fact that many phylogenetic studies show that many of the genera we commonly recognize are not monophyletic. This means that many of these will be combined into more natural groups.

Stability of classification: Why do they keep changing the damn names?

Most people want to have a name for an animal that doesn't change over time. People become frustrated when names change frequently, creating the impression that systematic biologists do not have their act together. Systematic biologists generally do not alter the name of a species without a sound scientific basis. There is a published set of rules a systematic biologist must follow called the <u>International</u> <u>Code</u> of <u>Zoological</u> <u>Nomenclature</u>. This code describes how a **taxonomist,** a person that describes species, must construct and publish a new species with its new scientific name. One of the fundamental rules of nomenclature is the law of **priority.** This law says that a name that is correctly published before other names, is the one that must be used to apply to the species. Often times a scientist will find an older name buried in the scientific literature. This name must be the valid name for that group, unless a special petition is filed, stating that a change of name would seriously disrupt the stability of a long-used name. In most cases the International Commission of Zoological Nomenclature will elect to use the oldest name, even if it disrupts stability. Scientific names may also change if it is determined that the species is better placed in a different genus. This only changes the first half of the scientific name, and the species name remains the same. With the changes that are occurring in systematic biology that are described above, many changes in classification will be made over the next several decades. We are in the midst of the greatest revolution in the classification of organisms that has occurred since Linnaeus first established the system of classification of animals in 1758. While this is exciting for systematic biologists, it is frustrating to individuals who desire a stable name for the plants and animals they recognize. The only comfort that systematic biologists can offer, is that we have now developed far more objective ways of determining monophyletic groups and this will eventually lead to much greater stability of names. In the meantime, people will hopefully be patient with systematic biologists, and recognize that changes are made for sound scientific reasons, not just to create confusion among the public.

Common names: Why can't they make it simple?

Most people like common names because they are easy to use and familiar to us. People feel more comfortable speaking their native language than a foreign one. Common names are plagued with problems. The group of tropical trees in the genus *Casuarina,* which are often found on beaches in the Indo-Pacific tropics, are known by different names in different geographical regions. The are known as ironwoods (Hawai'i), Australian pines (Florida) and she oaks (Australia). Oftentimes a particular common name may apply to several different organisms. A periwinkle can be a flowering plant of the genus *Vinca,* a littorinid snail (Europe and North America) or a trochid snail (South Africa). This kind of ambiguity is what has led systematic biologists to settle on a single scientific name for a species. Most people know and use far more scientific names than they realize. Most people, including small children, have no difficulty using scientific names like *Eucalyptus*, *Hippopotamus* and *Tyrannosaurus rex.* The other major problem with common names, is that most species do not have a common name. For the purposes of this book, we have given common names, when there is one in common usage. We have not made up common names, when there isn't one, as we feel this serves no useful purpose.

Islands, Reefs, and Lagoons: Biotic Communities of the Indo-Pacific

The Indo-Pacific is a vast region encompassing the tropical Indian and Pacific Oceans from Africa in the west to Hawai'i and French Polynesia in the east. This area represents the largest marine biogeographic region in the world. Many Indo-Pacific coral reefs develop along the margins of the African, Asian, and Australian continents but most are associated with islands. Examples of extensive continental coral reefs include Sodwana and Kosi Bays (South Africa), Inhaca Island (Mozambique), Eilat and Ras Muhammad (Red Sea), Phuket (Thailand), and the Great Barrier Reef (Australia).

Innumerable islands dot the Indian and Pacific Oceans, creating tropical shallow-water conditions conducive for coral reef growth. Island archipelagos of the Indian Ocean include the Comoros, Seychelles, Mascarene, Lakshadweep (Laccadive) and Maldives, Chagos, Andaman, and Nicobar Islands. Coral reefs have developed on these island chains. However, the greatest number of islands in the world are found in the tropical western and central Pacific. These are included in the modern nations of Indonesia, the Philippines, Papua New Guinea, Belau, Federated States of Micronesia, Northern Mariana Islands, Marshall Islands, Nauru, Solomon Islands, Vanuatu, New Caledonia, Kiribati, Tuvalu, Fiji, Tonga, Samoa, Cook Islands, and French Polynesia.

The western and central Pacific Ocean, the region of the highest marine biodiversity, can be divided into three sub-regions based on cultural geography as well as geology: Polynesia, Micronesia, and Melanesia. The three points of the Polynesian triangle are Hawai'i in the north, Easter Island in the southeast, and New Zealand in the southwest. Polynesia was inhabited originally by people migrating from southeast Asia. Most of Polynesia is made up of widely dispersed larger volcanic islands. The name "Polynesia" means "many islands." Micronesia, on the other hand, is situated between the Mariana Trench in the west and the Line Islands in the east. The name "Micronesia" means "tiny islands." This region is composed of thousands of small islands, mostly atolls, of the central Pacific. Micronesian peoples are thought to be related to islanders of the Philippines, the probable region of their ancient origin. Melanesia is a vast region of volcanic islands including New Guinea, the Solomon Islands, Vanuatu, Fiji, and New Caledonia. The name "Melanesia" means "black islands," referring to the dark complected people inhabiting these islands. Throughout the centuries, intermixing of coastal peoples from Polynesia, Micronesia, and Melanesia has taken place in areas of geographic overlap.

Geologically, the western Pacific is composed of a complex mosaic of tectonic plates, principally the Pacific, Eurasian, and Australian Plates. The smaller Philippine Plate produces an extensive boundary region between the Eurasian and Pacific Plates. Similarly, the Caroline Plate, and the Bismarck, Solomon, and Fiji Microplates create a boundary region separating the Pacific and Australian Plates. The entire region is part of the Ring of Fire or Pacific Rim and is consequently extremely active geologically. Periodic volcanic and seismic activity have produced high mountain ranges and deep-water trenches at the margins of all of these plates and microplates. This constant and often violent geological activity has created thousands of islands with extensive regions of shallow-water habitat suitable for coral reef growth.

The shallow subtidal regions of virtually all of these thousands of islands provide optimal conditions for coral reef growth and development. Depending on local conditions including currents, wave action, turbidity, temperature, and salinity, the expression of biotic communities may vary. Four primary types of biotic communities have developed in the shallow-water tropical Indo-Pacific, depending on differing physical parameters. These are coral reefs, seagrass beds, mangrove habitat, and sand flats. Coral reefs are treated in detail in a following introductory section.

Seagrass Beds: Seagrass meadows are considered an important ecosystem of shallow-water tropical regions. They occur in protected areas such as bays and lagoons. The seagrass ecosystem is remarkable for its high rates of primary productivity. Thick "forests" of rapidly growing seagrasses provide a protective and productive habitat for many animals that live in or on the sandy or muddy bottom (benthic forms), on the plants themselves (epiphytic forms), and in the water surrounding the plants (epibenthic or pelagic forms).

Despite the name, seagrasses are not true grasses but belong to two other families of flowering plants–the Potamogetonaceae and the Hydrocharitaceae. In the tropics, several species often grow together in any particular seagrass meadow. Seventeen species in eight genera–*Zostera, Halodule, Cymodocea, Syringodium, Thalassodendron, Enhalus, Thalassia,* and *Halophila* comprise the seagrass communities in the tropical Indo-Pacific. This represents about 35% of the world species total. The other 31 species are found mainly in monospecific stands in colder temperate regions such as southern Australia and New Zealand, the North Pacific, and parts of the Atlantic Ocean. Green algae comprise an important component of the tropical seagrass ecosystem–especially species in the genera *Caulerpa, Udotea, Codium, Acetabularia, Avrainvillea,* and *Halimeda.* They often grow as epiphytes on the seagrasses. Brown algae of seagrass beds include species of the genera *Padina* and *Sargassum.* Seagrasses are actually consumed by only a few animals including dugongs, sea turtles, and some mollusks and urchins. Many animals, however, graze on the green algae and diatom films that grow on the seagrass blades. For detailed information and identification of seagrasses consult Phillips and Meñez (1988), and Meñez, Phillips, and Calumpong (1983).

Mangrove Habitat: Mangroves are sometimes situated between coral reefs and a gently-sloping shoreline, often in protected bays and lagoons with limited circulation (Figure 5C). Mangroves have adapted to live in saltwater-saturated soils of highly saline condition. The mangrove ecosystem contributes significantly to the productivity of tropical shallow-water regions. The numerous aerial roots produced by some mangroves provide a sheltered habitat for a diverse array of marine life. The various flowering plants referred to as mangrove are actually unrelated species but share similar growth habits under similar environmental conditions. These include the red mangroves of the family Rhizophoraceae such as *Rhizophora* (pantropical), *Bruguiera* and *Ceriops* (tropical Asia and Africa), and *Kandelia* (southeast Asia); the black mangroves *Avicennia* (family Avicenniaceae); and the white mangrove *Lumnitzera* (family Combretaceae) from East Africa, Asia, and Australia (Lugo, 1990).

Sand Flats: Extensive sandy areas are found between patch reefs, or in depressions and gullies on the reef proper, or in deeper areas below or beyond a reef. Sand flats are often textured with ripple marks due to the action of strong bottom currents. Sand flats may seem barren by day, but at night the exploration of such areas can startle the diver with a surprising variety of animal life. Common night animals (including the sea pens *Veretillum* and *Virgularia*, the mollusks *Pleurobranchus* and *Coriocella,* and various cuttlefish, lobsters, crabs, and urchins) are active on sand flats and are invisible or highly cryptic during the day.

Biology of Corals and Coral Reefs

What are Corals?

The terms "coral" and "coral reef" often produce images of emerald blue waters at bathtub temperatures, calm shallow lagoons, and beaches of sparkling white sand lined with coconut palms. It is true that the majority of coral species are found on tropical coral reefs, but the term coral refers to a vast array of organisms that are found throughout the world's seas from freezing polar regions to equatorial reefs, and at all depths from the intertidal zone to the bottoms of the deepest **hadal** trenches (Williams, 1986).

The word "coral" is derived from the ancient Greek word "**korallion**," which referred to the precious red coral of the Mediterranean, known to us today as *Corallium rubrum* (Williams, 1993). The diverse assemblage of organisms known as corals are actually animals belonging to the Coelenterata–along with such things as hydroids, jellyfish, box jellies, and sea anemones. All corals are coelenterates, but some are more closely related to other coelenterates than to other corals. For example, hydrocorals are more closely related to hydroids than they are to other corals. Therefore corals are not a **monophyletic** group, but refer to coelenterates that produce a skeleton.

The life history of a coelenterate is characterized by having alternating life styles–either as a **polyp** (the attached stationary stage) or a **medusa** (the swimming or floating jellyfish-like stage). A coral is characterized by having some form of hard skeleton, composed of **calcium carbonate** or a tough fibrous protein known as **horn**, or a combination of these two. Most corals are colonial, but a few are composed of a solitary polyp throughout their entire life span. The polyp is the living individual of a coral colony, made up of two tissue layers surrounding a thin gelatinous matrix. Each polyp has a ring of tentacles surrounding a central mouth. The tentacles contain specialized stinging cells known as **nematocysts**, resembling miniature poison darts, which are important in defense and prey capture.

For the most part, corals are **polytrophic** feeders (Schlichter, 1982), that is, they obtain nutrition in a variety of ways. They are all micropredators and ingest plankton and particulate matter from the surrounding water medium, but they can also directly absorb dissolved organic matter from sea water through the epidermal tissues. In addition, many species can utilize the products of algal symbiosis via photosynthesis.

Many reef corals have one-celled algae called **zooxanthellae** living in their internal tissues. This symbiotic relationship allows for the production of enough calcium carbonate for coral reefs to originate and grow. Corals that take part in this relationship are called zooxanthellate corals and include the fire corals and blue corals, soft corals such as *Sarcophyton, Sinularia,* and *Lobophytum,* some sea fans such as *Rumphella,* a few sea pens such as *Virgularia* and *Cavernularia,* and all the reef-building hard corals. Zooxanthellate corals often have a golden-brown coloration and are not brighty colored. The coral provides a protected habitat for the algal cells and at the same time utilizes products of algal photosynthesis to produce more calcium carbonate than it could without the algae. This excess production of calcium carbonate is what builds coral reefs. The building of reefs is therefore a bipartisan effort between coral host and algal tenant and this close working relationship explains why coral reefs are restricted to the warm, clear, sunlit waters of the shallow-water tropics. Corals that do not contain zooxanthellae are called **aposymbiotic**.

Corals can reproduce either sexually or asexually. Sexual reproduction involves internally fertilized eggs which are brooded on the inside or outside of the parent polyps, or externally fertilized eggs that either develop into planktonic larvae and are dispersed in water currents, or the larvae develop and settle in the vicinity of the parent coral. Asexual reproduction is common is some reef corals and involves cloning by budding or fragmentation, which either originates within the body of the coral itself, or is due to external causes (Hughes, 1983).

Corals have a long fossil record dating back 450-500 million years to the Ordovician Period of the Paleozoic Era. Three groups of early corals–the heterocorals, the tabulate corals, and rugose corals–are now all extinct, having died out by the end of the Paleozoic. Four other groups of corals, which developed during the Mesozoic and Cenozoic Eras, survive to the present day–these are the hydrocorals, the black corals, the hard corals, and the octocorals. All four of these types inhabit Indo-Pacific coral reefs, but it is primarily certain hard coral species that actually build reefs as a result of the deposition of calcium carbonate onto the surface of the reef by the living tissues of the corals themselves. Coral reefs are actually **biogenic** geologic structures of limestone–having been created by countless generations of living corals. Corals can be classifed as **hermatypic** (reef-building) or **ahermatypic** (non-reef-building) (Schumacher & Zibrowius, 1985). Hermatypic corals are for the most part hard corals (scleractinians), but also include the octocoral *Heliopora* (blue coral) and the hydrocoral *Millepora* (fire coral). All hermatypic corals are zooxanthellate but not all zooxanthellate corals are hermatypic. Some ahermatypic zooxanthellate corals include the mushroom coral (*Fungia*) and Neptune's cap (*Halomitra*).

The Four Kinds of Corals

Hydrocorals belong to the Class Hydrozoa. All other corals are anthozoans. Members of the Class Anthozoa are exclusively polypoid, having lost the medusoid stage, while most hydrozoans retain both polypoid and medusoid stages in their life cycles. Hydrocorals include both the milleporine and stylasterine corals. Milleporine corals are also known as the "fire corals" or "stinging corals," representing a dozen or so valid species of the single genus *Millepora*. Found in both the Caribbean and Indo-Pacific, fire corals probably account for more toxic stings received by divers than any other coelenterates (Auerbach & Geehr, 1989: 955).

Stylasterine corals, also known as lace corals, include delicate and colorful species belonging to the genera *Stylaster* and *Distichopora*, both commonly found on Indo-Pacific reefs. All hydrocorals are characterized by a massive and relatively brittle calcium carbonate skeleton with numerous pinpoint-sized pores from which emanate two kinds of hydroid-like polyps, which are often finger-shaped with knob-like tentacles. The two kinds of polyps have a defensive function (**dactylozooids**) or a feeding function (**gastrozooids**).

Antipatharians are the black or thorny corals, characterized by having internal axes of dark horn; calcium carbonate is absent. The axis is covered with minute thorny or spiny projections, much like the branches of a rose bush (Figure 3D). Two genera of black corals, *Cirripathes* and *Antipathes*, are commonly encountered on Indo-Pacific reefs. Very thin tissues overlay the axis of a living black coral, usually resulting in a bright yellow or greenish-yellow appearance. Each black coral polyp has six finger-like tentacles surrounding the mouth.

The scleractinians or hard corals (also called stony corals) comprise most of the framework of a living coral reef. Hard corals have massive calcium carbonate skeletons with relatively large polyps (> 5mm in diameter), each containing internal radiating ribs called **septa**. Many important hermatypic species of Indo-Pacific reefs are colonial hard corals of genera such as *Acropora*, *Pocillopora*, *Goniopora*, and *Turbinaria*. Common ahermatypic forms include *Fungia*, *Tubastraea*, and *Dendrophyllia*. Several ahermatypic reef-inhabiting hard corals such as the mushroom coral, *Fungia*, are solitary (only one polyp is present) and do not form colonies.

Lastly, the octocorals include the soft corals, sea fans, sea whips, and sea pens. All octocorals are easily identified by the eight feather-like tentacles that surround the mouth of each polyp. Soft corals are important members of Indo-Pacific reef communities–their abundance, diversity, and biomass rivals or exceeds that of the hard corals in some regions. The blue coral (*Heliopora coerulea*, #65) has a massive **arago-nitic** skeleton and is an important reef builder in some areas. Other octocorals are ahermatypic and have skeletons of **calcitic** spicules known as **sclerites** (Figure 3E). (See Figure 3 for scanning electron micrographs of ultrastructural elements in various marine animals including soft coral sclerites.) In addition to sclerites, the gorgonians (sea fans and sea whips) also have internal axes composed of horn and/or calcium carbonate. The axis is always smooth, never thorny as in black corals. The flexible internal skeletons of sea fans and sea whips allow them to bend and sway in the currents and bottom surges like the branches of a tree in gusty winds. Soft corals flourish on shallow-water reef flats, while gorgonians inhabit vertical walls and deeper reef areas. Sea pens or pennatulaceans are adapted for life in soft substrata, are encountered mostly at night, and are restricted to sandy regions adjacent to reefs or in sandy depressions or gullies on the reef proper.

What are Coral Reefs?

Biogenic limestone reefs are geologic structures built over time by living organisms. The two most important types are algal reefs and coral reefs. Algal reefs are formed primarily by lime-secreting green and red algae such as certain species of *Halimeda* and *Lithothamnion*, while coral reefs are formed primarily by various species of hermatypic corals. Coral reefs represent the accumulated remains of the skeletons of lime-secreting organisms, primarily hermatypic corals. The thin living veneer of tissue lives on and builds upon the skeletal remains of past generations of corals below.

Coral reefs are home to an indeterminable number of species of organisms. The variety and abundance of marine life on coral reefs is overwhelming (Figure 5A). Virtually all of the 30 or so major animal groups are represented on coral reefs and many species no doubt have yet to be discovered and described in the scientific literature.

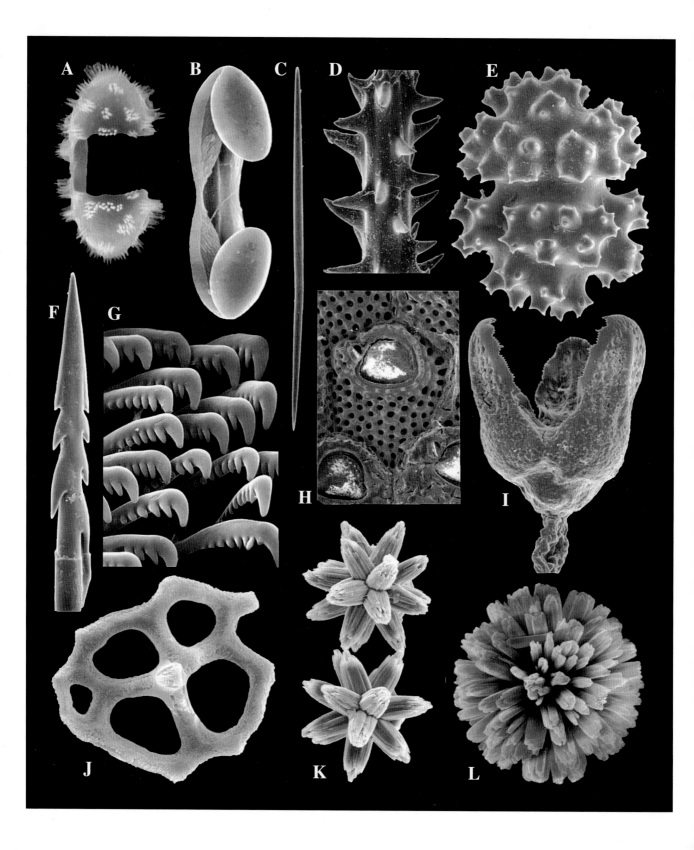

The importance of conserving coral reefs cannot be over-stressed. They are the world's most diverse marine communities representing banks of biological diversity. They are indicators of environmental stress such as pollution, sedimentation, and sea temperature fluctuations. They are also sources of pharmaceutically important compounds such as prostaglandins and anti-cancer agents.

Coral reefs are distributed in a circumtropical band mostly between 20° North Latitude and 20° South Latitude. The western Atlantic and the Indo-Pacific are the two main coral reef regions in the world (Wells, 1988). The western Atlantic is the region of tropical and subtropical America between Bermuda in the north and Brazil to the south, including the Gulf of Mexico and the Caribbean Sea. The Indo-Pacific covers the vast region from East Africa and the Red Sea to the Hawaiian and Tuamotu Archipelagos. In a comparison of biodiversity, the Indo-Pacific is roughly ten times more diverse than the western Atlantic. For example, there are approximately 60 species of hermatypic corals inhabiting the coral reefs of the western Atlantic compared with an estimated 500-600 species in the Indo-Pacific. Coral reefs are rare or absent from the tropical Atlantic of South America and Africa due mainly to the great influx and circulation of fresh water and silt from the Amazon and Zaire River systems.

Coral reefs need warm, clear, relatively quiet water for optimal growth. The distribution of coral reefs at any given point in time is determined by various limiting factors. The most significant of these are water temperature, depth and light intensity, salinity, water turbulence, and sedimentation. The optimum temperature for the growth of hermatypic corals and the development of coral reefs is 20-28°C. At temperatures below 18-20°C coral growth is limited or ceases. Temperatures above 28° frequently induce coral bleaching—the evacuation and depletion of zooxanthellae from coral tissues (Figure 5B). Some researchers feel that global warming will lead to widespread bleaching of corals, and

could have a serious negative impact on marine biodiversity (Pearce, 1994). Reef growth ceases with a widespread breakdown in the algae/coral symbiotic relationship. Water depth is important as it determines the intensity of light reaching the coral tissues for photosynthesis to occur. Most reefs grow best in depths of less than 25 meters. Light intensity, particularly ultraviolet radiation, may limit coral growth at the surface or in very shallow water, and in depths greater than 25 meters the diminution of light inhibits photosynthesis. Although many hermatypic corals can tolerate fluctuations in salinity as low as 18 parts/thousand and as high as 70 parts/thousand, around 35 parts/thousand is optimal. The amount of constant or periodic water turbulence can limit the extent of coral growth and reef development. Wave action and surge can lead to physical breakage, as well as limit growth and select for certain morphological types (i.e. robust and mound-like colonies vs. delicately branched forms). Lastly, the amount of sediment suspended in the ambient water can be a strong determinant of reef growth. Turbid water cuts down on light intensity and can also result in direct physical stress to coral colonies.

Natural threats to coral reefs and coral reef organisms include cyclones and hurricanes, periodic population explosions of echinoderms such as the crown-of-thorns starfish (*Acanthaster planci*, #964), periodic ocean warming events (El Niño), and the actions of earthquakes and volcanoes. Man-made threats include chemical and nutrient pollution, sedimentation from land clearing and coastal development, overfishing and collecting for the international aquarium, jewelry and sea shell trades, recreational use (ship anchor damage and tourism impact), and destructive fishing techniques including the use of dynamite and cyanide.

The evolutionary history of coral faunas and coral reefs is an ancient one extending back in geologic time to the early Paleozoic when the first corals appeared. During the breakup of the supercontinent Pangaea into Laurasia and Gondwanaland during the early Mesozoic, the Tethys Sea became circumglobal. A uniform worldwide coral fauna presumably flourished until the various continents began to split up. The splitting up of the two great land masses into various continents gave rise to a diversity of localized faunas. Approximately 1-6 million years ago, the Isthmus of Panama formed a complete closure between the Atlantic and Pacific. Independent evolution of corals then took place in isolation, leading to the separate faunas of the western Atlantic and the Indo-Pacific that we enounter today.

The Formation of Coral Reefs
Most coral reefs can be classified as fringing reefs, barrier reefs, coral atolls, table reefs, or patch reefs. Fringing reefs

Figure 3. Ultrastructural elements of marine animals. A-C. Sponge spicules, *Guitarra abbotti* (California) A. Magnification 5000X, B. 2000X, C. 300X. D. Axis of black coral, *Antipathes abies* (New Guinea), 120X. E. Sclerite of soft coral, *Minabea acronocephala* (New Guinea), 700X. F. Seta of polychaete, *Laetmonice* cf. *moluccana* (Philippines), 300X. G. Radular teeth of nudibranch, *Chromodoris hamiltoni* (South Africa), 700X. H. Zooid with avicularium of cheilostome bryozoan (Pakistan), 50X I. Pedicillaria of sea urchin, *Echinometra mathaei*, 200X. J. Sea cucumber ossicle, *Holothuria perviax* (Hawai'i), 2000X. K., L. Spicules of two didemnid tunicates (Belau). K. 1000X. L. 4000X.

are the most common type of reef in the Indo-Pacific as well as the western Atlantic. These reefs project seaward directly from the shore and form a fringe of stony coral around an island or along part of the shore of a large land mass (Figure 5C). Barrier reefs are located further from shore than fringing reefs and are separated from adjacent land by a lagoon (Figure 5D). The largest barrier reefs in the world are the Great Barrier Reef off Queensland, Australia and the Belize Barrier Reef off Belize in the Caribbean Sea. Atolls are low, ring-shaped, limestone islands with a central lagoon, encountered predominantly in Micronesia, Polynesia, and parts of the Indian Ocean. The word "atoll" is derived from "atolu," a native name in the Maldive Islands. Table reefs are small open ocean reefs with no central islands or lagoons, confined to the tops of guyots or seamounts. Lastly, patch reefs rise from the floor of lagoons and represent discrete units surrounded by sand or other non-reef substratum.

The modern theory of coral atoll formation was originated by Charles Darwin as a result of his observations at Tuamotu Archipelago and Cocos Keeling Island made between November 1835 and April 1836, during the voyage of the "Beagle." His book, The Structure and Distribution of Coral Reefs, was first published in 1842 as Part I of the Geology of the Voyage of the 'Beagle.'

Darwin believed that the fringing reefs, barrier reefs, and atolls of volcanic islands represented a successional series through geologic time. He hypothesized that the transition from fringing to barrier reef to atoll could result from the upward growth of coral on the edge of a gradually sinking volcano. Gradual subsidence and continuous reef growth were fundamental to his theory. He believed that barrier reefs represented an intermediate stage between fringing reefs and atolls, and that the ring-like appearance of an atoll with a central lagoon resulted from the total submergence of the summit of a volcano. The sequence of events in atoll form ation can be summarized as follows: (1) an emergent oceanic volcano that is no longer active, is colonized by reef-building corals; (2) initial coral growth forms a fringing reef around the island. As the magma chamber of the volcano is depleted the island begins to sink, but coral growth continues, building upon past generations of corals; (3) as the volcano continues to subside, a barrier reef is formed with a lagoon between the island and the reef; (4) finally, the volcano completely disappears below sea level, leaving an atoll composed of low coral islets in a ring with a lagoon in the center (Figure 4).

Darwin was not the only naturalist interested in the origin of coral islands. The geologist James Dwight Dana, and the conchologist Joseph Couthouy, as members of the United

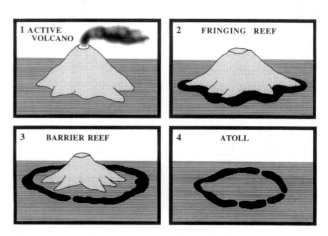

Figure 4. Atoll formation in the Indo-Pacific.

States Exploring Expedition between 1838 and 1842, in Fiji made observations similar to those of Darwin. They contributed significantly to the developing coral island theory initiated by Darwin, by recognizing that sea temperature can restrict coral growth and hence can help explain the distribution of coral reefs (Couthouy, 1842). They also found indirect evidence to support the idea that subsidence of some oceanic volcanos actually does take place. Dana believed that in islands such as Tahiti, a deeply embayed and irregular coastline coupled with an extremely eroded and dissected topography, was evidence for partial island submergence since the sea was not capable of eroding the shore in such a way (Dana, 1853, 1872). Neither of these principles were recognized by Darwin (Appleman, 1985).

It was not until the acceptance of the Theory of Glaciation in the latter part of the nineteenth century that we had a full and vigorous explanation for coral island formation. The Swiss naturalist Louis Agassiz came to the United States to a professorship at Harvard in 1837. He began to espouse the concepts of the newly developing Theory of Glaciation and

Figure 5. A. Coral reef community at Planet Rock (Madang, Papua New Guinea). B. Coral bleaching in a faviid coral (Batangas, Philippines). C. Fringing reef and mangrove communities (New Georgia Group, Solomon Islands). D. Barrier reef, Wistari and Heron Islands (Australian Great Barrier Reef). E. Exposed coral reef forming a coastal terrace (Mborokua Island, Solomon Islands). F. Exposed limestone formation showing notching by sea erosion at different levels (New Georgia Group, Solomon Islands). G. Karst topography (Naru Hills, Papua New Guinea). H. Karst landscape (Chocolate Hills, Bohol, Philpines).

G.C. Williams **A**

G.C. Williams **B**

G.C. Williams **C**

Leslie Newman & Andrew Flowers **D**

G.C. Williams **E**

Roy Eisenhardt **F**

G.C. Williams **G**

G.C. Williams **H**

11

the "Great Ice Age"–a theory that can be traced back to the observations and speculations by various Swiss geologists in the Alps since the early 1800s. It was not until the 1870s that the Theory of Glaciation became fully accepted by the scientific community (Larson & Birkeland, 1982). Between 1910 and 1948, the American geologist Reginald Daly developed the "Glacial Control Theory" to explain the many elevated or submerged notches and erosion terraces found on the coastlines of many coral islands, as well as recently exposed Pleistocene reefs (Figure 5E). The theory can be summarized as follows: (1) an island emerges above the surface of the sea; (2) sea levels drop significantly during an Ice Age; (3) horizontal terraces and ledges are cut by erosion during the period of low sea level; (4) as the Ice Age ends the sea level rises; (5) coral reef growth subsequently takes place on the newly created submerged platforms (Daly, 1915; Davis, 1928; and Kuenen, 1950).

The modern explanation of coral reef development should correctly be called the Darwin/Dana/Daly Theory of Coral Reef Formation since only the work of these three men taken together can fully explain the characteristics observed on atolls today.

Experimental verification of Darwin's and Dana's part of the theory dealing with subsidence came about in the early 1950s during studies made in the Marshall Islands prior to atomic and hydrogen bomb testing. As part of an environmental assessment of the region, the U.S. Navy drilled a series of deep holes at Bikini and Enewetak Atolls between 1947 and 1952. The earlier experiments at Bikini reached 767 meters below the surface of the atoll and yielded only coralline rock. In 1951-52 deeper holes were drilled at Enewetak and at 1266-1389 meters the volcanic rock basalt was encountered. The drilling passed entirely through the 1200 meter limestone cap composed of shallow-water coral reef rock. Fossil corals from the base of the cap were dated from the Eocene Epoch (Larson & Birkeland, 1982). It was during this period, while Bikini Atoll was in the news, that the marketing name was coined for the newly miniaturized two-piece swimsuit.

This study not only provided direct evidence for the subsidence of volcanos during atoll formation, but also showed that coral reef growth, which allowed for the formation of Enewetak Atoll, has been occurring for the past 60 million years. The rate of subsidence has not been constant, but has averaged only a fraction of a millimeter per year.

Coral Reef Landscapes

Volcanic subsidence has occurred mainly in the central Pacific regions of Micronesia and Polynesia–areas confined to the Pacific Tectonic Plate. This is the region where most

of the world's coral atolls are found on isolated hot-spot volcanos (Dana, 1849). The western Pacific, on the other hand, being part of the Pacific Rim's "Ring of Fire," is consistently very active geologically. The region is reticulated with crustal plate margins, where earthquakes, uplift, and volcanic activity are commonplace occurrences. It is probable that both geologic uplift as well as sea level fluctuations are responsible for the elevated notches and terraces in the coastlines of certain areas such as New Guinea, the Solomon Islands, the Philippines, and Indonesia (Figure 5F). The combined effect of changing sea levels and geologic uplift is also reponsible for the remarkable limestone landscapes known as **karst**. Karst theory states that sea level changes and uplift combined with terrestrial erosion and air exposure of biogenic reef regions have given rise to hummocky landscapes often impregnated with sinkholes and caves. Examples of such striking karst topography include the Blue Holes of Andros in the Bahamas, the limestone towers of Kwangsi Province in China, the Naru Hills of Papua New Guinea, and the Chocolate Hills of Bohol in the Philippines (Figures 5G & 5H).

Biogeography: the distribution of animals and plants

Each of the many thousands of marine animals present in the Indo-Pacific tropics has its own distributional limits. These limits reflect aspects of the evolutionary history, prey specificity, larval and adult dispersal capabilities and adaptability to changing environments. A species' distributional limits are dynamic and will likely change with time, depending on a variety of physical and biotic factors, including air and ocean temperatures, changes in oceanic current patterns and changes in predator populations.

As in the case of taxonomy, biogeography is poorly understood for most species present on Indo-Pacific reefs and other ecosystems. A great number of species from the Indo-Pacific are known only from their original description or from a few specimens that may have been found a century or more after they were first named. For example, the nudibranch, *Flabellina macassarana* (#621), was named from Indonesia in 1905 and was not seen again until single specimens were found in the Philippines and Tanzania in 1994. The fact that this species has been found only from these three places, probably indicates that its range is poorly known and the species is more widespread, rather than it being restricted to three, widely separated localities. Similarly, the heart urchin, *Eurypatagus ovalis* (#1019) and the sea star, *Ferdina sadhensis* (#943) were known only from the original descriptions and are reported for a second time here in this book. On the other hand, the limpet, *Cellana sandwicensis* (#424) is common in the Hawaiian Islands, but has not been found elsewhere. This suggests that it is endemic, or restricted geographically, to the Hawaiian Islands. Furthermore, it is restricted to the rocky intertidal zone and is unlikely to be overlooked. In contrast, *Flabellina macassarana* has been found only from depths exceeding 30 meters and appears to be associated with a single species or genus of hydroids. These factors indicate that it is far more likely to be overlooked than a common, intertidal limpet.

Owing to our incomplete knowledge of the distributions of the vast majority of species found in the Indo-Pacific tropics, it is difficult to know whether a species is found throughout the Indo-Pacific or is restricted to only a portion of the vast area. Many texts list the distribution of species as "Indo-Pacific" or "Indo-West Pacific." Stating the distribution in this manner is ambiguous, as it may mean that the species has been found within a single locality in the Indo-Pacific region or it may have been found from the entire region from the east coast of Africa to Hawai'i. Owing to this ambiguity, we have tried to specify the known geographic ranges of the species treated here, where sufficient information is available. If species are known from only a few isolated localities, we have listed those with the hope that users of the book will provide us with further detailed locality information to be included in future editions.

Modern biogeographers are concerned with **vicariance biogeography**, in addition to the classical interest in distribution patterns of species. Vicariance biogeography's focus is in comparing distribution patterns of closely related species. Vicariance biogeography is predicated on the fact that speciation requires geographical separation of populations to restrict gene flow and permit speciation to occur. In instances where two sister species (determined by detailed phylogenetic analysis) are geographically separated, the present distributions may indicate what geographical or other physical barriers may have isolated populations originally, thus permitting speciation to occur. Vicariance biogeographers try to look for repetition of these patterns to determine major isolating mechanisms, such as the closing of the Isthmus of Panama, which may explain many speciation events. In practice, many, if not most, Indo-Pacific species are often found in the same places as their closest relatives, thus indicating that dispersal has occurred after speciation. This dispersal often completely masks the original geographical isolation that permitted speciation to occur.

Biodiversity

It is estimated that about 3/4 of all species of living organisms are animals. Of animal species, only about 7% are vertebrates. On land about 3/4 of all animal species are arthropods (insects and arachnids), but in the marine environment mollusks are the largest animal group and the arthropods are represented mostly by crustaceans. Figure 6 shows relative estimated species diversity for the marine animal groups covered in this book.

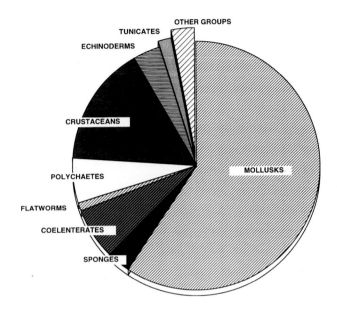

Figure 6. Marine animal species diversity (excluding vertebrates).

Although the vast Indo-Pacific is the most diverse biogeographical region in the marine realm, it is actually comprised of a complex mosaic of faunal subregions, each with differing species compositions and endemics. The Red Sea and the western Indian Ocean as well as the western Pacific are three such areas with very high diversity. The region with by far the highest diversity is defined by a geographic triangle formed by the Philippines in the north, Indonesia to the southwest, and New Guinea to the southeast.

A composite theory to explain tropical high diversity includes these elements: greater energy input near the equator together with great spatial heterogeneity and favorable physical factors punctuated with periodic disturbances.

How To Use This Book

This book focuses on benthic and epibenthic animals–those species that inhabit the sea bottom or the water column just above the sea bottom. **Pelagic–or open water species–have not been included since, strictly-speaking, they do not inhabit the coral reef proper.**

So little is known about Indo-Pacific marine life that this book can only be considered as covering "the tip of the iceberg." New discoveries are to be found on virtually every dive. Contrary to what you might think, only a small percentage of species are known well or even described sufficiently in the scientific and popular literature. Taxonomic revisions of certain groups are necessary before positive identifications can even be made. Much of the published information is inaccurate or based on limited observations. In addition, the extent of variation within a particular species, geographic distributions, and aspects of natural history, are virtually unknown for most species that are encountered on the reefs.

We have organized the text into chapters, each containing a major taxon or phylum. Some care must be given in selecting the correct phylum to find the species you are interested in. A short discussion can be found at the beginning of each group to assist you in making this determination. You may experience some difficulty in distinguishing sponges from tunicates and hydroids from erect bryozoans, for example. We recommend that you follow the text carefully, and not rely solely on the photographs presented, in making your identifications.

The taxonomic regime followed here for each group is based on that which we believe is currently the most widely accepted, for a particular taxon. In several of these groups the systematics below the level of class is still somewhat unstable, and is reported differently in the literature, by different researchers. In these cases we have followed the most recently published review of the species of that group.

Determining how to list the species in this book in a convenient order has been a dilemma. Alphabetical listings eliminate any hope of grouping similar species. Classical color categories used in botanical guides are also inappropriate. After much consideration, the approach taken here is to group the species, within each of the sixteen animal phyla presented, by major rank (order and suborder). Within these major taxonomic categories species are then separated by family, or groups of similar families, in an effort to be more friendly to the reader. This method groups species by general body shape and morphology. A systematic list is presented on page 306.

Our preference for the use of the scientific, Latin and Greek, names over those of common names has already been discussed above. We encourage the reader not to be scared off by the strange names. They will come easier than you think. If a two year old can say "Triceratops," then so too can adult divers!

In numerous instances we were unable to identify a given animal to species level, and list them here with the name of the genus to which they belong only, in *italics*, followed by "sp." In instances where "spp." follows the genus name, this indicates that the photographs refer to several different species within that genus. In the few instances where the photograph is representative of a family, the name is not italicized. The abbreviation "cf." stands for the Latin "conferre," to compare, and is used for tentative identifications–to compare with a given species.

The name of the author originally describing the species, and the date of original publication, follows the species name. The author and date of publication of each species are important in order to check the original source of the species name, as well as the original description and the source of type material.

In the *"Distribution"* section of each species description, we have included all known localities of the species. We have gone to much effort to include accurate distributional records for each species. Many of the localities reported are new to the literature and based on our own collections. Some represent extensions of previously known distributions. **The locality in parentheses is where the photograph was taken.**

New Discoveries and Future Editions

The contents of a book such as this one are dynamic, not static, as nomenclatural changes are recognized, identifications are corrected, species are added, and new information regarding natural history and distribution are recorded. Future editions will reflect this only if keen eyed observers and researchers contribute their findings for the good of science!

The authors would greatly appreciate the help of fieldworkers by receiving high quality color photographs or preserved specimens with written observations. Please send your contributions regarding the following groups to the appropriate author c/o Department of Invertebrate Zoology, California Academy of Sciences, Golden Gate Park, San Francisco, California 94118 U.S.A.: Sponges, coelenterates, ctenophores, and tunicates (Gary Williams); Flatworms and crustaceans (Dave Behrens); other worms, mollusks, lophophorates, and echinoderms (Terry Gosliner).

SPECIES IDENTIFICATIONS

Porifera - Sponges

Sponges are asymmetrical (or radially symmetrical) benthic animals, often strikingly colored. They represent a major component of reef communities.

The sponges are certainly one of the most poorly known groups of all reef animals found in the Indo-Pacific. Most researchers believe that only about half of the 2500-3000 species of sponges have been described. Much of the existing literature dates back to the early part of the century, a time when sponge taxonomy was based largely on dried specimens and upon microscopic analysis of their skeletal structures. Little consideration was given to the considerable variation in form that we find is often caused by differences in environmental conditions.

Because of the difficulty of field identification, we have limited our coverage to a few selected shallow-water species.

Identifications based solely on photographs should be considered as nothing more than "best guesses" or tentative at best. We refer the reader to Colin and Arneson (1995) who portray color photographs of some 238 living species, each referenced by museum voucher collection numbers for species confirmation, to Hooper and Wiedenmayer (1994) for a comprehensive catalogue of Australian sponges, and to Laubenfels (1954) for a review of the sponges of the West-Central Pacific.

Calcarea - Calcareous Sponges

The body of calcareous sponges is formed by calcareous spicules (rather than siliceous spicules as in the demosponges) embedded in a matrix of spongy fibers. This group is represented by small species, usually less than 10 cm in height. Their body forms vary from vase-shaped to cushion-shaped to branching forms.

1. *Clathrina* sp. 1
Identification: The body of this sponge forms a delicate branching network of tubes. Large spicules are lacking and therefore these sponges display a soft appearance and smooth surface texture. Color is mostly pale yellow.
Natural History: Often abundant in shaded areas such as alcoves, crevices, ledges, and overhangs.
Distribution: Philippines (Batangas, Luzon, Philippines).

Lynn Funkhouser
Mike Severns

2. *Clathrina* sp. 2
Identification: The body of this species is a matrix of long white tubes, the surface of which is shiny in appearance. This species is mostly greyish-white in color.
Natural History: Often encountered in sheltered areas such as overhangs and rock crevices.
Distribution: Hawaiian Archipelago (Hawai'i).

3. *Leucetta chagosensis* **Dendy, 1913**

Identification: This sponge forms variable, globular-shaped colonies. The spicules and spongin are tightly packed giving an opaque appearance. Color is uniform bright yellow.

Natural History: This conspicuous sponge generally occurs on reef slopes.

Distribution: Red Sea and the central Indian Ocean to the western Pacific (Sipadan, Borneo).

Nicholas Galluzzi

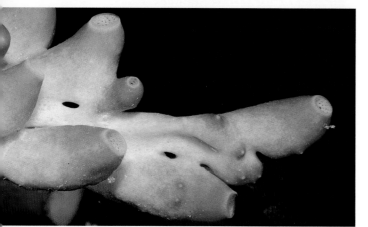

4. *Leucetta primigenia* **Haeckel, 1872**

Identification: The body of this species varies from a flattened encrusting species with raised ridges and apertures to the vase-shaped form shown here. Specimens are red in color near the apertures, and have a white band at their opening. One of the largest calcareous sponges, the spicules are large and very sharp.

Natural History: Species of the nudibranch genus *Notodoris* (#608-610) feed on this sponge. Depth: to 26 m.

Distribution: Eastern Africa; Australia; Indonesia; Borneo; Philippines; Pohnpei; Chuuk and Biak (Sulu Sea).

Lovell & Libby Langstroth

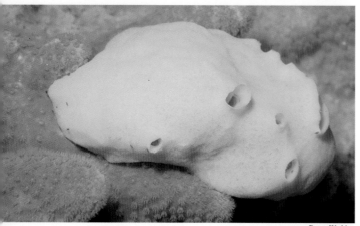

5. *Leucetta* **sp.**

Identification: Colonies of this species form a globular mass with several apertures. The body is dense and opaque, relatively compact to potato-shaped. Color is lemon yellow.

Natural History: This species often occurs on coral on reef flats.

Distribution: Melanesia (Solomon Islands).

Bruce Watkins
G.C. Williams

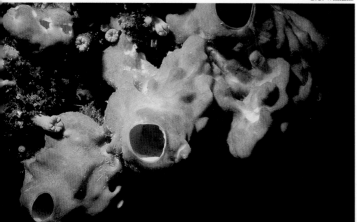

6. *Pericharax heterorhaphis* **Polejaeff, 1883**

Identification: A ridged and globose form, the exterior is olive to brown in color, while the interior is yellow to green. This species can reach 150 cm in size, and its shape is quite variable, depending on where it is growing.

Natural History: This species is known to occur to 18 m in depth.

Distribution: Andaman Islands; New Guinea; Philippines; Chuuk (Bohol, Philippines).

7. *Pericharax* sp.

Identification: A small, solitary, nipple-shaped or urn-shaped species. The sack-like structure is attached to the substrate by a short stalk. It varies from golden-brown to chocolate brown.

Natural History: This delicate and inconspicuous form is often encountered between coral heads on reef flats.

Distribution: Indonesia; Malaysia; Philippines (Sipadan, Borneo).

Nicholas Galluzzi

8. *Leuconia palaoensis* (Tanita, 1943)

Identification: A shiny-surfaced tubular colony. The colonies are soft and form a network of tubes. The color varies from pink to pale blue.

Natural History: This delicate sponge is found on walls and other areas exhibiting vertical relief.

Distribution: Australia; New Guinea; Belau and Federated States of Micronesia (Heron Island, Great Barrier Reef, Australia).

Leslie Newman & Andrew Flowers

9. *Leucosolenia* sp.

Identification: An encrusting gelatinous mat with numerous rounded and raised protuberances, many of which are slightly forked at the ends. Olive green in color with pink aperture walls.

Natural History: Often encountered on vertical surfaces, slopes, and ledges.

Distribution: Melanesia; Indonesia; New Guinea (Indonesia).

Demospongiae - Siliceous Sponges

The skeleton of demosponges is formed by siliceous spicules, spongin fibers or a combination of both. Some demosponges (known as sclerosponges) have calcareous spicules in addition to siliceous ones along with spongin fibers. Demosponges vary considerably in size and shape. They represent the most diverse and abundant sponge group, and are the most conspicuous reef sponges.

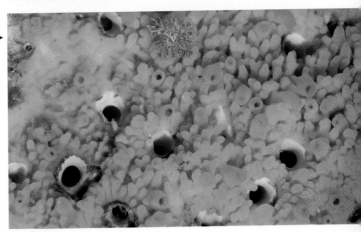
Bruce Watkins
Nicholas Galluzzi

10. *Plakortis* sp.

Chicken Liver Sponge

Identification: A pinkish-brown globose species with one to several osculae per colony. The exterior surface is smooth and fleshy, resulting from the lack of large spicules. Internal and external color is similar.

Natural History: Infrequently encountered on reef slopes. Depth: 5-20 m.

Distribution: Philippines and New Guinea (Batangas, Luzon, Philippines)

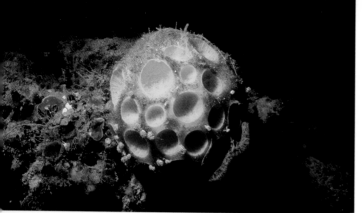

G.C. Williams

11. *Cinachyra* sp.

Golf Ball Sponge

Identification: Colonies of this species form orange puffy balls with large craters on the surface. The remaining surface between the craters is often covered with epizoic growth.

Natural History: The craters vary in color–golden yellow in this species and pink/magenta in a related form.

Distribution: Philippines and New Guinea (Bohol, Philippines).

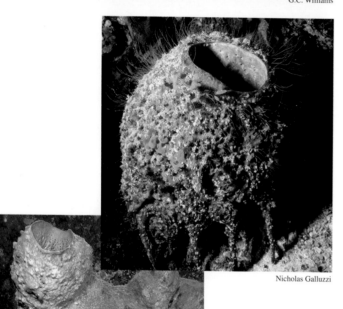

Nicholas Galluzzi

12. *Stelletinopsis isis* Laubenfels, 1954

Identification: A large brown barrel sponge, it is easily identified by the series of leg-like structures surrounding its base.

Natural History: The surface of this species provides a surface for numerous encrusting organisms such as hydroids, tunicates and other sponges.

Distribution: Indonesia; Malaysia and Micronesia (Sipadan, Borneo).

13. *Diacarnus spinipoculum* (Carter, 1879)

Ligament Sponge

Identification: A large globular sponge. The body wall is relatively thin for a sponge this size. The cavities are large and gaping.

Natural History: Often encountered on reef slopes at 20-25 m in depth.

Distribution: Australia; Fiji; Indonesia; Micronesia (Indonesia).

Terry Schuller

Bruce Watkins

14. *Atergia* sp.

Identification: A unique sponge colony. Colony varies from white to red. There is one large osculum and many smaller ostia (pores) which because of their ring of branched papillations around the orifice, give the appearance of octopus suckers.

Natural History: This species often occurs on dead coral and rock in protected areas.

Distribution: Belau and Philippines (Philippines).

15. *Theonella swinhoei* **Gray, 1868**

Identification: Robust, cigar-shaped tube sponge. Colonies may be slightly branched or appear as two or more fused individuals. The outer surface is densely spiculose, but smooth in appearance. The color is usually brown with ochre around the oscula.

Natural History: These sponges may be encrusted with algae and other epizoic growth.

Distribution: Nicobar Islands and Philippines (Bohol, Philippines).

T. M. Gosliner

16. *Coelocarteria singaporensis* **(Carter, 1883)**

Identification: A globose species, bright yellow in color. The surface is covered with densely-set, long and finger-like papillations.

Natural History: The surface papillations create habitat for numerous small fish and other benthic organisms.

Distribution: Australia; Singapore; New Guinea (Madang, Papua New Guinea).

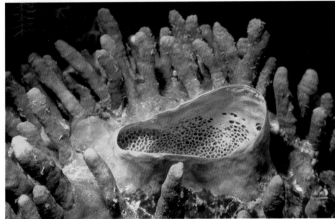

T. M. Gosliner

17. *Monanchora ungiculata* **(Dendy, 1921)**

Identification: A bright-red branching tube sponge. The colony shown in this photo is highly inflated. It can shrink to a thin layer when disturbed.

Natural History: This species prefers silty reef habitat up to 20 m in depth.

Distribution: Indian Ocean; New Guinea; Malaysia (Sipadan, Borneo).

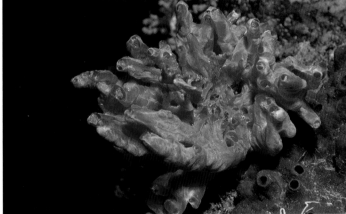

Nicholas Galluzzi
T. M. Gosliner

18. *Desmacella* **sp.**

Identification: Colonies of this species are thin and encrusting over relatively extensive areas of substratum, mostly on vertical surfaces. Color is bright orange-red.

Natural History: The tubular polyp stage of the jellyfish *Nausithoe punctata* (#306) lives in dense populations in species of *Desmacella* as well as other sponge genera such as *Suberites*, *Myxilla, Reneira,* and *Esperia.* The numerous white tentaculated mouths of the polyps can be seen clearly in this photograph.

Distribution: Philippines; Chuuk, Micronesia (Batangas, Luzon, Philippines).

G.C. Williams

19. *Liosina granularis* Borges & Bergquist, 1988

Identification: A distinctive tube sponge with robust stalk-like tubes, the surface of which is smooth with a matrix of irregular depressions. Color varies from tan to orange.

Natural History: This species is often found covered with silt. The specimen pictured here is covered with the small isopods *Santia* (#713).

Distribution: Philippines; Marshall Islands (Mindoro, Philippines).

Lovell & Libby Langstroth

20. *Clathria mima* (Laubenfels, 1954)

Identification: This is a handsome encrusting species, easily recognized by the white-veined pattern branching from the oscula.

Natural History: This sponge is recorded to 12 m in depth.

Distribution: Indonesia; Chuuk (Flores, Indonesia).

T. M. Gosliner
G.C. Williams

21. *Echinochalina* sp.

Identification: A flat encrusting species, the surface bears characteristic transparent canals leading to elevated oscula. Mostly brownish in color and finely speckled with salmon pink.

Natural History: This species is known to occur on reef slopes and ledges.

Distribution: Philippines (Batangas, Luzon, Philippines).

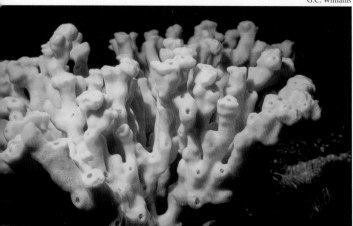

22. *Auletta* sp.

Identification: A firm branching colony with interconnected upright tubes arising from a common base. The branches often gradually taper from the tips toward the bases. Oscula are found toward ends of the branches. Mostly uniform bright yellow in color.

Natural History: Brittle stars commonly inhabit the branches of this sponge.

Distribution: Melanesia including New Guinea and the Solomon Islands (Mborokua Island, Solomon Islands).

23. *Axinellid* sp.

Identification: This undescribed, often fan-shaped, sponge is salmon in color. The apertures are transparent tubes distributed over the surface.

Natural History: Infrequently encountered on reef slopes and walls at about 15 m in depth.

Distribution: New Guinea; Philippines; Malaysia (Sipadan, Borneo).

Nicholas Galluzzi

24. *Axinyssa* sp.

Identification: A dark blue-black encrusting or lumpy sponge. The surface has a pattern of thin greyish curved lines between the oscula.

Natural History: Usually encountered at depths greater than 20 m.

Distribution: Red Sea to Melanesia and Micronesia (Red Sea).

Nicholas Galluzzi

25. *Callyspongia* sp. 1

Identification: Colonies are formed by several robust tubes. The outer surface is highly papillated and pinkish-red in color. This spiky-looking surface is highly irregular and fibrous in appearance.

Natural History: This species has been encountered on reef slopes.

Distribution: Philippines and Melanesia (Batangas, Luzon, Philippines).

Nicholas Galluzzi
G.C. Williams

26. *Callyspongia* sp. 2

Identification: A densely-spiculiferous encrusting species. The colonies have an irregular surface that varies in color from grey to pink.

Natural History: This species is found on vertical surfaces at 20-25 m in depth.

Distribution: Philippines (Batangas, Luzon, Philippines).

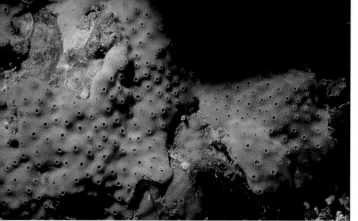

27. *Haliclona* sp. 1

Identification: A beautiful blue encrusting species with evenly distributed oscula that form numerous low volcano-like elevations.

Natural History: Colonies encrust exposed rock surfaces.

Distribution: Philippines (Mindoro, Philippines).

28. *Haliclona* sp. 2

Identification: Colonies of this species are composed of firm spiculiferous branching masses. Several oscula occur on each branch. The color is deep blue.

Natural History: Commonly encountered on exposed coral rock on reef flats and slopes. Fed upon by the dorid nudibranch *Jorunna funebris* (#564).

Distribution: Malaysia (Sipadan, Borneo).

29. *Nara nematifera* Laubenfels, 1954

Identification: A brilliantly-colored low-growing species, it is easily identified by the pattern of parallel white lines which almost camouflage the oscula. Color magenta/purple with conspicuous fine white thread-like striations.

Natural History: Encountered at depths to at least 18 m.

Distribution: New Guinea and Micronesia (Madang, Papua New Guinea).

30. *Kallypilidion* sp.

Tube Sponge

Identification: A slender tube sponge usually with numerous tubes arising adjacent to one another. The overall color is pink and the surface is covered with raised, densely-spiculated horns.

Natural History: This characteristic sponge is commonly encountered on reef slopes. It is the prey of an undescribed, cryptic nudibranch, *Rostanga* sp.

Distribution: Philippines (Batangas, Luzon, Philippines).

31. *Kallypilidion poseidon* Laubenfels, 1954
Vase Sponge

Identification: A large bluish-grey vase sponge with a smooth texture and a large tapering central cavity.
Natural History: The outer surface is often inhabited by small tube-forming polychaete worms. A locally common and characteristic sponge.
Distribution: New Guinea and Micronesia (Madang, Papua New Guinea).

T. M. Gosliner

32. *Amphimedon* sp.
Vase Sponge

Identification: A tapering blue to purple tube sponge with wide oscula and a smooth external surface.
Natural History: The outer surface of this species is often encrusted with tube worms.
Distribution: New Guinea (Madang, Papua New Guinea).

Leslie Newman & Andrew Flowers

33. *Cribrochalina* sp.
Identification: A stout tube sponge, with an undulating but smooth surface. The surface of this blue species is made up of tightly-packed spiculated tubercles.
Natural History: Often encountered on slopes and ledges.
Distribution: Indonesia (Bali, Indonesia).

Dan Gotshall
G.C. Williams

34. *Niphates callista* (Laubenfels, 1954)
Identification: A conspicuous and easily recognized sponge with paddle-shaped lobes. Numerous brittle fibers give this sponge a crunchy appearance. Color is violet or purple with white peaks and rings around the larger pores.
Natural History: A locally common and conspicuous species in deeper regions off reef flats. Depth: 25-35 m.
Distribution: Indonesia; New Guinea; Belau (Madang, Papua New Guinea).

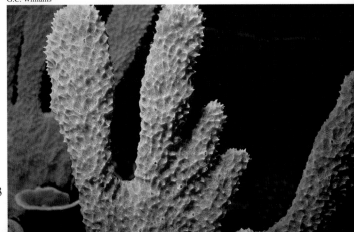

G.C. Williams

G.C. Williams

35. *Xestospongia testudinaria* (Lamarck, 1815)
Barrel Sponge

Identification: This is one of the largest and most easily recognized sponges in the western Pacific–a large barrel sponge that can attain over a meter in height. The brownish surface is very hard and deeply ridged.

Natural History: Found singularly or in small groups on coral flats and reef faces. The deeply-ridged surface provides substrate for many epizoic organisms. Various crustaceans are often found inside the barrel. This species is morphologically identical to the Giant Barrel Sponge, *Xestospongia muta*, of the Caribbean, but differs in chemical properties.

Distribution: Indonesia; Australia; New Guinea; Philippines (Batangas, Luzon, Philippines).

36. *Xestospongia* sp.

Barrel Sponge

Identification: A smaller barrel sponge usually less than 20 cm in height. The osculum varies in width and often is surrounded by finger-like protuberances. The outer wall of the sponge has numerous short ridges that do not align longitudinally as in the previous species. These ridges often have rounded protuberances on the outer margin. Color is pinkish-tan with lighter areas around the osculum.

Natural History: Encountered at 10-20 m in depth on reef slopes or ledges.

Distribution: Philippines (Batangas, Luzon, Philippines).

37. Petrosid sponge

Identification: Large colonies of robust tubes. The outer surface shows a reticulated pattern of raised ridges, which connect the individual tubes. A sponge of very firm and hard texture.

Natural History: This variable unidentified sponge forms upright masses on reef flats.

Distribution: Indian Ocean; Australia; New Guinea; Indonesia; Federated States of Micronesia; Belau (Indonesia).

Terry Schuller
Lovell & Libby Langstroth

38. *Oceanapia* sp.

Identification: A globular sponge having the central osculum surrounded by long slender finger-like extensions.

Natural History: The surface is often covered with green and red encrusting algae.

Distribution: New Guinea (Bismarck Sea, Papua New Guinea).

24

39. *Carteriospongia* cf. *contorta* (Bergquist, Ayling, & Wilkinson, 1988)

Identification: A unique encrusting sponge, the tall tubular apertures are connected by ridges which radiate and branch from one end of the colony.

Natural History: A distinctive sponge on reef slopes, often on vertical surfaces, 15-20 m in depth.

Distribution: Australia and Philippines (Batangas, Luzon, Philippines).

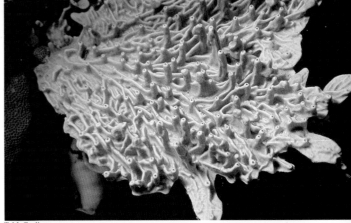
T. M. Gosliner

40. *Phyllospongia lamellosa* (Esper, 1799)
Fan Sponge

Identification: A tough leathery sponge, it grows in a plate-like, laminar-shape, or crescent-shaped form. The surface is covered with lighter-colored lumps, which presumably contain symbiotic algae.

Natural History: This conspicuous and distinctive sponge often occurs on the edges of walls and boulders. This species is superficially similar to other sponges of the genera *Carteriospongia* and *Strepsichordaia*. A species of nudibranch in the genus *Okenia* (#605) feeds on epizoic bryozoans growing on this species.

Distribution: Indian Ocean; Australia; Philippines and Japan (Batangas, Luzon, Philippines).

G.C. Williams

41. *Ircinia* sp.

Identification: Colonies are solitary pink balls with a large central cavity. Outer surface often heavily encrusted with epizoic organisms.

Natural History: A distinctive species often found attached to walls or other vertical surfaces on reef slopes.

Distribution: Malaysia (Sipadan, Borneo).

Nicholas Galluzzi
T. M. Gosliner

42. *Dysidea* cf. *herbacea* (Keller, 1889)

Identification: Colonies are tough and encrusting with long raised wall-like branches. Surface texture may be highly variable.

Natural History: The upright, flattened lobes of this species make this a distinctive sponge species; encountered on reef flats and slopes. Depth: 10-25 m. The opisthobranch *Sagaminopteron psychedelicum* (#534) feeds on this sponge.

Distribution: Tanzania; Indonesia; Philippines; Guam; Belau; Micronesia (Mindoro, Philippines).

25

G.C. Williams

43. *Ianthella basta* (Pallas, 1766)
Elephant Ear Sponge

Identification: A distinctive fan-shaped sponge. Specimens up to a meter in height have been observed. Color varies greatly–yellow, brown, green, blue, or purple.

Natural History: May occur in relatively dense, localized populations–probably the most common and widely distributed species in the genus. Depth: 10-20 m. Gobies of the genus *Pleurosicya* are often found on the surface of these sponges.

Distribution: Indonesia; Guam; Australia; New Guinea; Solomon Islands (Mborokua Island, Solomon Islands).

Coelenterata (Cnidaria) - Hydroids, Hydrocorals, Soft Corals, Sea Fans, Sea Pens, Anemones, Hard Corals, Black Corals and Jellyfish

The coelenterates are a large group of over 9000 species, mostly marine aminals. They are characterized by having an internal space for digestion called the gastrovascular cavity, with the same opening for the mouth and anus. They are radially symmetrical and have a ring of tentacles surrounding the mouth of each polyp. The tentacles are armed with microscopic stinging structures called nematocysts. These contain toxic mixtures of proteins and phenols. For many coelenterate species, there is an alternation of life cycle between an attached polyp and free-swimming medusa. There are two tissue layers, but no organ systems. Coelenterates are polytrophic feeders–deriving nutrition in several possible ways–as predators, through algal symbiosis, and/or by direct uptake of dissolved organic matter.

Hydroida - Hydroids

Hydroids are mostly benthic feather-like colonies, in which the polypoid generation predominates. Many species have a jellyfish-like sexual stage known as a hydromedusa. The colony often has a chitinous exoskeleton and the tentacles of each polyp are either capitate or filiform.

Stephen Smith
Kathy deWet

44. *Ralpharia* sp.

Identification: The slender polyps have very elongate and spreading outer tentacles, as well as shorter oral tentacles held close together. Grape-like clusters of gonophores are found between the two rings of tentacles.

Natural History: A low-growing, intertidal and shallow water species. Often inconspicuous, species of *Ralpharia* are frequently encountered with encrusting organisms such as sponges, bryozoans, and tunicates.

Distribution: Australia. A genus of widespread distribution in warm waters of the Atlantic and the Indo-Pacific including Indonesia (South Solitary Island, New South Wales, Australia).

45. *Tubularia* sp.

Identification: Similar to the previous species, but the polyps are stouter, with more robust stems; usually not more than 1.5 cm in height.

Natural History: As in other species of *Tubularia*, often found littoral to shallow subtidal, growing amongst dense growths of encrusting organisms.

Distribution: New Guinea (Mapia, Irian Jaya).

46. *Pennaria disticha* Goldfuss, 1820

Identification: Delicate, feather-like hydroids, generally less than 12 cm in height, with side branches mostly opposite one another. The white urn-shaped polyps may have very short stalks and are upright, being sparsely arranged and restricted to the upper side of each side branch.

Natural History: Fed on by *Caloria indica* (#626).

Distribution: Nearly cosmopolitan in tropical and temperate neritic waters, including the Philippines (Philippines).

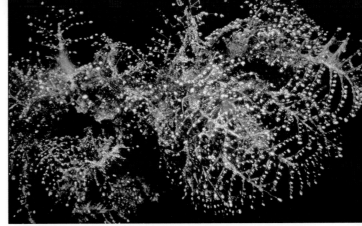
Lynn Funkhouser

47. *Solanderia* sp.

Identification: Colonies are tree-like and strongly-branched, with branching more-or less in one plane. An internal skeleton of chitin (a brown polysaccharide carbohydrate) is present, which is rough, spiny, or grooved. The polyps are club-shaped with numerous capitate tentacles scattered over the surface.

Natural History: Several species occur in reef areas, distinguished mainly by the nature of the skeleton, and number of tentacles per polyp. The nudibranch *Protaeolidia atra* preys upon this hydroid. Depth: 5-30 m.

Distribution: Philippines. The genus is widespread in the Indo-Pacific (Batangas, Luzon, Philippines).

Terry Schuller

Zanclea spp. (#'s 891, 897, 898 & 904)

Zancleid hydroids are symbiotically associated with phidoloporid and celleporariid bryozoans, such as *Iodictyum axillare* (#891), *Rhynchozoon larreyi* (#897), *Triphyllozoon inornatum* (#898) and *Celleporaria sibogae* (#904). The hydroid presumably benefits from feeding currents generated by the bryozoan. The bryozoan in turn benefits from the protection of the hydroid's stinging cells. Zancleids usually have filiform or clavate polyps with short capitate tentacles.

48. *Eudendrium* sp.

Identification: Colonies are composed of a mass of upright branching tubular stalks. The polyps are trumpet-shaped or funnel-shaped, each with one whorl of narrow tentacles. The stalks are brown, and the polyps are often milky-white with a purple central region.

Natural History: The aeolid nudibranch *Flabellina rubrolineata* (pictured here and #622) feeds on this hydroid.

Distribution: Philippines. The genus is worldwide in distribution with several Indo-Pacific species (Batangas, Luzon, Philippines).

G.C. Williams
T. M. Gosliner

49. Hydractiniid sp.

Identification: The crowded polyps grow directly from a common network of basal tubes that cover snail shells. The conical polyps taper toward the mouth and each has a single whorl of slender tentacles.

Natural History: Hydractinid colonies are epizoic on snail shells, such as *Nassarius fissilabrus* pictured here.

Distribution: Philippines. Two related genera, *Hydractinia* and *Stylactaria*, have worldwide distribution, with perhaps several species in the Indo-Pacific (Batangas, Luzon, Philippines).

G.C. Williams

50. *Zygophylax* sp.

Identification: Colonies are erect and feather-like, up to 100 mm in height. The side branches have two rows of polyps on the upper and lower surfaces. The polyps are elongate, each with a narrow basal stalk. Color is usually golden yellow.

Natural History: Infrequently encountered on shallow reef flats and slopes.

Distribution: Philippines. The genus is distributed worldwide with several species in the Indo-Pacific (Batangas, Luzon, Philippines).

G.C. Williams

51. *Aglaophenia cupressina* Lamouroux, 1816
Stinging Hydroid

Identification: Conspicuous and robust, feather-like hydroids up to one-third of meter in height. Colonies are repeatedly branched. Polyps are very closely-set on one side of the ultimate branches. Color usually white to tan throughout.

Natural History: This common and distinctive hydroid is dreaded for its powerful sting, often inflicted upon the unsuspecting diver. In spite of this capability, it is preyed upon by the nudibranch *Doto ussi* (#613). Depth: 10-30 m.

Distribution: East Africa; Madagascar; Comoros to Australia; Philippines; New Guinea and Sea of Okhotsk (Bohol, Philippines).

T. M. Gosliner
Leslie Newman & Andrew Flowers

52. *Gymnangium gracilicaule*
(Jäderholm, 1903)

Identification: Broad, feathery hydroids usually less than 15 cm in height. The colonies have many branches with minute and inconspicuous polyps. Color is a lustrous white with ends of ultimate branches and polyps sharply contrasting in dark brown.

Natural History: A considerable amount of morphological variation has been recorded in this species.

Distribution: East Africa; Red Sea to Indonesia; Philippines and China Sea (Mindoro, Philippines).

53. *Macrorhynchia philippina*
Kirchenpauer, 1872
Fire or Stinging Hydroid

Identification: Feathery hydroids up to 15 cm in height. Each branch is narrow and elongate with numerous side branches of equal length. Stems are usually dark brown, polyps are usually white and in striking contrast to the darker stems.

Natural History: Several stinging hydroids are common in the Indo-Pacific, but this species is perhaps the most virulent, capable of inflicting an intense sting. In spite of this, the nudibranch *Lomanotus vermiformis* feeds on this hydroid. Intertidal to 45 m in depth.

Distribution: Circumglobal in tropical to warm temperate regions (Heron Island, Great Barrier Reef, Australia).

Hydrocorals

Hydrocorals are common constituents of coral reefs. They are characterized by having a brittle and massive calcareous skeleton with minute polyp pores (mostly less than one mm in diameter). The golden-brown fire corals are often associated with reef flats while the brightly-colored lace corals are mostly restricted to caves and vertical walls.

Milleporidae - Fire corals

Fire corals are often very brittle with extremely diverse growth forms. The colonies can be encrusting and spreading, or upright and multiply-branched, lobe-like, or plate-like. This great morphological variation has produced a great deal of taxonomic confusion. Because of this, the number of valid species is at present undetermined, although approximately 50 species have been described worldwide.

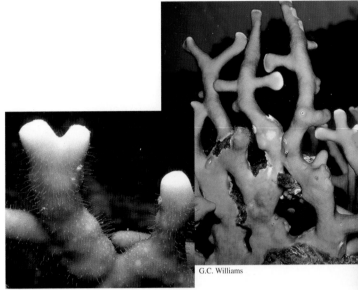

G.C. Williams

G.C. Williams

54. *Millepora* sp. 1

Branching Fire Coral

Identification: Upright branching colonies. Upright forms are mostly aligned in a plane perpendicular to the prevailing currents. The branches are often forked at the ends. The minute pores containing the polyps are of two kinds–larger holes containing the feeding polyps are surrounded by more numerous smaller ones containing the defensive polyps. The left photo shows the polyps fully extended.

Natural History: Due to toxins injected by the stinging cells, the aptly-named fire coral is capable of inflicting very irritating stings to scantily-dressed divers.

Distribution: Melanesia. The genus is distributed throughout the tropical reefs of the world and very common throughout the Indo-Pacific (Madang, Papua New Guinea; New Georgia Group, Solomon Islands).

55. *Millepora* sp. 2

Encrusting Fire Coral

Identification: Colonies are usually low-growing, encrusting and spreading with finger-like knobs or lobes.

Natural History: Severe storms can break off the upright portions of fire corals. Regrowth takes place from the encrusting base. Fire corals along with the blue corals are the only non-scleractinian corals that are reef-builders (hermatypic).

Distribution: Melanesia (Madang, Papua New Guinea).

G.C. Williams
G.C. Williams

56. *Millepora* sp. 3

Plate Fire Coral

Identification: Colonies with thickened and upright plate-like growth. Color is golden-brown due to symbiotic alage in the coral tissues, while the growing tips of the branches or lobes are usually white.

Natural History: The circular protuberances seen here result from pyrgomatid coral barnacles (#694) that often infest fire corals, as do a variety of tube worms and other creatures.

Distribution: Melanesia (New Georgia Group, Solomon Islands).

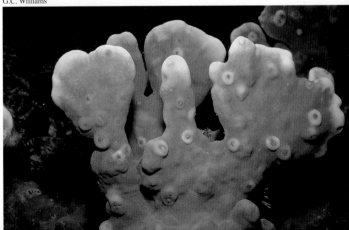

Leslie Newman & Andrew Flowers

Stylasteridae - Lace Corals

Unlike the fire corals, the stylasterines do not contain symbiotic algae and are mostly brightly-colored. The taxonomy of the shallow-water stylasterines of the Indo-Pacific is at present unsynthesized and unworkable, and hence it is not possible to make accurate species identifications. The genus *Stylaster* contains approximately 71 species worldwide, while *Distichopora* comprises about 19 species. *Stylaster sanguineus* Milne Edward and Haime, 1850, and *D. violacea* (Pallas, 1766) are two common and widespread species.

57. *Stylaster* sp. 1   Lace Coral
Identification: A beautiful, delicate and finely-branched coral, with brittle calcareous skeletons. The numerous branches are slender and pointed at the tips. The polyps are represented by minute holes with scalloped margins. Color is usually pink or magenta, often with white-tipped branches.
Natural History: This species is one of the most frequently encountered stylasterids on western Pacific coral reefs.
Distribution: Western Pacific (Lizard Island, Great Barrier Reef, Australia).

G.C. Williams

58. *Stylaster* sp. 2   Lace Coral
Identification: Fan-shaped lace corals, up to 10 cm in height. Branches are less slender, more sinuous, and less openly branched than *Stylaster* sp. 1. Color is usually red or wine-red with white branch tips.
Natural History: Encountered on the ceilings of caves and on overhangs.
Distribution: Melanesia (New Georgia Group, Solomon Islands).

59. *Stylaster* sp. 3   Lace Coral
Identification: A delicate lace coral with open branching and slender branches, usually less than 8 cm in height. Color is tan to orange with white tipped branches.
Natural History: Encountered on overhangs and vertical surfaces.
Distribution: Western Pacific (Batangas, Luzon, Philippines).

G.C. Williams
Marc Chamberlain

60. *Stylaster* sp. 4   Lace Coral
Identification: A copiously-branched stylasterid with slender ultimate branches. Color is white or pinkish-tan and white.
Natural History: Commonly encountered in shaded locations.
Distribution: Melanesia (Milne Bay, Papua New Guinea).

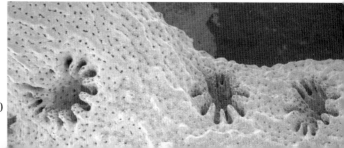

61. *Distichopora* sp. 1

Lace Coral

Identification: Somewhat robust coral colonies composed mainly of brittle and brightly-colored calcium carbonate. Branches fan out in one plane. Branch tips are blunt and rounded. Polyps are restricted to a single groove along the outer margin of the coral. Color is usually orange with the branches sometimes tipped with white or yellow.

Natural History: Commonly found on the ceilings or walls of caves and overhangs.

Distribution: Melanesia (New Georgia Group, Solomon Islands).

G.C. Williams

62. *Distichopora* sp. 2

Lace Coral

Identification: Stylasterid corals with very stout and compressed main branches and short side branches. Often strikingly bicolored: blood red with white-tipped branches.

Natural History: Encountered in shaded areas off of reef flats, inhabiting caves and alcoves.

Distribution: Marshall Islands (Enewetak, Marshall Islands).

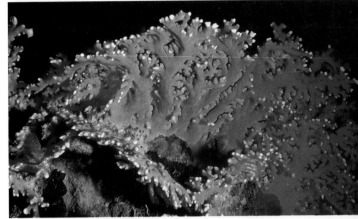
Jack Randall

63. *Distichopora* sp. 3

Lace Coral

Identification: Stout coral colonies with short and rounded side branches that are few in number. The main branches are robust and mostly cylindrical. Color violet often with white-tipped branches.

Natural History: Frequently encountered in caves and shaded vertical surfaces.

Distribution: Micronesia (Belau).

Marc Chamberlain
Marc Chamberlain

64. *Distichopora* sp. 4

Lace Coral

Identification: A robust, fan-shaped species with stout and rounded branches. Color is mostly yellow with branches tipped white.

Natural History: Encountered on vertical surfaces away from shallow reef flats.

Distribution: New Guinea (Milne Bay, Papua New Guinea).

Anthozoa - Anemones and Corals

Anthozoans have only the polyp stage; a medusa is absent.

Octocorallia - Octocorals

Octocorals are conspicuous and important members of coral reef communities, as well as being some of the most beautiful and colorful of all reef animals. They contribute indirectly to the building of coral reefs by their contribution of countless spicules (sclerites) to the unconsolidated calcareous sediments of reef communities, that over time become incorporated into the overall limestone reef structure. Octocorals are an extremely diverse group and include the blue corals, stoloniferans, soft corals, sea fans and sea whips, and sea pens. All species characteristically have eight feathery tentacles surrounding the mouth of each polyp–hence the name "octocoral."

Helioporacea - Blue corals

The blue corals are unique among the octocorals in that they have a massive and brittle calcium carbonate skeleton, much like the hydrocorals or hard corals. However, they differ from these by having eight feathery tentacles surrounding the mouth of each polyp.

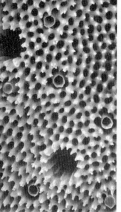

G.C. Williams

65. *Heliopora coerulea* (Pallas, 1766)
Blue Coral

Identification: Variable in appearance, but often forming many upright flattened plates or lobes. The plates are brittle due to the massive non-spicular calcium carbonate skeleton. The deep brown surface coloration of the tissues contrasts strongly with the distinctive powder-blue color of the internal skeleton. Distal tips of the lobes or plates often have conspicuously white tips. Veron (1986: 614) provides a descriptive account of this species.
Natural History: Common with patchy distribution, in the warmer shallow waters, low tide to 18 m in depth. Coloration of the tissues results from the presence of zooxanthellae. Blue coral is probably the only reef-building octocoral of significance. Eaten by the arminacean nudibranch *Doridimorpha gardineri.*
Distribution: East Africa; Red Sea to Philippines; New Caledonia; Ryukyu Islands; Samoa (Batangas, Luzon, Philippines).

G.C. Williams
G.C. Williams

Alcyonacea - Stoloniferans, Soft Corals and Gorgonians

These octocorals have a skeleton in the form of free or fused spicules imbedded in the tissues of the coral (sclerites). Some also have an internal axis of a dark horn-like protein and/or calcium carbonate. Included here are the stolon-bearing octocorals, soft corals, sea fans, and sea whips.

66. *Clavularia* sp. 1   Tree Fern or Palm Coral

Identification: Large polyps attached by a common stolon that adheres to the substratum - often forming large sheets. The polyp bodies are 10-20 mm in height and 4-10 mm in width. The diameter of the extended circle of tentacles may exceed 20 mm. The pinnules of the tentacles are long and conspicuous. Polyp tentacles green with yellow green centers.
Natural History: Several species of this large and widespread genus assume a similar large growth habit, including *Clavularia inflata* and *C. kuekenthali.* Depth: 3-30 m.
Distribution: Melanesia (Madang, Papua New Guinea).

67. *Clavularia* sp. 2
Tree Fern or Palm Coral

Identification: This species is similar and may be synonymous with the previous species, but differs in having grey to brownish tentacles with white centers.

Natural History: Common and abundant in some areas, covering extensive areas of up to a meter or more in diameter.

Distribution: Melanesia (Mborokua, Solomon Islands).

G.C. Williams

68. *Clavularia* sp. 3
Tree Fern or Palm Coral

Identification: This species differs from the previous two by having very fine pinnules. The tentacles are pale yellow with whitish centers. The polyps are densely-set and colonies may cover large areas of substratum.

Natural History: Locally common on ledges and slopes.

Distribution: Melanesia (Madang, Papua New Guinea).

T. M. Gosliner

69. *Carijoa* sp. 1

Identification: These stolon-bearing octocorals often form tangled mats of long cylindrical branches, each with many bulbous polyps. The polyps can completely retract into the calyces. Color is white throughout. The related and similar genus *Telesto*, containing species that are red or orange, is recorded throughout the Indo-Pacific from eastern Africa to Hawai'i, as well as the western Atlantic.

Natural History: Species of *Carijoa* are white in color, but often have a thin dark-red sponge growing on the surface of the colony between the polyps, which results in a superficially reddish appearance. Preyed upon by the nudibranch *Phyllodesmium poindimiei*.

Distribution: Philippines. The genus is found throughout much of the Indian and Pacific Oceans from eastern Africa to New Guinea and Hawai'i (Batangas, Luzon, Philippines).

G.C. Williams
G.C. Williams

70. *Carijoa* sp. 2

Identification: The colonies are large and very bushy, up to a meter in height. The polyps can fully retract into the tubular calyces. Color cream-white to light tan.

Natural History: Locally abundant, often forming dense forest-like stands. Fed upon by the ovulid snail, *Primovula* sp. (#483) and the nudibranch, *Tritonia* sp. (#616). Depth: 10 m.

Distribution: Philippines (Mindoro, Philippines).

G.C. Williams

G.C. Williams

G.C. Williams

G.C. Williams

David K. Mulliner
G.C. Williams

71. *Pachyclavularia violacea* (Quoy & Gaimard, 1833)

Identification: Colonies are membranous, about 2 mm thick with tall, deep purple, cylindrical polyp calyces to 10 mm in height. The crowded polyps have slender green tentacles with inconspicuous pinnules surrounding the white oral discs. A similar genus is *Briareum* (#137), which can also be membranous or encrusting (~10 mm thick) with calyces mostly less than 5 mm in height. Several species of both genera inhabit Indo-Pacific reefs.

Natural History: Often forming an extensive carpet on ledges and slopes; 5-25 m depth. This species is preyed upon by the nudibranch *Phyllodesmium briareus* (#627).

Distribution: Australia; Philippines; Indonesia; New Guinea; Solomon Islands (Madang, Papua New Guinea).

72. *Tubipora musica* Linnaeus, 1758
Organ Pipe Coral

Identification: The soft polyps retract into hard calcareous tubes. The densely-set tubes are joined by calcareous shelves that often form multi-level platforms. The pinnules of the tentacles are short and inconspicuous. Colonies form rounded clumps 20-300 mm in diameter. Tubes and platforms are deep reddish-purple, while the tentacles are greenish, golden-brown, or gray. Veron (1986: 612) provides a descriptive account of this coral.

Natural History: Shallow-water to about 15 m in depth.

Distribution: East Africa; Red Sea to New Caledonia; Solomon Islands; southern Japan; Marshall Islands (Mborokua, Solomon Islands).

73. *Eleutherobia grayi* (Thomson & Dean, 1931)

Identification: Colonies of this striking soft coral are elongate and finger-shaped, often curved and tapering toward the tip, up to 45 mm in length. Polyps are of one kind and are present on the entire surface of the colony. Orange or red, often with yellow polyp bases. About 14 species of *Eleutherobia* are known from the Indo-West Pacific. Verseveldt & Bayer (1988) provide a key.

Natural History: Encountered on rubble. Depth: 30-73 m. Fed upon by the nudibranch *Dermatobranchus gonatophora* (#617).

Distribution: Indonesia to Ryukyu Islands (Okinawa).

74. *Eleutherobia* sp.

Identification: Globular, bilobed, or finger-shaped soft corals, often somewhat curved, usually 10-60 mm in length. Expanded polyps are tubular and elongate, translucent or milky white. Retracted polyps form low, rounded mounds or flat eight-parted discs on the surface of the coral. Color is brick-red or deep reddish-purple, sometimes mottled with orange or yellow.

Natural History: Inhabits the ceilings of caves and overhangs, often forming dense populations. At night, the expanded polyps are strikingly bioluminescent. Epizoites such as small tunicates are often found growing on its surface. Depth: 10-20 m.

Distribution: Philippines and Solomon Islands (Mborokua, Solomon Islands).

75. *Minabea aldersladei* Williams, 1992

Identification: Finger-shaped colonies 10-50 mm in length. The polyps are of two kinds and cover most of the surface of each colony; they retract completely into the colony and do not form mounds. The larger polyps are extended to feed at night, usually tightly retracted during daylight. Uniform red-orange or yellow. See Williams (1992) for a description and key.

Natural History: Common on the ceilings or walls of caves and alcoves, or on barrier walls. Depth: 2-25 m.

Distribution: Western Pacific: New Guinea; Australia; Malaysian Borneo; Belau; Philippines; Solomon Islands (Belau and Madang, Papua New Guinea).

Roger Hess   Marc Chamberlain

76. *Minabea acronocephala* Williams, 1992
Button Soft Coral

Identification: Colonies are dome-shaped, less than 15 mm long. The polyps are restricted to the very tip of the colony. Color can be yellow, orange, or red, or some combination of these. Williams (1992) provides a full description and key.

Natural History: Often found near *Minabea aldersladei* in caves and vertical walls. Depth: 5-21 m.

Distribution: Fiji; New Guinea; Solomon Islands (Madang, Papua New Guinea).

G.C. Williams

77. *Lobophytum* sp. 1   Ridged Leather Coral

Identification: This species is often encrusting or low and spreading, circular or irregular in shape up to a meter or more in diameter, with the upper surface showing radiating or parallel ridges and crests. The folds or ridges at the margins are closed and fused. Two kinds of polyps are the larger tubular feeding polyps with tentacles, and smaller, crowded mound-like polyps without tentacles. Golden-brown to grey or greenish-brown. Of the 47 species of the genus *Lobophytum*, many are conspicuously ridged, at least in mature colonies. Verseveldt (1983) provides a revision of the genus.

Natural History: Common on reef flats. Depth: 5-30 m.

Distribution: Melanesia. This growth form is found from East Africa; Red Sea; Micronesia and the Tuamotu Archipelago (New Georgia Group, Solomon Islands).

G.C. Williams
G.C. Williams

78. *Lobophytum* sp. 2   Ridged Leather Coral

Identification: In this species, the ridges or plications may be tall and wall-like. The colonies are often roundish in outline. Although common constituents of shallow-water reef flats, species of *Lobophytum* are usually not as commonly encountered as those of the other leather coral genera *Sarcophyton* and *Sinularia*. Younger colonies of many species of *Lobophytum*, as well as mature colonies of some species, may exhibit finger-like lobes rather than distinct ridges or crests. Some larger colonies may have both kinds of processes.

Natural History: The coloration of these soft corals results from the symbiotic algal cells living in the tissues of the coral.

Distribution: Melanesia (New Georgia Group, Solomon Islands).

G.C. Williams

G.C. Williams

G.C. Williams
G.C. Williams

79. *Lobophytum* sp. 3
Lobed Leather Coral

Identification: Colonies usually low and spreading or encrusting. Often dish-shaped or bowl-shaped, or upright with a distinct stalk. The upper surface is mostly lobed, the lobes being rounded or finger-like. The lobes on the edge of the colony are formed by folds of the margin that are closed and fused. As in *Sarcophyton* two kinds of polyps are evident. The polyps can retract completely into the body of the coral.

Natural History: Intertidal to 30 m in depth, on reef flats.

Distribution: Melanesia. Species of this growth form are widespread in the Indo-Pacific from East Africa; Red Sea to Ryukyu, Marshall, and Tuamotu Archipelagos (New Georgia Group, Solomon Islands).

80. *Lobophytum* sp. 4
Lobed Leather Coral

Identification: This species has numerous, closely-set, robust, finger-like lobes. Color yellowish brown.

Natural History: Frequently encountered, but not as common as the large ridged forms of the genus.

Distribution: Melanesia (Madang, Papua New Guinea).

81. *Sarcophyton* sp. 1
Mushroom Leather Coral

Identification: Mushroom-, funnel-, or cup-shaped colones up to 0.5 m or more in diameter, often with a folded margin, in which the folds are open and do not fuse. The polyp-bearing capitulum arises from a smooth basal stalk. Two types of polyps are present: larger tubular autozooids and minute siphonozooids, which are crowded on the colony surface between the larger polyps. When the larger polyps are retracted, the colony takes on a very smooth, leathery appearance.

Natural History: Fed on by the nudibranch *Phyllodesmium magnum* (#630), the ovulid snails *Calpurnus lacteus* (#464) and *Ovula ovum* (#476). The species shown here is growing on top of a coral head of *Porites*. Intertidal to at least 30 m in depth.

Distribution: Melanesia. The genus occurs from East Africa; Red Sea to Micronesia; New Caledonia; Japan (New Georgia Group, Solomon Islands).

82. *Sarcophyton* sp. 2
Mushroom Leather Coral

Identification: The polyp-bearing upper portion is spreading and flattened, and deeply scalloped along the margin. Color yellowish-green. At least 36 species of the genus *Sarcophyton* are common members of shallow-water reef flat communities, of which the most common and widespread include *S. ehrenbergi*, *S. glaucum*, and *S. trocheliophorum*. Verseveldt (1982) provides a revision for the genus.

Natural History: This species reproduces extensively by asexual reproduction by cloning via budding or fragmentation.

Distribution: Melanesia (New Georgia Group, Solomon Islands).

83. *Sarcophyton* sp. 3
Mushroom Leather Coral

Identification: This small mushroom-shaped colony has a rounded capitulum. The smaller siphonozooids are clearly seen on the surface of the capitulum below the larger feeding polyps. Color white or greyish-white.

Natural History: The polyps seen here are fully extended in feeding mode.

Distribution: Western Pacific (Flores, Indonesia).

Ben Tetzner

84. *Sarcophyton* sp. 4
Mushroom Leather Coral

Identification: This species has the capitulum deeply folded. The colony in the foreground is tightly contracted with the polyps totally withdrawn, while the colonies in the background are more extended with the polyp tentacles contracted.

Natural History: Many soft corals in the genus *Sarcophyton* have recently been found to produce toxic terpenoid compounds.

Distribution: Melanesia (Madang, Papua New Guinea).

G.C. Williams

85. *Sarcophyton* sp. 5
Mushroom Leather Coral

Identification: This species is erect with a well-developed white stalk. The greyish capitulum is strongly folded along the margin, and ranges up to approximately 25 cm in diameter.

Natural History: A locally common and a conspicuous species in open areas of reef flats and slopes.

Distribution: Melanesia (New Georgia Group, Solomon Islands).

G.C. Williams

Nicholas Galluzzi

86. *Sinularia dura* (Pratt, 1903)
Flat Leather Coral

Identification: Spreading and plate-like or upright and funnel-shaped, in which the central part of the polyp-bearing portion of the colony is free of lobes or ridges and is thus relatively plain or featureless. The margin can be thin and undulating or variously lobed. The polyps can be very sparse or densely set. Alderslade and Baxter (1987) list 8 other species that share this unusual growth form.

Natural History: Found on shallow reef slopes.

Distribution: Maldives to Micronesia and Melanesia (Sipadan, Borneo).

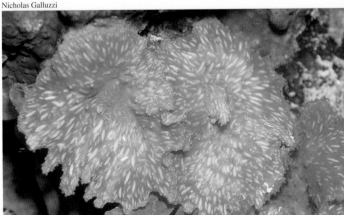

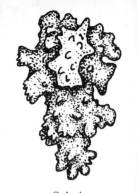

Sclerite

Mike Miller

87. *Sinularia flexibilis* (Quoy & Gaimard, 1833) Slimy Leather Coral

Identification: Distinguished by the very long, slender, tube-like lobes, which can be extremely slimy and flabby, often seen undulating in areas of current or surge. Colony color is white, greyish, or tan to brownish. This species is similar in appearance to *Sinularia procera* Verseveldt, 1977 and *S. sandensis* Verseveldt, 1977. The three species can only be distinguished positively by comparing sclerites.

Natural History: Common on flats. Depth: 3-15 m.

Distribution: Australia; New Guinea; Philippines; Indonesia; Belau; New Caledonia; Fiji; Samoa; Vietnam; southern Japan (Philippines).

G.C. Williams

88. *Sinularia* sp. 1   Digitate Leather Coral

Identification: These soft corals can be low, flat, and encrusting with an inconspicuous stalk, or taller and erect with well-developed stalks. The upper surface has many finger-like lobes that contain the numerous polyps. The polyps are all of one type and of equal size. The colonies may be soft and flaccid or have a tough leathery texture. Colors are mostly tan to brownish, cream to yellowish green, or greyish white. Of the more than 110 species of the genus *Sinularia*, many share this distinctive digitate growth form. See Verseveldt (1980) for a revision of the genus.

Natural History: Intertidal to at least 25 m in depth.

Distribution: Melanesia (New Georgia Group, Solomon Islands).

89. *Sinularia* sp. 2   Digitate Leather Coral

Identification: In this species, the fleshy colonies are composed of numerous elongated finger-like lobes that may be bifurcated or multiply-branched. Usually light brown with yellowish polyps.

Natural History: Locally common. Chemical products such as various cytotoxic diterpenes have recently been isolated from species of *Sinularia*.

Distribution: Melanesia (Njapuana, Solomon Islands).

G.C. Williams
G.C. Williams

90. *Sinularia* sp. 3   Knobby Leather Coral

Identification: A generally broad stalk gives rise to an upper portion covered with short lobes or rounded knob-like processes. The polyps are of one kind and are densely-set on the lobes. Many species of *Sinularia* share this growth form, having low rounded knobs or short lobes. Species of *Sinularia* can only be positively identified by microscopic examination of the skeletal elements (sclerites).

Natural History: Intertidal to 25 m in depth, on reef flats.

Distribution: Melanesia. Species of this growth form are found from the Red Sea and Madagascar to the Society Islands (New Georgia Group, Solomon Islands).

91. *Cladiella australis* (Macfadyen, 1936)

Identification: Colonies are generally thick and encrusting with a lumpy appearance. The upper surface is bumpy or has many round or finger-like lobes. The retracted polyps are only of one kind and form conspicuous spots darker on the light colored colony. Color of the colonies is white with chocolate brown or greenish brown polyps.

Natural History: Single colonies, 10-30 cm in diameter, are occasionally encountered on reef flats or ledges. Depth: 10-20 m.

Distribution: South and East Africa; Andaman Islands; Nicobar Islands; Australia (Natal, South Africa).

T. M. Gosliner

92. *Cladiella* sp. 1

Identification: Colonies can be somewhat encrusting or more commonly arise from a broad and short stalk with a lumpy or knobby upper surface. The many small polyps are all of one size and can quickly withdraw into the body of the coral when disturbed. Colonies with retracted polyps are usually bright white, while the polyps are often cream, light green, or brown in color. Encrusting colonies of *Cladiella* can superficially resemble the membranous xeniid soft coral *Sympodium coeruleum* (#136).

Natural History: Intertidal to 30 m in depth.

Distribution: Melanesia. The genus is distributed from the Red Sea and East Africa to the Solomon Islands and Tahiti (New Georgia Group, Solomon Islands).

G.C. Williams

93. *Cladiella* sp. 2

Identification: Several knobby lobes emanate from the base of this soft coral, approximately 10 cm in diameter. Color is greyish white with brownish polyps. At least 40 species of *Cladiella* occur throughout much of the Indo-West Pacific. Commonly occurring western Pacific species include *C. pachyclados*, *C. sphaerophora*, and *C. digitulata*.

Natural History: Locally common on ledges and slopes, often inconspicuous in shady or protected areas.

Distribution: Philippines (Batangas, Luzon, Philippines).

G.C. Williams
G.C. Williams

94. *Capnella imbricata*
(Quoy & Gaimard, 1833)

Identification: Colonies are compact in appearance with several short branches that arise from the top of a basal stalk. The crowded polyps do not retract into the branches and often appear to partly overlap other polyps. The polyps are restricted to the small terminal branches, and are smooth without supporting bundles of sclerites. Color is cream to tan or brown. Verseveldt (1977: 184) provides a key to the 17 species, which are distributed from Africa to the western Pacific.

Natural History: A locally common inhabitant of fringing reef flats and slopes. Depth: 5-20 m.

Distribution: New Guinea and Philippines (Batangas, Luzon, Philippines).

39

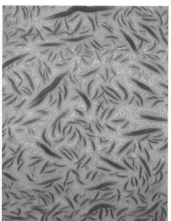

G.C. Williams

Ben Tetzner

Bruce Watkins

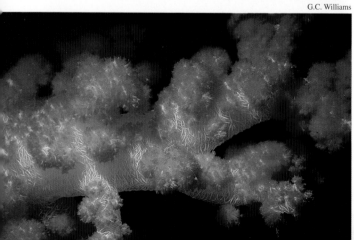

G.C. Williams
G.C. Williams

95. *Dendronephthya* (*Roxasia*) sp. 1
Divaricate Tree Coral

Identification: Bushy or tree-like soft corals that are profusely branched with long and slender branches. The colonies have a prickly appearance due to sharp supporting bundles of sclerites on each polyp. Adjacent polyps are united into distinct bundles that are conspicuously separated, and each polyp is incapable of totally withdrawing into the branches of the coral. Vividly colored–yellow and pale orange. *Dendronephthya* is a large genus of at least 250 described species and is in desperate need of revision–species identification at present is therefore extremely difficult, or virtually impossible.

Natural History: Common inhabitants of Indo-Pacific reefs. Intertidal to at least 50 m in depth.

Distribution: Melanesia. The genus ranges throughout the Indo-West Pacific, East Africa and the Red Sea to Japan, Philippines, Indonesia, New Guinea, Australia, Micronesia and parts of Polynesia (New Georgia Group, Solomon Islands).

96. *Dendronephthya* (*Roxasia*) sp. 2
Divaricate Tree Coral

Identification: This species is superficially similar to the previous one, but has white branches with red to magenta polyp clusters.

Natural History: Locally common, often growing upright from horizontal surfaces such as rocky shelves or ledges.

Distribution: New Guinea (Milne Bay, Papua New Guinea; Indonesia).

97. *Dendronephthya* (*Spongodes*) sp. 1
Glomerate Tree Coral

Identification: Bushy and prickly colonies sparsely branched with many bundles of polyps crowded together forming rounded bunches.

Natural History: The genus *Dendronephthya* is a favorite subject of underwater photographers.

Distribution: Melanesia. The genus is distributed throughout the Indo-West Pacific (Florida Group, Solomon Islands).

98. *Dendronephthya* (*Spongodes*) sp. 2
Glomerate Tree Coral

Identification: Large robust colonies with ivory white branches and vivid red clusters of spiny polyps. The rounded clumps of polyps are often ball-like and vary considerably in size.

Natural History: A magnificent species attaining a height of almost one meter.

Distribution: Melanesia (Madang, Papua New Guinea).

99. *Dendronephthya (Morchellana)* sp. 1
Umbellate Tree Coral

Identification: Tree-like prickly colonies with bundles of polyps grouped together forming corymb-like or umbrella-like aggregations. The polyps are restricted to the outer surface of the colony. The genus *Dendronephthya* has some of the most variably and brightly colored species of any reef soft corals.

Natural History: A common inhabitant of reefs and adjacent areas.

Distribution: Southeastern Africa (Natal, South Africa).

Sclerite

G.C. Williams

100. *Dendronephthya (Morchellana)* sp. 2
Umbellate Tree Coral

Identification: A tricolored species with white branches, magenta or white sclerites, and yellowish polyps; ranging in height up to about 0.5 m. The polyp clusters form numerous umbrella-like groupings. The large magenta sclerites are mostly restricted to the extremities of the branches.

Natural History: An infrequently encountered species.

Distribution: Philippines (Batangas, Luzon, Philippines).

G.C. Williams

101. *Lemnalia cervicornis* (May, 1898)

Identification: Slender tree-like soft corals with long thin branches that arise from a tall and delicate-looking stalk. The small globular to cylindrical polyps are loosely scattered on the branchlets and twigs. They are smooth and cannot withdraw completely into the body of the coral. Color cream or tan. These are slender and delicate-looking soft corals.

Natural History: On reef flats and slopes. Depth: 5-20 m.

Distribution: East Africa to Australia; Philippines; Indonesia; New Guinea; southern Japan (Bohol, Philippines).

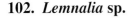
Sclerite

G.C. Williams

102. *Lemnalia* sp.
Cauliflower Soft Coral

Identification: Cauliflower-like, copiously branched soft corals with well-defined elongate stalks. The small globular polyps are scattered on the distal branches and twigs, and most polyps are without supporting bundles. Usually cream to tan. Numerous species of three superficially similar genera (*Lemnalia*, *Litophyton*, and *Nephthea*) inhabit Indo-Pacific reefs.

Natural History: Common and conspicuous on reef flats.

Distribution: Melanesia. The genus ranges throughout much of the Indo-Pacific from East Africa to parts of Polynesia (Mborokua Island, Solomon Islands).

G.C. Williams

41

G.C. Williams

G.C. Williams

103. *Litophyton* sp.

Identification: Colonies are elongate and tree-like with numerous slender branches arising form a tall stalk. The polyps cannot withdraw into the branches and are restricted to small terminal twigs and branches. The polyps are mostly smooth, usually without a supporting bundle of sclerites. Color is usually cream. This species, along with others in the genus, are very difficult to distinguish from the genus *Nephthea*.

Natural History: A common and often very abundant member of flats and slopes of fringing and patch reefs. Depth: 5-20 m.

Distribution: Philippines. The genus ranges throughout much of the Indo-Pacific; Philippines; New Guinea; Solomon Islands (Batangas, Luzon, Philippines).

104. *Nephthea* sp.

Identification: Tree-like soft corals with numerous often slender branches arising from a tall stalk. Some polyps may have supporting bundles of sclerites and are incapable of withdrawing into the body of the coral. The polyps are contained in short lobes or catkins on the branchlets of the abundantly branched colonies. Cream with tan or brown polyps. Roxas (1933b: 408) provides a key to 21 Philippine species. This genus is impossible to distinguish underwater from *Litophyton* and certain species of *Capnella*.

Natural History: Intertidal to 30 m in depth, on reef flats or slopes.

Distribution: Philippines. The genus ranges from eastern Africa to southern Japan, Philippines, Indonesia, New Guinea, Australia, Solomon Islands (Batangas, Luzon, Philippines).

G.C. Williams
G.C. Williams

105. *Paralemnalia* cf. *clavata* Verseveldt, 1969

Identification: Somewhat similar to *Paralemnalia thysoides* but with thicker more robust lobes emanating from a thickly encrusting and spreading base. The polyps are restricted to the end portions of the lobes and are cylindrical or tubular, and presumably do not completely withdraw into the lobes. Color white, cream, grey, tan or brown. A locally abundant species forming spreading mats up to 20 cm in diameter with numerous finger-like lobes arising from the base.

Natural History: Commonly encountered on horizontal ledges and shelves. Depth: 5-25 m.

Distribution: Madagascar and Melanesia (Mborokua, Solomon Islands).

106. *Paralemnalia thyrsoides* (Ehrenberg, 1834)

Identification: This soft coral produces numerous and elongate, slender, fingerlike lobes that arise from a common base. The polyps are capable of withdrawing completely into the lobes and are scattered on the distal portions of these lobes. Tan or grey.

Natural History: Locally common on shallow-water fringing reefs, on reef flats and slopes. Depth: 3-30 m.

Distribution: Australia; Philippines; New Guinea; Solomon Islands; southern Japan (Mborokua, Solomon Islands).

107. *Paralemnalia* cf. *thyrsoides*
(Ehrenberg, 1834)

Identification: This species is similar to *Paralemnalia thyrsoides* but has thicker, more robust branches. The numerous polyps, as seen here, are totally retractable into the lobe-like branches, and are restricted to the terminal portions of the branches. Color is uniform light brown.
Natural History: Locally common on reef flats and slopes up to 30 m in depth.
Distribution: Melanesia (New Georgia Group, Solomon Islands).

G.C. Williams

108. *Scleronephthya* sp. 1

Identification: Contracted colonies have a fleshy and lobular appearance due to the presence of a few thick branches that arise from a very short and inconspicuous stalk. The polyps are not armed with supporting bundles of spicules and are scattered on the twigs and branches. Boldly and variably colored, this species varies from orange to pink.
Natural History: Often encountered on ledges, overhangs, or vertical walls. Depth: 10-30 m.
Distribution: Philippines. The genus ranges in the Indo-West Pacific from the Red Sea to New Guinea, Philippines, Solomon Islands and Micronesia (Sabang, Mindoro, Philippines).

G.C. Williams

109. *Scleronephthya* sp. 2

Identification: A handsome lobular species of soft coral with pale yellow branches and bright orange polyp clusters.
Natural History: Infrequently encountered on ledges and slopes.
Distribution: Melanesia (New Georgia Group, Solomon Islands).

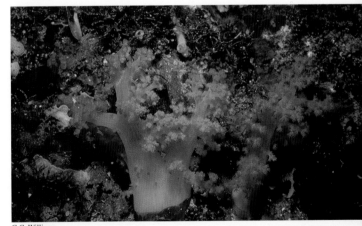

G.C. Williams
G.C. Williams

110. *Scleronephthya* sp. 3

Identification: A flesh-colored species with apricot-colored polyp centers. A robust and upright soft coral, copiously branched, but with thick main branches.
Natural History: Locally common on vertical surfaces in areas with appreciable current. This species is commonly preyed upon by the solenogaster, *Epimenia australis* (#418) (Scheltema & Jebb, 1994).
Distribution: Melanesia (Madang, Papua New Guinea).

43

G.C. Williams

G.C. Williams

Robert Bolland
Marc Chamberlain

G.C. Williams

111. *Scleronephthya* sp. 4

Identification: Some colonies when fully expanded are not lobular, but appear so only upon contraction. It is yellow orange with oral regions of the polyps orange to red.

Natural History: Locally common on vertical surfaces such as walls or ceilings of alcoves.

Distribution: Melanesia (New Georgia Group, Solomon Islands).

112. *Stereonephthya* sp.

Identification: This soft coral forms stiff spiny colonies usually less than 10 cm in height. The polyps are armed with supporting bundles of spicules. They are scattered along the branchlets of the sparsely divided colonies, and do not appear in bundles. Colonies are usually not brightly colored–mostly tan, grey, or creme to white. Numerous species have been described.

Natural History: Occasionally encountered on reef flats, ledges, or walls. Depth: 15-20 m.

Distribution: Melanesia. The genus is distributed from Zanzibar to New Guinea, Philippines, Solomon Islands, Belau, and Australia (New Georgia Group, Solomon Islands).

113. Nephtheid sp.
Slender-Stalked Soft Coral

Identification: An unusual soft coral resembling in general appearance the deep-water sea pen *Umbellula*. A slender elongated stalk gives rise to a flower-like cluster of soft branches, which radiate out from the end of the stalk, resembling an umbel. The polyps are sparsely distributed on the upper surface of the branches. Color of stalk is light red while the umbel is salmon orange, often with a greenish sheen; sclerites and polyps are white.

Natural History: A recently discovered species infrequently encountered in deeper water near coral reefs; 52-59 m in depth.

Distribution: Ryukyu Islands (Okinawa).

114. *Nidalia simpsoni*
(Thomson & Dean, 1931)
Dandelion Coral or Daisy Coral

Identification: Mushroom-shaped soft corals composed of a hemispherical capitulum (10-30 mm in diameter), atop a stalk 10-60 mm in length. The polyps retract into permanent cup-like calyces on top of the capitulum. Color varies from pale orange to rust brown or reddish brown. Verseveldt and Bayer (1988) provide a key to the twelve described species of *Nidalia*. Faulkner and Chesher (1979, as *N. lampas*) provide an illustrated description of this species. This species is similar in appearance to *N. occidentalis* from the West Indies.

Natural History: This species frequently inhabits caves and alcoves of vertical walls in areas of subdued light, and is less frequently found on the outer face of such walls. Depth: 6-24 m. The polyps are seen expanded in feeding mode at night.

Distribution: Indonesia; Belau; Philippines; Solomon Islands; New Guinea (Solomon Islands).

115. *Siphonogorgia godeffroyi* Kölliker, 1874

Identification: Soft corals with a gorgonian-like appearance. The polyps are clustered at the ends of the terminal branchlets and are absent from the trunk and main branches. They are capable of completely withdrawing into the branches. Colonies are bicolored: vivid red or wine-red with white to yellow polyps.
Natural History: A frequently encountered inhabitant of reef slopes, vertical surfaces, and overhangs. Depth: 10-20 m.
Distribution: Belau; Philippines and Solomon Islands (Mindoro, Philippines).

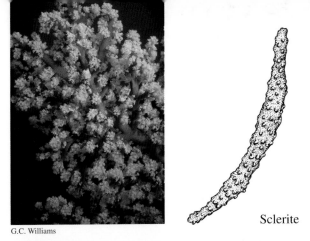
G.C. Williams

Sclerite

116. *Siphonogorgia* cf. *godeffroyi* Kölliker, 1874

Identification: This soft coral species is superficially similar to and may be synonymous with *Siphonogorgia godeffroyi*, but differs in its uniform pale orange coloration.
Natural History: This species forms magnificent fan-like colonies up to almost a meter in height, on vertical surfaces in deep reef regions.
Distribution: Melanesia (New Georgia Group, Solomon Islands).

G.C. Williams

117. *Siphonogorgia* sp.

Identification: Similar to *Siphonogorgia godeffroyi* except that the polyps are found on the main branches as well as the terminal branches. Color orange with white or cream polyps.
Natural History: Commonly found on ledges, walls, and reef slopes. Depth: 10-20 m.
Distribution: Australia; Philippines; New Guinea; Solomon Islands (Madang, Papua New Guinea).

G.C. Williams
G.C. Williams

118. *Chironephthya* cf. *macrospiculata* Thomson & Henderson, 1906

Identification: Striking soft corals with a few thick branches arising from an erect basal stalk. The drooping branches are thickly-set with very large visible sclerites. Polyps form distinct spiny calyces. Color vivid red with white sclerites.
Natural History: Occasionally encountered on ledges, overhangs, and vertical walls. Depth: 6-20 m.
Distribution: India and Chagos Archipelago to New Guinea and Australia (Madang, Papua New Guinea).

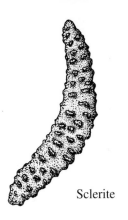
Sclerite

45

G.C. Williams

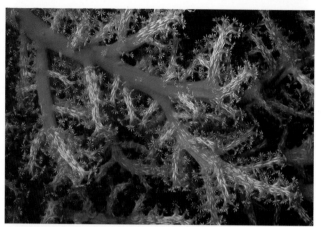

Bruce Watkins

G.C. Williams
G.C. Williams

119. *Chironephthya* sp. 1

Identification: These are gorgonian-like soft corals but, unlike the sea fans, there is no internal axis running throughout the colony. They are similar to species of *Siphonogorgia*, but differ in that each polyp retracts to form a distinct calyx. Polyps are most numerous toward the terminal portions of the branches. The terminal branches are often thick and taper little. This species is upright and bushy, pale yellow in color with pinkish polyps. Numerous species of *Siphonogorgia* have been described from the Indo-Pacific–from deeper water as well as coral reefs. Many of these are probably assignable to the related genera *Chironephthya* or *Nephthyigorgia*.

Natural History: On slopes and walls. Depth: 10-15 m.

Distribution: Melanesia. The genus extends from the East Africa to southern Japan, Philippines, New Guinea, Indonesia, Solomon Islands, Australia, and Micronesia (New Georgia Group, Solomon Islands).

120. *Chironephthya* sp. 2

Identification: This species is copiously-branched and tree-like with wine-red main branches and yellow terminal branches due to the presence of large yellowish spindle-like sclerites.

Natural History: Often forming large delicate colonies in the deeper, more protected regions of coral reefs.

Distribution: Melanesia. The genus is distributed throughout much of the Indo-West Pacific (Fiji).

121. *Chironephthya* sp. 3

Identification: Some species of *Chironephthya*, such as this one, are more drooping and less upright.

Natural History: Infrequently encountered in protected areas, such as vertical surfaces or alcoves, in deeper reef areas or margin of reefs. A xanthid crab (*Quadrella* sp.) is found living in pairs on colonies of *Chironephthya* species.

Distribution: Melanesia (New Georgia Group, Solomon Islands).

122. *Anthelia flava* (May, 1899)

Identification: The striking electric blue color of this soft coral is diagnostic. The individual polyps are crowded and arise from a common membranous base.

Natural History: This species often forms flat carpet-like sheets on rocks or dead coral fragments in shallow water. Intertidal to 18 m in depth.

Distribution: Zanzibar; South Africa; Mauritius; Philippines (Natal, South Africa).

123. *Anthelia glauca* Lamarck, 1816

Identification: This soft coral is comprised of many polyps arising from a common membranous base. The slender tentacles of the polyps have numerous thin pinnules. Color slate grey to cream with golden brown or yellowish tentacles. This is an extremely variable species throughout its range.
Natural History: Mostly on reef flats. Depth: 6-20 m.
Distribution: South Africa; Red Sea to Indonesia and Philippines (Batangas, Luzon, Philippines).

T. M. Gosliner

124. *Anthelia* sp.

Identification: Robust polyps arise from a common membranous base. Pinnules are stubby and widely separated. Color is a uniform chalk white to very pale grey. At least 22 species of *Anthelia* are recorded in the Indo-Pacific from eastern Africa and the Red Sea to Hawai'i. Roxas (1933a: 62) provides a key to most of these.
Natural History: A tritoniid nudibranch feeds on this species, and mimics the host soft coral remarkably well. Depth: 6-15m.
Distribution: Philippines and New Guinea (Madang, Papua New Guinea).

G.C. Williams

125. *Xenia* cf. *actuosa* Verseveldt & Tursch, 1979

Identification: Colonies are large with many elongate tubular polyps. The colonies are a lustrous white with brown pinnules.
Natural History: The polyps characteristically pulsate rapidly– the tentacles of a particular polyp repeatedly close and open in unison during the day time. Depth: 5-20 m.
Distribution: Philippines and New Guinea (Batangas, Luzon, Philippines).

T. M. Gosliner

G.C. Williams

126. *Xenia* cf. *mucosa* Verseveldt & Tursch, 1979

Identification: Very slimy soft corals with thick stalks that may branch to form 2-3 trunks. The numerous polyps are contained on the capitulum that terminates the stalk, and are incapable of withdrawing into the body of the coral. Color is lustrous white with brown polyps.
Natural History: This species gives off copious amounts of mucus when handled. Depth: 5-20 m.
Distribution: Philippines and New Guinea (Batangas, Luzon, Philippines).

G.C. Williams

127. *Xenia* sp. 1

Identification: The polyps are restricted to a well-defined head-shaped swelling (capitulum) at the end of a stalk. The stalk may be branched at the base. The polyps cannot retract into the capitulum. The tentacles of each polyp have several rows of pinnules along each side. In this species the stalks are smooth and white and the polyps are mottled. Over forty species of *Xenia* occur throughout much of the Indo-Pacific. Roxas (1933a: 78) provides a key to the Philippine species.

Natural History: Nudibranchs of the genus *Phyllodesmium* feed on *Xenia.* Intertidal to at least 15 m in depth.

Distribution: Philippines. The genus ranges from the Red Sea and East Africa to southern Japan, Australia, Philippines, New Guinea, Micronesia, and southern Polynesia (Batangas, Luzon, Philippines).

128. *Xenia* sp. 2

Identification: A *Xenia* species with a relatively short stalk, large feathery tentacles, and cream to white coloration.

Natural History: Frequently encountered on reef flats. The pinnules contain zooxanthellae and expand fully for maximum exposure to light.

Distribution: Melanesia (New Georgia Group, Solomon Islands).

G.C. Williams

129. *Xenia* sp. 3

Identification: The stalks of this species are relatively elongate and narrow, white in color with bluish tubular polyps. Roxas (1933a: 78) records 22 species of *Xenia* from the Philippines.

Natural History: Commonly encountered on Philippine coral reefs.

Distribution: Philippines (Batangas, Luzon, Philippines).

G.C. Williams

130. *Heteroxenia* sp. 1

Identification: Virtually identical to *Xenia*, except that species of *Heteroxenia* have two kinds of polyps (at least during breeding periods): in addition to the large feeding polyps (autozooids), smaller polyps (siphonozooids) occupy the surface of the capitulum between the bases of the autozooids. This particular species is relatively robust and greyish-white in color. Approximately ten Indo-Pacific species. Roxas (1933a: 96) provides a key to the Philippine species.

Natural History: Intertidal to at least 15 m in depth.

Distribution: Melanesia. The genus ranges from the Red Sea and South Africa to the Philippines, New Guinea, Australia, Micronesia, and southern Polynesia (New Georgia Group, Solomon Islands).

131. *Heteroxenia* sp. 2

Identification: This species has robust club-like stalks that are rust-orange in color and the polyps are pale greenish.
Natural History: The nudibranch *Phyllodesmium kabiranum* (#628) feeds on this species.
Distribution: Philippines (Batangas, Luzon, Philippines).

T. M. Gosliner

132. *Heteroxenia* sp. 3

Identification: Large robust colonies with conspicuous stalks, cream-white in color with brownish pinnules.
Natural History: Symbiotic algae (zooxanthellae) inhabit the tissues of the pinnules and produce food for the soft coral through photosynthesis.
Distribution: Melanesia (New Georgia Group, Solomon Islands).

G.C. Williams

133. *Cespitularia* sp. 1

Identification: These are fleshy, branching, tree-like soft corals in which the polyps do not retract into the body of the coral. Branching may occur on any part of the colony and is not restricted to the base. Polyps occur mostly on the branches. This species is strikingly colored tan with bright green polyps. See Roxas (1933a: 104) for a key to the species. Fifteen species have been described, and three have of these are recorded from the Philippines–*C. coerulea* May, 1898; *C. hypotentaculata* Roxas, 1933; and *C. quadriserta* Roxas, 1933.
Natural History: Often very slimy to the touch due to the production of copious amounts of mucus. Usually 5-15 m in depth.
Distribution: Philippines. The genus ranges from Red Sea and Madagascar to Indonesia, Philippines, Australia (Batangas, Luzon, Philippines).

G.C. Williams

G.C. Williams

134. *Cespitularia* sp. 2

Identification: This species is cream-white with blue tentacles, sometimes with a greenish sheen on the branches. Colonies are often highly branched.
Natural History: Encountered on protected reef flats. This species as well as others in the genus, can be locally very abundant in warm shallow areas.
Distribtution: Philippines (Batangas, Luzon, Philippines).

49

G.C. Williams

135. *Cespitularia* sp. 3

Identification: A large fleshy species with a smooth slimy surface. Color is pale blue or mottled blue-violet and white with brownish polyps.

Natural History: This species, along with others, can dominate extensive areas of reef flats in parts of the Philippines, forming dense, soft coral "forests."

Distribution: Philippines (Batangas, Luzon, Philippines).

G.C. Williams

136. *Sympodium coeruleum* Ehrenberg, 1834

Identification: The colonies appear as spreading membranous sheets on hard surfaces. The numerous polyps are capable of completely withdrawing into the basal sheet. Color white to pale blue, with brownish polyps.

Natural History: This unusual soft coral forms thin irregularly-shaped sheets that are either smooth or have a distinctly lumpy appearance. Depth: 4-18 m.

Distribution: South Africa; Red Sea to Australia; Philippines; New Guinea; Solomon Islands (Batangas, Luzon, Philippines).

G.C. Williams

137. *Briareum* cf. *stechei* (Kükenthal, 1908)

Identification: Thick encrusting mats give rise to crowded polyps with slender grass green tentacles. Polyp calyces less than 5 mm in height. The encrusting mat is tan to violet in color.

Natural History: Often covering extensive horizontal or vertical surfaces, up to one square meter or more in area. Common on ledges and slopes. Depth: 5-20 m.

Distribution: Belau and Philippines (Bohol, Philippines).

G.C. Williams

G.C. Williams

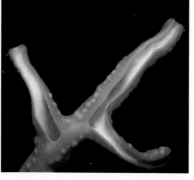

138. *Semperina* cf. *rubra* Kölliker, 1870

Identification: Sea fans intricately-branched in one plane, up to one-half meter in height. Many of the flattened or spatulate terminal branches have furrowed or fistulose tips. Color deep orange or red to purple or purplish-brown, with white fistulose branch tips. Kükenthal (1924: 20) provides a key to the 5 species that have been reported to occur in the western Pacific.

Natural History: On slopes and deeper portions of reefs. Depth: 20-30 m.

Distribution: Originally described from the Philippines but presumably inhabiting a broader range in the western Pacific (Batangas, Luzon, Philippines).

50

139. *Semperina* sp.

Identification: Colonies of this sea fan are upright, intricately-branched, and planar. The branch tips have white fistulose furrows. The polyps are mostly biserial. Colony color is often deep reddish-brown. Five species of *Semperina* have been described: *S. rubra, S. australis, S. brunnea, S. koellikeri, S. macrocalyx.*

Natural History: On slopes and deeper regions of reefs; 20-30 m in depth.

Distribution: Melanesia. The genus is distributed in the western Pacific from northwestern Australia and north of New Zealand to Indonesia, Philippines, New Guinea, and Solomon Islands (New Georgia Group, Solomon Islands).

G.C. Williams

140. *Subergorgia appressa* Nutting, 1911

Identification: Sea fans with lateral branching. A medial longitudinal groove runs along the surface of the branches. The branches are somewhat flattened. The polyps are biserial. Color varies from red to reddish brown.

Natural History: Infrequently encountered in deeper regions of reef margins, mostly on slopes.

Distribution: Indonesia and Melanesia (New Georgia Group, Solomon Islands).

G.C. Williams

141. *Subergorgia mollis* (Nutting, 1910)

Identification: Colonies are fan-like and intricately branched in one plane. The branches form a netlike, reticulated pattern with rectangular meshes that are 10-60 mm in length. The distal branches are thin, usually 1-3 mm in diameter. These large sea fans may reach 2-3 meters in height. Color in life is cream to pale pinkish orange, but sometimes scarlet red due to a very thin and uniform coating of epizoic growth, presumably algal in nature. Bayer (1949) provides a nicely illustrated description of this species from Bikini Atoll.

Natural History: Often encountered on vertical walls, not on shallow water reef flats; a characteristic species of the western Pacific. Depth: 9-32 m.

Distribution: Keeling Island; Okinawa; Indonesia; Philippines; New Guinea; Solomon Islands; Marshall Islands (Solomon Islands; Philippines; Solomon Islands).

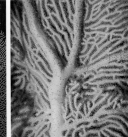

G.C. Williams

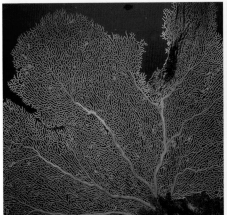

G.C. Williams
G.C. Williams

142. *Subergorgia suberosa* (Pallas, 1766)

Identification: Sea fans with dichotomous branching. A medial longitudinal groove is evident on the surface of the branches. The polyps are mostly biserial–two rows of polyps arranged opposite one another. Color rust orange or reddish brown.

Natural History: Often encountered on vertical walls, slopes, and ridges. Depth: 15-30 m.

Distribution: Mauritius to Philippines; New Guinea; and Solomon Islands (Bohol, Philippines).

G.C. Williams

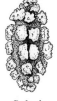

Sclerite

G.C. Williams

Sclerite

Fred McConnaughey

G.C. Williams

G.C. Williams

143. *Acabaria* sp. 1

Identification: Sea fans that are delicately-branched in one to several planes, planar or bushy, usually 8-20 cm in height. The branching points are swollen, representing flexible nodes of the internal axis. The slender branches have numerous hemispherical to short cylindrical polyp calyces, into which the polyps can withdraw. Color often vivid and variable, even within the same specimen - white, pink, yellow, orange, or red. At least 36 species have been described. Several of them are common inhabitants of Indo-Pacific coral reefs. Ofwegen (1987) provides a review of the Indian Ocean species.

Natural History: On slopes and ledges. Depth: 10-50 m.

Distribution: Philippines. The genus is distributed throughout the Indo-Pacific from the East Africa to Hawai'i (Batangas, Luzon, Philippines).

144. *Acabaria* sp. 2

Identification: A red orange sea fan intricately-branched in one plane. As in other related species, the branches are discontinuous, being composed of calcareous internodes and often swollen nodes of horn.

Natural History: Infrequently encountered on ledges and walls away from reef flats.

Distribution: Micronesia (Belau).

145. *Acabaria* sp. 3

Identification: A tall planar sea fan with biserial polyps. The colonies are bicolored–the branches are rust orange with yellow polyps. A relatively large species up to 30 cm in height.

Natural History: Infrequently encountered on reef flats.

Distribution: Philippines (Bohol, Philippines).

146. *Acabaria* sp. 4

Identification: Colonies up to 50 cm in height. A species of bushy colonies in which anastomosis (the fusing of adjacent branch tips) takes place frequently. Anastomosis may even take place between two adjacent but differently-colored colonies. Color varies from red to pale pink.

Natural History: Encountered on reef slopes, walls, and ledges.

Distribution: Philippines (Mindoro, Philippines).

52

147. *Melithaea* sp. 1

Identification: Large sea fans up to 1-2 meters in height, intricately branched in one plane with robust trunk and main stems. Swollen nodes at the branching points are noticeable throughout the colony. Color variable, sometimes bicolored–mostly yellow, orange, or red. Ofwegen (1987) provides a good review of two common Indo-West Pacific forms.

Natural History: A commonly encountered and very photogenic sea fan of vertical walls and steep slopes. Depth: 10-50 m. Preyed upon by the ovulid *Prosimnia semperi* (#484).

Distribution: Melanesia. The genus ranges throughout the Indo-West Pacific: East Africa to Japan, Singapore, Indonesia, Philippines, New Guinea, Solomon Islands, Fiji, and Australia (Solomon Islands).

Marc Chamberlain

148. *Melithaea* sp. 2

Identification: Large colorful and picturesque sea fans up to one meter or more in height. They are intricately-branched and are often bicolored with thick yellow main branches and slender orange secondary and terminal branches. At least 29 species have been described.

Natural History: *Melithaea ochracea* (Linnaeus, 1758) is perhaps the most well-known species. On slopes and vertical surfaces. Depth: 20-40 m.

Distribution: Melanesia (Vanuatu).

Terry Schuller

149. *Acalycigorgia* sp. 1

Identification: These arborescent gorgonians are readily distinguished by the many prominent and elongate polyp calyces, which are non-retractable and cylindrical to tubular in shape. Branches often sinuous. Colonies are often brightly colored: mostly scarlet red, yellow, rust orange, or deep purple. This particular species is bicolored–red branches with yellow tips. Kükenthal (1924: 237) provides a key and diagnoses to the five described species known at that time from Japan and Taiwan.

Natural History: Frequently encountered in deeper parts of reefs, below the shallow reef flats, often on vertical walls; mostly at depths greater than 10 m.

Distribution: Philippines. The genus inhabits the western Pacific: Japan, Taiwan, Belau, Philippines, Indonesia, Australia, Papua New Guinea, Solomon Islands (Batangas, Luzon, Philippines).

150. *Acalycigorgia* sp. 2

Identification: A lemon-yellow species with white tentacles. The tubular polyp calyces are up to eight mm in length. The colonies are highly branched.

Natural History: This is a beautiful species, infrequently observed on slopes and walls away from reef flats.

Distribution: Indonesia (Indonesia).

G.C. Williams

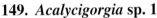

Sclerite

Bruce Watkins

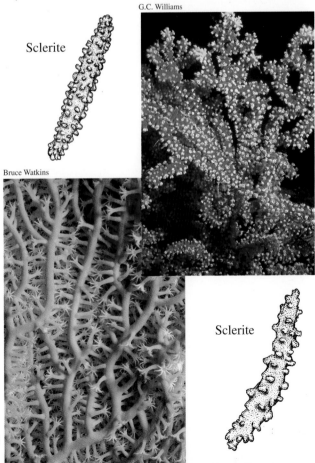

Sclerite

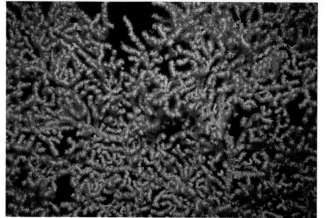

G.C. Williams

151. *Muricella* sp.

Identification: Large intricately-branched sea fans, up to over a meter in height. The numerous polyps form low rounded mounds on the branches when contracted. The scarlet-red polyps contrast markedly with the white to greyish white branches.
Natural History: Locally common on vertical walls. The plane of the fans is aligned perpendicular to the prevailing current.
Distribution: Indonesia; Philippines; New Guinea; China (Batangas, Luzon, Philippines).

G.C. Williams

152. *Echinogorgia* (*Menella*) sp.

Identification: Sea fans having elongate terminal branches, either smooth or with a distinctive bristly or prickly texture due to the presence of numerous spiny sclerites that surround the polyps. Color often red to wine red, reddish-brown, or orange. Numerous species have been described and allocated to this genus, which is in need of major revision. A few species (including the one pictured), allocated to a separate genus *Menella* by some authors, have distinctive sclerites.
Natural History: On slopes and ledges. Depth: 10-40 m.
Distribution: Philippines. The genus ranges throughout much of the Indo-West Pacific from East Africa to Indonesia, Philippines, New Guinea and Australia (Batangas, Luzon, Philippines).

Sclerite

G.C. Williams
Jack Randall

153. *Rumphella* sp.

Identification: Sea fans with relatively thick and very flexible branches that taper little and often have slightly club-shaped tips. Color grey or brownish. Several species in the Indo-Pacific; these are often encountered on or adjacent to shallow-water coral reefs.
Natural History: Unlike most other gorgonians, at least one species of *Rumphella* contains zooxanthellae and inhabits sandy depressions or gullies on reef flats, adjacent to the larger hard and soft corals that dominate this community; 4-36 m in depth.
Distribution: Philippines. The genus ranges throughout much of the Indo-West Pacific; South Africa, Australia, Philippines, Indonesia, New Guinea, Solomon Islands (Bohol, Philippines).

154. *Ctenocella* (*Ctenocella*) *pectinata*
(Pallas, 1766)

Identification: Sea fans with very distinctive pectinate or comb-like growth form. Many parallel vertical branches arise from a Y-shaped base that is formed by the basal trunk and two main branches. Color variable: red, yellow, or white.
Natural History: A very distinctive but infrequently encountered species. Depth: 20-30 m.
Distribution: Thailand; Indonesia; Australia (Phuket, Thailand).

155. *Ctenocella (Ellisella)* sp.

Identification: Brightly colored, very slender, repeatedly-branched sea fans with dichotomous branching. The terminal branches are often extremely long. The retracted polyps often form mound-like bumps on the surface of the branches, or the branches can be relatively smooth. Bright red or orange.

Natural History: A common species, often on barrier walls and other vertical surfaces. Depth: 9-32 m.

Distribution: Melanesia. The subgenus is circumtropical: western Atlantic, eastern Pacific, and Indo-West Pacific from East Africa to Philippines, Indonesia, New Guinea, Australia, and the Solomon Islands (New Georgia Group, Solomon Islands).

Sclerite

G.C. Williams

156. *Ctenocella (Viminella)* sp.

Sea Whip

Identification: Colonies are unbranched and whip-like, or with one or a few elongate whip-like branches. The polyps are often biserially arranged. Color is often orange or reddish brown. The systematics of the ellisellid genera have recently been revised by Bayer and Grasshoff (1994).

Natural History: Frequently encountered on reef slopes and deeper portions of reefs.

Distribution: Melanesia. The subgenus is distributed in tropics and warm temperate regions worldwide (Madang, Papua New Guinea).

G.C. Williams
G.C. Williams

157. *Junceella (Junceella)* sp. 1

Sea Whip

Identification: These gorgonians are mostly unbranched sea whips. The polyps often form crowded mound-like or finger-like calyces on the colony surface. Color is cream or white to tan or pale orange.

Natural History: Localized populations often found in relatively dense aggregations, in deeper portions of reef communities.

Distribution: Melanesia. The subgenus is found throughout the Indo-West Pacific from eastern Africa to the Philippines, Indonesia, New Guinea, and Australia (New Georgia Group, Solomon Islands).

G.C. Williams

158. *Junceella (Junceella)* sp. 2

Sea Whip

Identification: Mostly unbranched and whip-like up to a meter or more in length. The stem is usually smooth when preserved, as the polyps do not form pronounced calyces. Color varies from pure white to light orange or tan.

Natural History: Often locally common on barrier walls or on deeper reef flats in sandy areas, forming "undersea forests" in some areas. Intertidal to 30 m depth.

Distribution: Philippines (Bohol, Philippines).

Sclerite

Pennatulacea - Sea Pens

Sea pens are highly specialized octocorals that are well adapted for life in soft sediments such as sand or mud. They are composed of two parts–an "above-ground" rachis that contains the feeding polyps and a "below-ground" muscular peduncle that anchors the colony into the sea bottom. Most sea pens have a straight, unbranched, rod-like axis composed mainly of calcium carbonate.

Sclerite

G.C. Williams
Robert Bolland

G.C. Williams

Many sea pens are nocturnal, that is, the rachis emerges only after dusk. Primitive forms are sausage- or club-shaped with the polyps arising directly from the rachis, while the more derived forms have elaborate polyp leaves containing the numerous polyps. See Williams (1995) for a review of the pennatulaceans.

159. *Cavernularia* cf. *obesa* Valenciennes *in* Milne Edwards & Haime, 1850

Identification: Cylindrical and sausage-shaped or club-shaped sea pens. The large polyps surround the entire upper portion (the rachis) and are not contained on polyp leaves. Rachis cream-white to tan with brown, white or colorless polyps; peduncle may be differently colored–sometimes orange. Williams (1989: 308) provides a key to the thirteen valid species. The related genera *Cavernulina* and *Lituaria* have approximately 14 species and range from India to the eastern Pacific. The veretillid sea pens can only be positively differentiated by microscopic examination of sclerites.

Natural History: The species shown here is often seen in full expansion during the day on sandy flats adjacent to reefs, the brown polyps are inhabited by symbiotic algae for photosynthesis. Other species are nocturnal.

Distribution: India; New Guinea; Marquesas Islands (Madang, Papua New Guinea).

160. *Cavernularia* sp.

Identification: Sausage-shaped colonies up to 15 cm in height. The numerous polyps are radially arranged and cover the surface of the rachis.

Natural History: Like most shallow-water sea pens, this species is found on sand adjacent to reefs. Depth: to 320 m.

Distribution: Ryukyu Islands. The genus ranges from southern Europe and the Mediterranean, East Africa to Japan, China, southeast Asia, New Guinea, Australia, New Caledonia, and the Marquesas Islands (Okinawa).

161. *Veretillum* sp. 1

Identification: Flexible, elongated sausage-shaped sea pens with two kinds of polyps distributed all around the cylindrical rachis. Color varies from purplish brown to light brown–expanded polyps are colorless transparent. There are at least 7 valid species in the genus.

Natural History: As nocturnal feeders, these unusual sea pens are encountered mainly at night on sand flats or sand-filled depressions on reef flats–during the day, they are for the most part completely retracted beneath the surface of the sand. Often very common locally, with a density as high as two or three per square meter. *Porcellanella picta* (#826) lives on this sea pen.

Distribution: Philippines. The genus ranges from the southern Europe and the eastern Atlantic to East Africa, the Indian Ocean to the western Pacific (Batangas, Luzon, Philippines).

162. *Veretillum* sp. 2

Identification: A distinctly bicolored species with fewer polyps than the preceding species. Colonies are cylindrical with polyps radiating out on all sides of the rachis.
Natural History: Encountered at night on sand flats.
Distribution: Philippines (Batangas, Luzon, Philippines).

G.C. Williams

163. *Veretillum* sp. 3

Identification: A robust, bicolored species with purplish brown tentacles. The smaller polyps (siphonozooids) are numerous and appear in longitudinal rows along the surface of the rachis.
Natural History: A nocturnal species on sand flats adjacent to reefs.
Distribution: Philippines (Batangas, Luzon, Philippines).

Sclerite

Mike Miller

Robert Bolland

164. *Sclerobelemnon burgeri* (Herklots, 1858)

Identification: Dart-shaped or club-shaped sea pens with 2-4 irregular longitudinal rows of polyps along the rachis. Color is reddish-brown to dark brown.
Natural History: Infrequently encountered on sandy flats adjacent to reefs or between patch reefs. Depth: 22-91 m.
Distribution: Philippines; Indonesia; Japan and Ryukyu Islands (Okinawa).

G.C. Williams

165. *Sclerobelemnon* sp. 1

Identification: Sea pens that vary from slender, delicate, and transparent, to more robust and opaque–mostly dart-shaped or club-shaped, usually less than 12 cm in height. The calcareous axis can be seen clearly in the middle of the colony pictured here. Polyps are mostly sparsely distributed in 1-4 opposite longitudinal rows along the rachis–they do not occur on all sides of the rachis. Color is brown to tan, whitish and transparent or colorless. The genus has eight described and apparently valid species.
Natural History: This species has been observed only at night. Depth: 10-25 m.
Distribution: Melanesia. The genus is distributed in the western Atlantic and Indo-West Pacific: Red Sea to Australia, New Guinea, Indonesia, Philippines and Japan (Madang, Papua New Guinea).

Sclerite

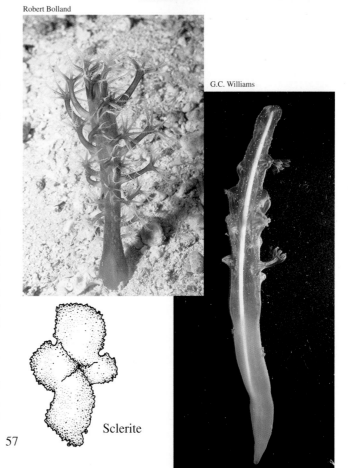

57

166. *Sclerobelemnon* sp. 2

Identification: The rachis has bilateral symmetry as the polyps are in two opposite longitudinal rows. The polyps in this species are large, robust, and bicolored. The emergent portion of the rachis is usually less than 5 cm in height.

Natural History: Observed only at night on sand flats.

Distribution: Philippines (Batangas, Luzon, Philippines).

T. M. Gosliner
Bert Hoeksema

167. *Scytalium* cf. *sarsii* Herklots, 1858

Identification: Sea pens that are similar in appearance to *Virgularia* but have plate-like spicules in the polyp leaves, and the leaves are usually less dense and more open. Mostly brightly-colored, varying from yellow, orange, brick red, or purple.

Natural History: Infrequently encountered in sandy areas of deeper reef margins. Depth: 20-50 m.

Distribution: Indian Ocean; Red Sea to Philippines; Indonesia and parts of Micronesia (Spermonde Archipelago, Indonesia).

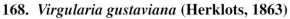

168. *Virgularia gustaviana* (Herklots, 1863)

Identification: Feathery sea pens that are robust and firm with numerous thin but conspicuous and fleshy polyp leaves. Each leaf with over 100 polyps. Color cream or yellowish to rose, red, violet, or reddish-purple.

Natural History: Commonly encountered intertidally or in subtidal sandy areas. Depth: up to 100 m.

Distribution: South Africa to Indonesia; Japan; China (Japan).

Bruce Watkins

Marc Chamberlain

169. *Virgularia* sp. 1

Identification: Flexible, fleshy, feather-like sea pens with soft polyp leaves in two opposite rows, usually 8-60 cm in height. This particular species has relatively long polyp leaves, and the calcareous axis is clearly seen running throughout the length of the rachis. Bicolored white and purple. Numerous species of *Virgularia* have been described, of which perhaps 20 are valid.

Natural History: Several species are commonly seen in sandy areas near coral reefs at night, although at least one species is diurnal and brown in color due to the presence of symbiotic algae. Intertidal to over 30 m in depth.

Distribution: Philippines. The genus has worldwide distribution; in the Indo-Pacific from Africa and the Red Sea to Japan, Indonesia, Philippines, New Guinea, Australia, and Hawai'i (Batangas, Luzon, Philippines).

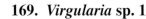

170. *Virgularia* sp. 2

Identification: An elongated feather-like sea pen with relatively short and uniform polyp leaves.

Natural History: Infrequently encountered at night on deeper sand flats adjacent to reefs. Depth: 35 m. A solenogaster mollusk, *Epimenia* sp., has recently been found living in association with a similar species of *Virgularia* in the Ryukyu Islands.

Distribution: Philippines (Batangas, Luzon, Philippines).

171. *Virgularia* sp. 3

Identification: A broad feather-like sea pen, mottled with cream white, red, and orange. The exposed rachis is about 10 cm in height.

Natural History: Encountered only at night on sandy slopes or flats next to coral reefs.

Distribution: This species is common in Philippines (Batangas, Luzon, Philippines).

Jack Randall

Marc Chamberlain
G.C. Williams

172. *Pteroeides* sp. 1

Identification: Stiff feather-like sea pens often with a prickly or bristly appearance due to the presence of projecting rays in the polyp leaves composed of radiating rows of needle-like sclerites. The large, stiff polyp leaves are arranged in two opposite rows along the visible portion of the rachis. Color is variable. This particular species has a robust, fleshy, lemon yellow rachis, which may be as much as 60 cm in height. Numerous species have been described, of which at least 25 are valid.

Natural History: The pea crab *Porcellanella picta* (#826) often inhabits the spaces between the polyp leaves of some species. Depth: 9-320 m.

Distribution: Philippines. The genus is distributed in Europe and the Mediterranean; Africa to Japan, Indonesia, Philippines, New Guinea, Australia, and New Zealand (Batangas, Luzon, Philippines).

T. M. Gosliner

173. *Pteroeides* sp. 2

Identification: A tall, beautiful species with an elongate rachis and thin bicolored polyp leaves that have an undulating margin. These are chocolate brown, edged with orange.

Natural History: Infrequently encountered in sandy or rubble areas.

Distribution: Philippines (Bohol, Philippines).

Robert Bolland

Mike Severns

174. *Pteroeides* sp. 3
Identification: A slender feather-like species. The polyp leaves are straight and rigid. Color rose-pink.
Natural History: Infreqently encountered in deeper sandy areas near reefs. Depth: 50 m.
Distribution: Ryukyu Islands (Okinawa).

175. *Pteroeides* sp. 4
Identification: A robust species with thick, fleshy, densely-set polyp leaves. The calcareous rays often project beyond the margins of the polyp leaves. Color is greyish white with some brown polyps.
Natural History: Infrequently encountered on sandy slopes and flats near reefs.
Distribution: Indonesia (Manado, Indonesia).

176. *Pteroeides* sp. 5
Identification: A relatively spiny species with the white calcareous needle-like rays projecting well beyond the polyp leaf margins.
Natural History: Encountered in deeper sandy areas off coral reefs. Depth: 45 m.
Distribtution: Ryukyu Islands (Okinawa).

Hexacorallia (Zoantharia)

Actininaria - Sea Anemones

Sea anemones are characteristic and conspicuous members of the coral reef. Anemones and hard corals are anatomically similar, differing mainly in the presence of a calcium carbonate skeleton in the latter. At least ten species of Indo-Pacific anemones in the genera *Cryptodendrum, Entacmaea, Heteractis, Macrodactyla,* and *Stichodactyla,* have symbiotic associations with 29 species of anemonefishes–27 of these in the genus *Amphiprion.* Fautin and Allen (1992) provide an excellent account of these animals and the nature of their relationships.

177. *Boloceroides* sp.
Identification: Small anemones usually less than 20 mm in diameter, with numerous slender and elongated tentacles, which sometimes partly hide the small oral disc. The column is often short and inconspicuous. The tentacles are variably-colored or clear to mottled. The mouth region is pinkish tan to brownish.
Natural History: Several species are commonly encountered in sandy areas. The nudibranch, *Berghia major* (#632) preys upon species of *Boloceroides* (see Gosliner, 1980).
Distribution: Philippines. The genus is widespread in the Indo-Pacific (Moalboal, Philippines).

Robert Bolland
Jerry Allen

60

178. *Bunodeopsis / Triactis* sp.
Crab Claw Anemone or Pom-Pom Anemone

Identification: Hemispherical, pom-pom-like anemones about
1 cm in diameter. The many tentacles usually hide the oral disc.
Color is usually cream, tan, or yellowish.

Natural History: A single pom-pom anemone is commonly
held in each claw of the crab, *Lybia tessellata* (#872). This represents a mutualistic association, beneficial to both partners.

Distribution: Mozambique; Mauritius; Seychelles to Hawai'i
(Seychelles).

T. M. Gosliner

179. *Alicia* cf. *sansibarensis* Carlgren, 1900
Tuberculate Night Anemone

Identification: Anemones with tentacles that are extremely long,
thin, and string-like. The column is tall and cylindrical. Grape-
like clusters of vesicles are present on the surface of the column.
Color is mostly transparent white with orange in the column.

Natural History: This unusual anemone is observed at night.

Distribution: East Africa to Philippines (Batangas, Luzon, Philippines).

D.W. Behrens

180. *Phyllodiscus semoni* Kwietniewski, 1897
Night Anemone

Identification: White central column and tentacles, emerging
from the center of a very dark green to blackish mass of intricately-branched vesicles that contain symbiotic algae.

Natural History: A strange and highly unusual anemone, seen
fully emerged in feeding mode at night in shallow areas.

Distribution: Western Pacific: Indonesia and Philippines
(Batangas, Luzon, Philippines).

T. M. Gosliner

Robert Bolland

181. *Triactis producta* Klunzinger, 1877

Identification: Numerous slender tentacles surround the smooth
oral disc atop an upright column. Many branching tree-like
vesicles are present near the top of the column and just below
the tentacles. Color is mostly green and white.

Natural History: Infrequently encountered in deeper water
adjacent to reefs. Depth: 30 m.

Distribution: Red Sea to Ryukyu Islands (Okinawa).

Ben Tetzner

G.C. Williams

182. *Entacmaea quadricolor*
(Rüppell & Leuckart, 1828)
Bulb Tentacle Sea Anemone

Identification: The tentacles are inflated and bulb-like, the bulb usually situated just below the tips. The column is smooth and brown in color–sometimes reddish or green. The oral disc is the same color as the tentacles, usually golden-brown or greenish-brown.

Natural History: These anemones are usually found deep in crevices with only the bulb-like tentacles visible. Thirteen species of anemonefishes are known to occur in symbiosis with this anemone.

Distribution: Eastern Africa; Red Sea to Australia; Micronesia and Melanesia; Japan, (Fiji; Madang, Papua New Guinea).

G.C. Williams

183. *Macrodactyla* cf. *doreensis*
(Quoy & Gaimard, 1833)
Corkscrew Tentacle Sea Anemone

Identification: Large anemones with the column often buried in sediment. The oral disc is often flared and flattened, up to one-half meter in diameter. The tentacles are sparse, long and slender, up to 18 cm in length.

Natural History: Only 3 species of symbiotic fish are associated with this anemone: the Mauritian, Pink, and Clark's anemonefishes.

Distribution: Ryukyu Islands; Philippines; Indonesia; New Guinea and Coral Sea (Batangas, Luzon, Philippines).

Marc Chamberlain
G.C. Williams

184. cf. *Condylactis* sp.

Identification: A beautifully colored anemone with smooth, slender tube-like tentacles that are green near the oral disc and blue toward the ends. The mouth is atop a cone-like oral projection.

Natural History: Locally common in open rubble areas or depressions in dead coral heads.

Distribution: Micronesia (Belau).

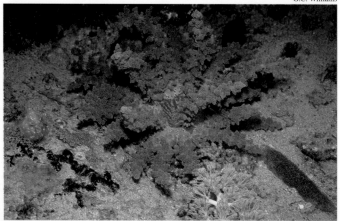

185. *Actinodendron* cf. *glomeratum*
Haddon, 1898
Branching Anemone

Identification: Bushy, low-growing anemones, often somewhat flattened. The tentacles are large and are covered with grape-like clusters of short knobby vesicles. Color is usually light green.

Natural History: Frequently encountered on sand, fully emergent in daylight.

Distribution: Western Pacific: Australia and Philippines (Batangas, Luzon, Philippines).

186. *Actinodendron* cf. *plumosum*
Haddon, 1898
Branching Anemone

Identification: Tree-like anemones, tall and upright. The tentacles are large, with clusters of finger-like vesicles. Color is golden brown or tan.
Natural History: Often seen emerging from sand at night.
Distribution: Western Pacific: Australia and Philippines (Batangas, Luzon, Philippines).

G.C. Williams

187. *Actinostephanus haeckeli*
Kwietniewski, 1897

Identification: Anemones with 12 or more large snake-like tentacles that radiate outward from a flattened oral disc. Conspicuous tubercles are present on the tentacles. Color wine-red with translucent tentacles, sometimes greyish green.
Natural History: Mostly active at night in shallow rubble areas with mud or sand.
Distribution: Philippines and Indonesia (Batangas, Luzon, Philippines).

T. M. Gosliner

188. *Actineria* sp.

Identification: Broad anemones often with a convoluted margin. The tentacles are numerous, relatively short, and grass green or brownish with distinctive grape-like clusters of white or magenta vesicles.
Natural History: This striking anemone is infrequently encountered on reef flats.
Distribution: Melanesia (Solomon Islands).

Roy Eisenhardt
Terry Schuller

189. *Cryptodendrum adhaesivum*
Klunzinger, 1877
Adhesive Sea Anemone

Identification: These anemones form flat and undulating discs up to a third of a meter in diameter. The numerous and densely-set tentacles are very short (< 5 mm long) and extremely sticky. Tentacles are of two forms (bulb-like near the edge of the disc, and glove-like toward the center). These are usually differently colored–with blue, grey, yellow, pink, green or brown.
Natural History: Only one symbiotic fish, Clark's anemonefish (*Amphiprion clarkii*) is associated with this anemone.
Distribution: Red Sea; India; Indonesia; Philippines; Melanesia; Micronesia and Polynesia (Manado, Indonesia).

63

Carol Buchanan

190. *Heteractis aurora*
(Quoy & Gaimard, 1833)
Beaded Sea Anemone

Identification: Large anemones up to one third meter in diameter. The tentacles have ring-like swellings at intervals along their 50 mm length giving them a bead-like appearance.
Natural History: Seven species of anemonefishes are recorded to be associated with this anemone.
Distribution: Red Sea; Mascarene Archipelago; India; Solomon Islands; Ryukyu Islands; Melanesia and Polynesia (Solomon Islands).

G.C. Williams

191. *Heteractis crispa* (Ehrenberg, 1834)
Leathery Sea Anemone

Identification: Large anemones sometimes over one-half meter in diameter. Tentacles are long, slender and sinuous (up to 100 mm in length), tapered to a point at the tips, often greenish or violet.
Natural History: No less than 14 species of anemonefishes are recorded as symbiotic associates with this anemone.
Distribution: Red Sea; India; Philippines; Japan; Micronesia and Polynesia (Batangas, Luzon, Philippines).

G.C. Williams
T. M. Gosliner

192. *Heteractis magnifica*
(Quoy & Gaimard, 1833)
Magnificent Sea Anemone

Identification: A beautifully colored anemone up to one half of a meter in diameter. The tentacles are the same color as the oral disc. The column is colored–usually magenta or purple but also blue, green, red, white or brown. Previously *Radianthus ritteri*.
Natural History: This flamboyant anemone is perhaps the most commonly photographed of all reef anemones. It prefers exposed locations with sufficient current or surge. It is the host for no fewer than twelve species of anemonefishes. Depth: 5-20 m.
Distribution: East Africa; Mozambique; Red Sea; Australia; Melanesia; Philippines; Ryukyu Islands and Tuamotu Archipelago (New Georgia Group, Solomon Islands).

193. *Heteractis malu*
(Haddon & Shackleton, 1893)
Delicate Sea Anemone

Identification: Oral disc up to 20 cm in diameter. The tentacles are robust and finger-like. Expanded column often tall and cylindrical, often with a mottled pattern.
Natural History: Only one fish, *Amphiprion clarkii,* has been recorded as a symbiotic associate with this anemone.
Distribution: Malaysia; Thailand; Australia; Indonesia; Philippines; Melanesia; Japan; Hawai'i (Batangas, Luzon, Philippines).

194. *Stichodactyla haddoni*
(Saville-Kent, 1893)
Haddon's Sea Anemone

Identification: Large anemones with the oral disc convoluted on the margin, up to four-fifths of a meter in diameter. The tentacles are short and sticky to the touch.

Natural History: Six species of symbiotic anemonefish associates have been recorded for this anemone in various parts of its range.

Distribution: East Africa; Red Sea; India to Australia; Philippines; Melanesia; southeast Asia and Japan, (New Georgia Group, Solomon Islands).

G.C. Williams

195. *Stichodactyla mertensii* (Brandt, 1835)
Merten's Sea Anemone

Identification: This can be a very large anemone having the oral disc flared and adhering close to the sea bottom, with convoluted margin. Perhaps the world's largest sea anemone with the oral disc often exceeding one meter. Previously *Stoichactis giganteum.*

Natural History: Twelve species of anemonefish symbionts are associated with this anemone.

Distribution: East Africa; Red Sea to Australia; Indonesia; Philippines; Melanesia; Micronesia; Polynesia; Ryukyu Islands (Batangas, Luzon, Philippines).

Nicholas Galluzzi

196. *Calliactis* sp.
Hermit Crab Anemone

Identification: The column is robust and often mottled, 1-4 cm in height–usually with slender and darkly-mottled tentacles. Individual anemones are attached to the shells of hermit crabs.

Natural History: These hormathiid anemones are symbiotic with hermit crabs in the genera *Dardanus* and *Eupagurus.*

Distribution: Philippines (Batangas, Luzon, Philippines).

D.W. Behrens
G.C. Williams

197. *Verrillactis paguri* (Verrill, 1869)
Hermit Crab Anemone

Identification: The column is robust and often mottled as in *Calliactis,* but the tentacles are usually shorter and uniformly whitish or transparent. Individual anemones are attached to the shells of hermit crabs.

Natural History: These anemones are often symbiotic with hermit crabs in the genus *Dardanus.*

Distribution: South Africa; China Sea; Japan (Natal, South Africa).

65

Marc Chamberlain

Kathy deWet

Kathy deWet

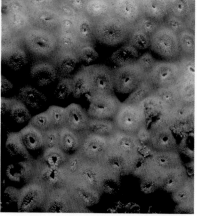

G.C. Williams
Nicholas Galluzzi

198. *Nemanthus annamensis* Carlgren, 1943
Gorgonian Wrapper

Identification: Basal part of column broad, mostly tapering toward the tip, 1-3 cm in height. The tentacles and oral disc can withdraw into the top of the column. Color highly variable–white to cream or orange to red.

Natural History: These anemones are often found attached to the axes of gorgonians, black corals, or other sessile benthic animals that have stalks or branches.

Distribution: Western Pacific: Indonesia and Philippines (Batangas, Luzon, Philippines).

Zoanthidea - Zoanthids

Zoanthids are small anemone-like animals. A single polyp is usually less than 20 mm in diameter. Some zoanthids are solitary but most are colonial, being connected by a basal stolon or a common membranous mat. Zoanthids are mostly tropical and are common inhabitants of intertidal or shallow-water reef communities.

199. *Parazoanthus* sp.

Identification: The columns of *Parazoanthus* are generally smooth and free of foreign particles such as sand grains. The common membranous mat often covers parts of other animals. Numerous differently-colored polyps emanate from the mat. Each polyp has a conspicuous stalk.

Natural History: Epizoic on other sessile animals–often associated with encrusting sponges that grow on the bare axes of sea fans and black corals.

Distribution: Indonesia. The genus ranges throughout much of the Indo-Pacific: South Africa to western Pacific (Indonesia).

200. *Palythoa caesia* Dana, 1848

Identification: Colonial animals in which the polyps are joined by common basal tissues, and display numerous flattened disc-like polyps up to 3 cm in diameter, with inconspicuous stalks. The polyp discs are broad and flat, dish-shaped and up to 20 mm in diameter. Tentacles often appear knob-like around the margin of the disc. Color dark brown to tan.

Natural History: Encountered on ledges and slopes.

Distribution: Western Pacific: Indonesia and Philippines (Indonesia; Batangas, Luzon; Philippines).

201. *Protopalythoa* sp. 1

Identification: Polyps have flat oral discs (1-2 cm in diameter) with knob-like tentacles surrounding the margin and conspicuous stalks. Color green or brown.

Natural History: Conspicuous and often encountered in densely-concentrated aggregations.

Distribution: Malaysia (Sipadan, Borneo).

66

202. *Protopalythoa* sp. 2

Identification: Olive green zoanthid with relatively long stalks (up to three cm), and knob-like white tentacles surrounding the oral disc.

Natural History: Aggregations are common on reef slopes.

Distribution: Micronesia (Sapwauhfik, Pohnpei).

Dave Zoutendyk

203. *Protopalythoa* sp. 3

Identification: Zoanthids with large flat discs up to 3 cm in diameter. The tentacles are often yellowish in color.

Natural History: These zoanthids live in dense concentrations forming brown mats.

Distribution: Australia (Great Barrier Reef, Australia).

Corallimorpharia - Corallimorpharians

Corallimorpharians are considered intermediate between sea anemones and hard corals–being virtually identical to hard corals anatomically, but without a skeleton as in anemones. Some have radially arranged capitate tentacles. In several species, numerous solitary polyps live crowded together, forming aggregations that may cover large areas of hard substratum.

Leslie Newman & Andrew Flowers

204. *Pseudocorynactis* sp.

Identification: Distinctive anemone-like animals. The oral disc is smooth with radiating lines, usually 5-10 cm in diameter. The numerous tentacles are mostly clear or tan and transparent with conspicuous golden to orange, ball-shaped tips. The brown basal column is conspicuous and upright, often up to 10 cm in height. This species is superficially similar to *Pseudocorynactis caribbeorum* Hartog, 1980, from the western Atlantic.

Natural History: Frequently encountered in crevices on reef slopes.

Distribution: Malaysia (Sipadan, Borneo).

Nicholas Galluzzi
Mike Miller

205. *Amplexidiscus fenestrafer*
Dunn & Hamner, 1980

Identification: Fully expanded animals form plate-like discs up to 30 cm or more in diameter, with a smooth, often light-colored ring near the margin. The central portion of the disc is covered with short cylindrical tentacles. Contracted individuals are rounded and bowl-like, with the column covering most of the disc. Color is cream to brown or greenish. This is the largest recorded species of corallimorpharian.

Natural History: Common on shallow reef flats.

Distribution: Australia; Indonesia; Belau; Guam; Philippines (Batangas, Luzon, Philippines).

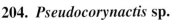

G.C. Williams

206. *Discosoma* cf. *rhodostoma* (Ehrenberg, 1834)

Identification: The expanded polyps form open discs mostly less than 15 cm in diameter. The usually magenta or pink mouth is surrounded by numerous branched or glove-like tentacles, giving a shaggy appearance to the surface of the disc. Color is greenish or brownish.

Natural History: Frequently encountered on reef flats or slopes.

Distribution: Throughout much of the western Pacific (Batangas, Luzon, Philippines).

207. *Discosoma* cf. *rhodostoma* (Ehrenberg, 1834)

Identification: This form shows the wall of a partially contracted individual. The wall is finely-lined with white striations.

Natural History: Locally common forming large mats in dense aggregations on vertical surfaces, slopes, and ledges.

Distribution: Philippines (Verde Island, Philippines).

207. *Discosoma* cf. *rhodostoma* (Ehrenberg, 1834)

Identification: The tentacles of this species are fully extended in this photograph and are tipped with pale yellow. The discs are up to approximately 10 cm in diameter.

Natural History: Dense aggregations occur on reef slopes and areas of vertical relief.

Distribution: Melanesia (Madang, Papua New Guinea).

209. *Discosoma* sp. 1

Identification: The aggregated polyps form flared discs usually less than 10 cm in diameter. Radiating lines of short knob-like or rounder tubercles often emanate from the central mouth. Color is variable - violet, brown, or green.

Natural History: Frequently encountered on reef flats.

Distribution: Australia. The genus is distributed throughout much of the western Pacific (Great Barrier Reef, Australia).

210. *Discosoma* sp. 2

Identification: A deeply-colored, greenish brown species with conspicuous radiating lines of knobby white tubercles.
Natural History: Dense aggregations of this species often cover large areas of rocky substrata.
Distribution: Philippines (Batangas, Luzon, Philippines).

Marc Chamberlain

211. *Discosoma* sp. 3

Identification: This corallimorpharian has numerous large mammiform to cylindrical tubercles on the face of the disc. Color greenish with pale pink tubercles.
Natural History: Infrequently encountered in cracks or depressions on ledges or slopes.
Distribution: Micronesia (Belau).

Marc Chamberlain

212. *Discosoma* sp. 4

Identification: This species is usually greyish white with rich brown branching tubercles on the face of the disc. The discs are often 8-10 cm in diameter.
Natural History: Infrequently encountered in crevices or depressions on ledges or slopes.
Distribution: Polynesia (Bora Bora, Society Islands).

Scleractinia - Hard corals

Hard corals are the builders of coral reefs. The actual material comprising the geologic structure of a coral reef is deposited by hard corals over time. Coral reef scleractinians (hard corals) are diverse, abundant, and morphologically variable. An estimated 500-600 species of reef-building hard corals inhabit the Indo-Pacific, about ten times that of the Caribbean region. Hard corals provide homes for countless numbers of reef organisms. See Veron (1986) for the best overall account of Indo-Pacific scleractinians.

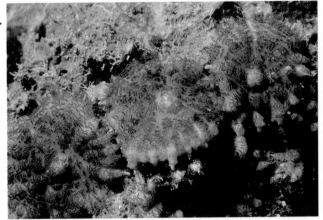
D.W. Behrens
G.C. Williams

213. *Pocillopora damicornis* (Linnaeus, 1758)

Identification: Colonies are mostly delicately-branched with slender cylindrical branches and large wart-like projections that are often difficult to distinguish from true branches, as the two intergrade. Color is cream to tan, green or pink.
Natural History: Commonly encountered in a wide range of shallow water habitats.
Distribution: East Africa; Red Sea to Mexico and Ecuador (Batangas, Luzon, Philippines).

69

214. *Pocillopora verrucosa*
(Ellis & Solander, 1786)

Identification: Compact colonies composed of congested upright branches. The wart-like projections (verrucae) that cover the branches are variable in size. Color is usually cream or tan to reddish-brown.

Natural History: Frequently encountered on fringing reefs.

Distribution: East Africa; Red Sea to Australia; Solomon Islands; Hawai'i (Mborokua, Solomon Islands).

215. *Pocillopora* cf. *meandrina* Dana, 1846

Identification: Low compact colonies composed of many congested sprawling branches. The wart-like projections that cover the branches are uniform in size. Cream to tan or pink.

Natural History: Occasionally encountered on exposed reefs.

Distribution: Australia to Hawaiian and Society Islands (Tuamotu Archipelago).

216. *Pocillopora eydouxi*
Milne Edwards & Haime, 1860

Identification: Colonies are robust with tall, upright, flattened branches that flare out toward the tips. Wart-like projections (verrucae) are conspicuous and are densely-set on the branch surfaces. Color is forest green or brown.

Natural History: Abundant in high energy environments often forming monospecific stands.

Distribution: East Africa; Red Sea to Australia; New Guinea; Solomon Islands to Hawai'i (Hawai'i).

217. *Seriatopora hystrix* Dana, 1846

Identification: Delicately-branched colonies with narrow branches that are conspicuously tapered to a sharp point. Some polyp openings appear in longitudinal rows along the branches. Color is cream to tan, bluish or pink.

Natural History: Commonly encountered in many shallow reef habitats.

Distribution: East Africa; Red Sea to Australia; New Guinea; Samoa; Ryukyu Islands (Madang, Papua New Guinea).

218. *Seriatopora caliendrum* Ehrenberg, 1834

Identification: Colonies have branches that are only somewhat tapered with blunt tips. Polyps do not appear in rows and often have projecting hoods above each polyp. Color is cream to brown.

Natural History: Infrequently encountered on shallow reef slopes.

Distribution: East Africa; Red Sea to Australia; Solomon Islands and New Caledonia (New Georgia Group, Solomon Islands).

G.C. Williams

219. *Stylophora pistillata* Esper, 1797

Identification: Mostly low-growing colonies with thick branches that are rounded or club-shaped with blunt ends. The skeletons of individual polyps have conspicuously projecting hoods. Color is cream to tan, green or pink.

Natural History: In low energy environments the colonies may be more delicately branched, with slender cylindrical branches.

Distribution: East Africa; Red Sea to southern Japan; Tuamotu Archipelago and Pitcairn Island (Bohol, Philippines; Mborokua, Solomon Islands).

T. M. Gosliner

G.C. Williams

220. *Montipora* sp. 1

Identification: Colonies can be encrusting, plate-like, scroll-shaped, or columnar. The polyps are small and often hidden in the lowest areas between prominent projecting points. Color cream to brown or brightly-colored. This is the second largest genus of hard corals (second only to *Acropora*) with about 40 species from Australia alone. The species pictured is similar to *M. undata* and several other species, in which the highly ornamented surface has numerous fused processes forming complex ridges.

Natural History: Frequently encountered on reef flats.

Distribution: Melanesia. The genus ranges from Africa; Red Sea to New Guinea; southern Japan; Hawai'i; Marquesas and Pitcairn Islands (Madang, Papua New Guinea).

G.C. Williams

G.C. Williams

221. *Montipora* sp. 2

Identification: The colonies form large thick plates with many sharp peaks and ridges. Brownish green with pink or magenta peaks and margin.

Natural History: Frequently encountered on reef flats and slopes.

Distribution: Philippines (Batangas, Luzon, Philippines).

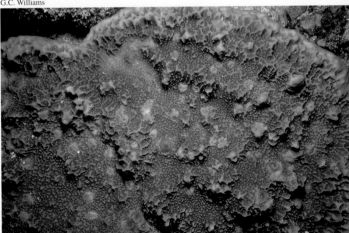

Jack Randall

222. *Anacropora forbesi* **Ridley, 1884**

Identification: Delicately-branched colonies having slender elongate tapering branches with blunt tips. The polyps form numerous mound-like bumps on the surface of the branches. Color is tan with white-tipped branches.

Natural History: Infrequently encountered–mostly restricted to areas of turbid or muddy water.

Distribution: Seychelles; Maldives to New Guinea; Fiji; Ryukyu Islands, Marshall Islands (Madang, Papua New Guinea).

223. *Acropora cerealis* **(Dana, 1846)**

Identification: Colonies are low-growing and cushion-shaped with crowded, slender branches. The skeletons of individual polyps are prominent and cylindrical to tubular. Color is tan or cream often with purple, pink, or blue.

Natural History: Frequently found on upper reef slopes.

Distribution: Australia; Indonesia; Philippines; Solomon Islands; Marshall Islands (Mborokua, Solomon Islands).

G.C. Williams

224. *Acropora clathrata* **(Brook, 1891)**

Identification: This species is very similar in appearance to *Acropora hyacinthus,* except that in *A. clathrata* the branches fuse to form a solid plate near the center of the colony, and the free branches near the edges do not conspicuously project upwards, but rather lie mostly horizontal.

Natural History: Commonly encountered on fringing reefs.

Distribution: Madagascar to Australia; New Guinea; Tuamotu Archipelago (Madang, Papua New Guinea).

G.C. Williams
G.C. Williams

225. *Acropora gemmifera* **(Brook, 1892)**

Identification: Colonies are composed of conical branches that are thick and taper to a blunt tip. Skeletons of individual polyps decrease in size toward branch tips, and sometimes form longitudinal rows. Color is cream, brown, green, pinkish or bluish.

Natural History: Frequently found on reef flats and upper slopes.

Distribution: Indian Ocean to Australia; Solomon Islands, Fiji, New Caledonia (New Georgia Group, Solomon Islands).

226. *Acropora* cf. *hyacinthus* (Dana, 1846)

Identification: Colonies form flat circular or scroll-shaped plates or tables. The numerous and congested branchlets project upward throughout most of the colony. Color is uniform cream to tan, brown, or green. The growth margin may be differently colored than the rest of the colony.

Natural History: A very abundant species of reef flats and upper slopes. Intertidal to shallow subtidal.

Distribution: Mascarene Islands to Australia; New Guinea; Society Islands (Madang, Papua New Guinea).

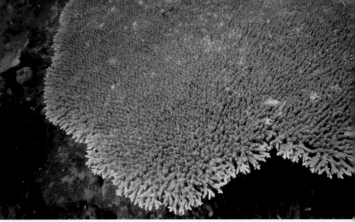

G.C. Williams

227. *Acropora robusta* (Dana, 1846)

Identification: Colonies have thick, elongate, conical branches standing vertically toward the center of the colony, with branches of the colony margin thinner, shorter, and mostly horizontal. Color is vivid green or brown with cream or pink branch tips.

Natural History: Common on high energy reef habitats. Intertidal to shallow subtidal.

Distribution: Central Indian Ocean to Solomon Islands and Society Islands (New Georgia Group, Solomon Islands).

G.C. Williams

228. *Acropora* cf. *tenuis* (Dana, 1846)

Identification: Colonies are composed of congested, cylindrical, upright branches. The skeletons of individual polyps have large, often crescent-shaped lower tips. Color is cream to blue, pink, or purple.

Natural History: Commonly encountered on upper reef slopes.

Distribution: Mauritius to Australia; New Guinea and Marshall Islands (Madang, Papua New Guinea).

G.C. Williams
G.C. Williams

229. *Astreopora gracilis* Bernard, 1896

Identification: Colonies are convex and shield-like or hemispherical in shape. The polyps form turret-like or conical to cylindrical projections. Color usually cream, pale yellow, or brownish. Similar in appearance to *Turbinaria frondens* (#286).

Natural History: Infrequently encountered on reef flats.

Distribution: Western Pacific: Philippines and Australia to Micronesia (Batangas, Luzon, Philippines).

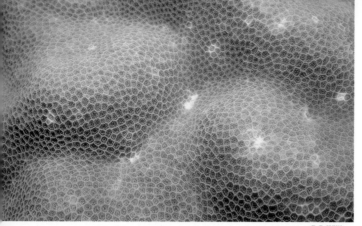

G.C. Williams

230. *Porites solida* (Forskål, 1775)

Identification: Colonies can be very large and encrusting, often with a lumpy surface appearance. Adjacent polyps share thin walls between them.

Natural History: Common and abundant throughout its range. Low tide line to shallow subtidal.

Distribution: Widespread in the Indo-Pacific: Red Sea to Australia; New Guinea and Hawai'i (Madang, Papua New Guinea).

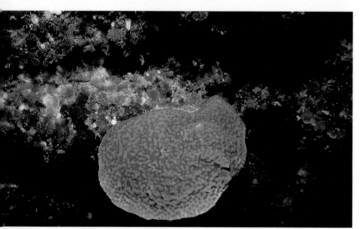

D.W. Behrens

231. *Porites* cf. *lobata* Dana, 1846

Identification: Colonies are hemispherical or dome-shaped often with a lumpy surface appearance. This species, as with others of the genus, is difficult to distinguish without examination of skeletal details.

Natural History: Common and abundant in lagoons and protected fringing reefs. Interidal and shallow subtidal.

Distribution: Widespread in the Indo-Pacific: central Indian Ocean to Hawai'i and the Galápagos Islands (Tuamotu Archipelago).

232. *Porites vaughani* Crossland, 1952

Identification: Colonies are often plate-like, but sometimes forming upright columns. The skeletons of individual polyps are contained in low areas between rounded ridges and tubercles that are of uniform height. Color is cream, brown, pink, green, or purple.

Natural History: Encountered in a wide range of reef habitats.

Distribution: Southeast Asia; Australia; Indonesia; Philippines; Solomon Islands (Mborokua, Solomon Islands).

G.C. Williams
G.C. Williams

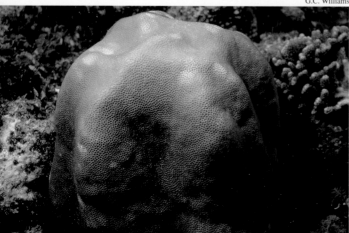

233. *Porites* sp.

Identification: Colonies are often encrusting or hemspherical to dome-shaped with a smooth surface, up to more than 5 meters in height or diameter, making them perhaps the largest of all coral colonies. The openings of the polyps are crowded and very small (ca. 2 mm in diameter) and share walls with adjacent polyps. Color is cream to brown or green, sometimes pink, blue or purple. There are at least 16 species in the western Pacific. Intertidal to shallow subtidal.

Natural History: Some large colonies of *Porites* have been estimated to be nearly 1000 years old.

Distribution: Melanesia. The genus is circumtropical: East Africa; Red Sea to Gulf of California and Ecuador; Caribbean (Madang, Papua New Guinea).

234. *Goniopora* cf. *djiboutiensis* Vaughan, 1907

Identification: Colonies are mound-like or form short columns. The skeletal openings of the polyps are about 5 mm in diameter. The extended polyps are long, slender, and tubular, with 24 tentacles surrounding the very prominent cone-like mouth. Color is brown or green with pink or white oral cones.

Natural History: Polyps are seen extended during daylight. Abundant colonies may cover extensive areas in turbid water habitats.

Distribution: Western Indian Ocean to Fiji; including Philippines and Australia (Cebu, Philippines).

Mike Miller

235. *Goniopora* spp.

Identification: Colonies can be encrusting, lobate, or columnar. The elongate polyps have 24 finger-like tentacles surrounding the oral cone that contains the central mouth. Color is variable– usually cream, tan, grey, light green or brown.

Natural History: At least 14 species of the genus inhabit the western Pacific, usually in protected and often turbid water habitats. A nudibranch, *Phestilla* sp. (#625) feeds on the polyps of *Goniopora*.

Distribution: The genus ranges from East Africa, Red Sea to Australia, Philippines, Solomon Islands, Tuamotu Archipelago, Marshall Islands, southern Japan (Batangas, Luzon, Philippines; New Georgia Group, Solomon Islands).

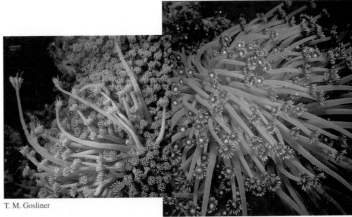

T. M. Gosliner

G.C. Williams

236. *Alveopora verrilliana* Dana, 1872

Identification: Colonies form knob-like branches. The polyp openings are less than 2 mm in diameter, with 12 slender tentacles surrounding the oral disc. Color is light green, brown, or grey.

Natural History: Infrequently encountered on reef flats and slopes.

Distribution: Pacific Ocean from Australia to Hawai'i (New Georgia Group, Solomon Islands).

G.C. Williams

T. M. Gosliner

237. *Alveopora* spp.

Identification: The polyps resemble those of *Goniopora*, but have 12 instead of 24 slender or knob-like tentacles surrounding the oral disc. Colony and polyp colors are similar to those of *Goniopora* species. About 16 species of the genus are recognized.

Natural History: Colonies are infrequently encountered in a wide variety of reef habitats.

Distribution: Philippines. The genus ranges throughout the Indo-Pacific; eastern Africa; Red Sea to Australia; Philippines; Hawai'i; Marquesas; Tuamotus; southern Japan (Batangas, Luzon, Philippines; Bohol, Philippines).

Marc Chamberlain

G.C. Williams

238. *Pavona cactus* (Forskål, 1775)

Identification: Colonies are composed of thin, undulating, upright leaf-like fans and plates. The inconspicuous openings of the polyps are aligned in concentric lines parallel to the fan margins. Color is tan to greenish-brown.

Natural History: Usually encountered in lagoons and slopes of protected fringing reefs, sometimes in turbid water habitats.

Distribution: Red Sea to Philippines; Australia; Marshall Islands (Bohol, Philippines).

D.W. Behrens

239. *Pavona decussata* (Dana, 1846)

Identification: Colonies are usually composed of upright, thickened plates that tend to flare toward the upper margin. The skeletons of individual polyps form shallow depressions, and are often situated in irregular lines parallel to the plate margins. Color cream to ochre, or light green to brown.

Natural History: This species may be enountered in a variety of reef habitats.

Distribution: Red Sea to Australia and Society Islands (Bora Bora, Society Islands).

240. *Pachyseris speciosa* (Dana, 1846)

Identification: Colonies are usually plate-like with numerous concentric ridges of uniform height. The polyps are contained in one or more rows between the ridges. Color is grey, tan, brown, or greenish-brown.

Natural History: Encountered in a wide range of reef habitats.

Distribution: Red Sea to Australia; Samoa; Solomon and Society Islands (Mborokua, Solomon Islands).

Fungiids - Mushroom corals

The mushroom corals represent over 40 species in 11 genera, inhabiting the tropical Indo-Pacific in shallow water. The six genera included here are *Fungia, Heliofungia, Ctenactis, Herpolitha, Polyphyllia,* and *Halomitra.* Mushroom corals are conspicuous and commonly encountered members of coral reef communities. Most species are free-living and unattached to the sea bottom, and exhibit a mushroom-like appearance. See Hoeksema (1989) for a comprehensive treatment and revision.

G.C. Williams
Marc Chamberlain

241. *Fungia (Cycloseris) costulata* Ortmann, 1889

Identification: Solitary corals that are circular in outline with a flattened undersurface. The radiating ridges (septa) nearest the mouth are conspicuously thickened in comparison to the other septa. The septa are straight throughout, not wavy. Closely related and difficult to distinguish from *Fungia tenuis.* The specimen pictured here has been placed on another fungiid coral *Podabacia motuporensis.*

Natural History: Inhabit sandy areas in the western Pacific.

Distribution: Red Sea to Australia; Indonesia; Belau; Ryukyu Islands; Marquesas (Belau).

76

242. *Fungia (Cycloseris) somervillei* Gardner, 1909

Identification: These mushroom corals are oval in shape. The tentacles of the polyp are green with a small white bead-like tip.

Natural History: As with other mushroom corals, this species is usually encountered semi-buried in sandy areas.

Distribution: Seychelles; Australia; New Guinea; Philippines; Indonesia; Guam (Lembeh Strait, Indonesia).

Mike Severns

243. *Fungia (Fungia) fungites* (Linnaeus, 1758) Shallow Water Form

Identification: About 25 valid species are recognized, and most of these are widespread in the Indo-Pacific. Solitary, the skeleton usually is ovoid with a strong central arch. Dark brown to ochre, sometimes mottled.

Natural History: Many species of *Fungia* are often abundant on reef slopes below areas of substantial wave action. In the juvenile stage, this coral is sometimes found attached to the sea bottom, like other mushroom corals. In nearshore turbid water (2-3 m), often on coral rubble.

Distribution: East Africa; Red Sea to Australia; Philippines and Society Islands (Darwin, Australia).

G.C. Williams

244. *Fungia (Fungia) fungites* (Linnaeus, 1758) Deep Water Form

Identification: Circular in shape, relatively flat, with a central mouth. The radiating ridges (septa) have rounded or pointed teeth on the upper margin.

Natural History: Mostly deep subtidal (10-20 m) on offshore reefs. Polyps are extended at night.

Distribution: East Africa; Red Sea to Australia; Philippines; Solomon Islands; Hawaiian and Society Islands (Mborokua Island, Solomon Islands).

G.C. Williams
T. M. Gosliner

245. *Fungia (Lobactis) scutaria* Lamarck, 1801

Identification: This beautiful ovoid mushroom coral shows the tentacles fully extended. The tentacles are translucent and pale green in color. Many such corals are strikingly pigmented or mottled with magenta around the mouth and margin of the coral polyp.

Natural History: Locally common and encountered semi-buried in shallow sandy areas.

Distribution: Philippines (Batangas, Luzon, Philippines).

77

G.C. Williams

246. *Fungia* spp.

Identification: Several species of mushroom corals are evident in this underwater landscape view. The corals are distributed relatively densely, and can often cover large areas of underwater landscapes.

Natural History: In areas such as the one pictured here, the numerous fungiids live in coral rubble as opposed to a sandy area.

Distribution: Philippines (Mindoro, Philippines).

Bruce Watkins

247. *Heliofungia actiniformis* (Quoy & Gaimard, 1833)

Identification: Large solitary hard corals with a circular disc-like skeleton, up to 21 cm in diameter. These corals live unattached to the sea bottom. The tentacles are slender and smoke grey to brown with cream-colored knob-like tips, which surround the single centrally located mouth. The tentacles of this species may be confused with those of the colonial hard coral *Euphyllia glabrescens* (#279) and some small individuals of the anemone *Stichodactyla mertensii* (#195).

Natural History: Often encountered in areas of rubble and unconsolidated sediments. Depth: 5-15 m.

Distribution: Australia; Philippines; Indonesia; New Guinea; Solomon Islands; New Caledonia; southern Japan (Philippines).

Mike Miller
G.C. Williams

248. *Ctenactis albitentaculata* Hoeksema, 1989

Identification: These free-living corals are oval or elliptical and elongated in shape. The radiating ridges (septa) appear as vertical walls with uniform tapering serrations on the upper margins. Color is brownish with conspicuous white tentacles. This species is easily recognized by the striking white pigmentation of the tentacles.

Natural History: Infrequently encountered on reef flats and slopes.

Distribution: Malaysia; Philippines; Indonesia; New Guinea; Australia; Micronesia and New Caledonia (Philippines).

249. *Herpolitha limax* (Esper, 1797)

Identification: Corals are elongate-ovoid with prominent raised central region. A medial longitudinal groove is evident. The septa have smooth margins. Color is often greenish brown.

Natural History: Infrequently encountered in protected areas of rubble or sandy bottom.

Distribution: Central Indian Ocean to Australia; Philippines; Belau (Batangas, Luzon, Philippines).

250. *Polyphyllia talpina* Lamarck, 1801

Identification: Adult corals are free-living (not attached to the sea bottom), elongate-elliptical in shape, with many mouths distributed over the upper surface. A medial longitudinal groove is often evident.

Natural History: Polyps are usually extended during the day.

Distribution: Madagascar to Philippines; Ryukyu Islands; Marshall Islands; Samoa and Tonga (Batangas, Luzon, Philippines).

G.C. Williams

251. *Halomitra pileus* (Linnaeus, 1758)
Neptune's Cap

Identification: Adult corals are unattached to the sea bottom, circular and dome-shaped, up to one-half meter in diameter. Color is light brown.

Natural History: Polyps are seen only at night. Infrequently encountered in lagoons and on rubble of lower reef slopes.

Distribution: Eastern Indian Ocean: Malaysia; Philippines; Ryukyu Islands; Micronesia and central Polynesia (Mindoro, Philippines).

G.C. Williams

252. *Galaxea astreata* (Lamarck, 1816)

Identification: Colonies are usually low and encrusting, sometimes upright and columnar. There are usually less than 12 radiating ridges (septa) that immediately surround the mouth of each circular polyp.

Natural History: Polyps are only rarely seen extended during the day.

Distribution: Red Sea to Australia; Indonesia and Melanesia (Mborokua, Solomon Islands; Fiji).

Ken Howard

G.C. Williams

253. *Galaxea fascicularis* (Linnaeus, 1767)

Identification: Colonies often form cushion or dome-shaped mounds. Some larger colonies may be lobed or slightly branched. There are 12 or more radiating ridges (septa) that immediately surround the mouth of each circular to oval polyp.

Natural History: Polyps are often seen extended during the day.

Distribution: Red Sea to Australia; Melanesia to Samoa (New Georgia Group, Solomon Islands; Mapia, Irian Jaya, New Guinea).

Kathy deWet

G.C. Williams

G.C. Williams

254. cf. *Oxypora lacera* (Verrill, 1864)

Identification: Colonies form thin scroll-like plates or may be encrusting. Conspicuous tooth-like tubercles are present surrounding the mouths of the polyps. Many beaded parallel lines are evident running between the polyps and the plate margin. Color is golden brown to greenish brown. This species is difficult to distinguish underwater from the superficially similar hard corals *Echinophyllia aspera* and *Echinopora lamellosa* (#276).

Natural History: Infrequently encountered on reef flats.

Distribution: Red Sea to Melanesia and Micronesia (Mborokua, Solomon Islands).

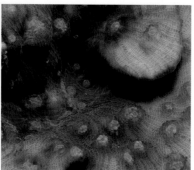

G.C. Williams   Marc Chamberlain

255. *Mycedium elephantotus* (Pallas, 1766)

Identification: Colonies are similar in appearance to *Oxypora* except that the polyps are periscope-shaped (facing outward toward the margin of the colony rather than facing upward), and the tubercles on the upper surface of the colony are either small and inconspicuous or absent. Color is brown, greyish blue, green or pinkish.

Natural History: Often encountered in a variety of reef habitats.

Distribution: Red Sea to Australia; New Guinea; Philippines; Micronesia; and Society Islands (Batangas, Luzon, Philippines; Belau).

G.C. Williams
G.C. Williams

256 *Pectinia lactuca* (Pallas, 1766)
Carnation or Hibiscus Coral

Identification: These hard corals form colonies that are more-or-less disc-shaped up to a meter in diameter, with tall, thin, often undulating walls covering the upper surface. Color usually brown to light green or grey.

Natural History: Occasionally encountered on reef slopes; intertidal and shallow subtidal. This species can tolerate more turbid water conditions than many other corals.

Distribution: Madagascar to Australia; Solomon Islands and Fiji (New Georgia Group, Solomon Islands).

257. *Pectinia paeonia* (Dana, 1846)
Carnation or Hibiscus Coral

Identification: Colonial hard corals with colonies up to 30 cm in diameter, in which the upper surface is covered with clusters of tall fluted projections and spine-like spires that point upward. Color dull brown or smoke grey.

Natural History: Often found in turbid conditions. Depth: 5-15 m.

Distribution: Central Indian Ocean to Australia; Fiji; Solomon Islands; New Caledonia (New Georgia Group, Solomon Islands).

258. *Cynarina lacrymalis* (Milne Edwards & Haime, 1848)

Identification: Solitary and free-living (semi-buried) corals that are circular to oval in shape. Many of the radiating ridges (septa) around the mouth are very thick with prominent teeth on the outer margins. Puffy, bubble-like swellings often form around the thickest septa. Color is highly variable.

Natural History: Inhabits a variety of reef habitats, although inconspicuous and infrequently seen.

Distribution: Madagascar; Red Sea to Vanuatu; Kermadec Islands; southern Japan (Philippines).

Mike Severns

259. *Lobophyllia hemprichii* (Ehrenberg, 1834) Young Stage

Identification: Anemone-like and saucer-shaped. Tentacles are short with knob-like ends. Color is variable and often mottled– green, brown, or reddish-brown.

Natural History: Infrequently encountered in a variety of reef habitats.

Distribution: East Africa to Polynesia (Batangas, Luzon, Philippines).

G.C. Williams

260. *Lobophyllia hemprichii* (Ehrenberg, 1834) Mature Stage

Identification: Colonies are flat or form hemispherical mounds. Retracted polyps are thick and fleshy giving the colony an inflated or billowy appearance, which hides the underlying growth form that is necessary for species identification. This species is very difficult to distinguish underwater from *Symphyllia agaricia* Milne Edwards & Haime, 1849.

Natural History: Both species are encountered on upper reef slopes.

Distribution: East Africa; Red Sea to Australia; New Guinea; and Tuamotu Archipelago (Madang, Papua New Guinea).

G.C. Williams
T. M. Gosliner

261. *Lobophyllia corymbosa* (Forskål, 1775)

Identification: Mostly hemispherical colonies. The teeth of the radiating ridges (septa) are elongate and blunt-ended, thus giving the polyp mounds a conspicuously spiny appearance. Color greyish brown or greenish brown.

Natural History: Often encountered on upper reef slopes. Extended polyps are observved only at night.

Distribution: East Africa; Red Sea to Australia; Philippines; Society Islands and Tuamotu Archipelagos (Batangas, Luzon, Philippines).

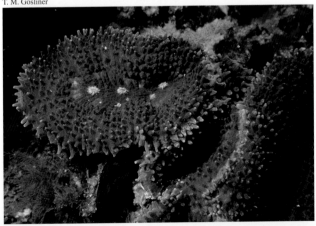

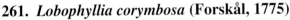

262. *Lobophyllia* cf. *pachysepta* Chevalier, 1975

Identification: These corals form small colonies that are flat or hemispherical. The main radial ridges of the skeleton (septa) are thick and have several lobed teeth on their outer margins. Color is dark green with yellow septal teeth.

Natural History: Infrequently found in protected reef habitats.

Distribution: Central Indian Ocean to Australia; Philippines and Coral Sea (Batangas, Luzon, Philippines).

Jerry Allen

263. *Symphyllia* sp.

Identification: Colonies of these corals are flat with pronounced lobes that radiate out from the center. Large teeth of the septa give the lobes a spiny appearance. Color is mottled with light green or grey and dark green or greenish brown. Superficially similar to *Lobophyllia hataii*.

Natural History: A strikingly beautiful species of various reef habitats, but infrequently encountered.

Distribution: Philippines (Batangas, Luzon, Philippines).

G.C. Williams

264. *Favia* cf. *favus* (Forskål, 1775)

Identification: Colonies are flat or hemispherical and rounded. The skeletons of individual polyps are 15-20 mm in diameter and their radiating ridges (septa) are not uniform in size. Color is often mottled light or dark green to brown. This species is very similar in superficial appearance to *Barabattoia amicorum*, which has septa of uniform size, and more protuberant polyp mounds.

Natural History: Often encountered on reef flats.

Distribution: Red Sea to Australia; Melanesia; Micronesia and Samoa (New Georgia Group, Solomon Islands).

265. *Favia* cf. *pallida* (Dana, 1846)

Identification: Colonies are hemispherical and rounded with crowded polyps, each about 10 mm in diameter. The radial ridges (septa) of the skeletons of individual polyps are uniform in size. Usually pale green with brown polyp mounds. This species is very difficult to distinguish underwater from the related hard coral *Montastrea magnistellata*.

Natural History: The colony in the photo exhibits: (1) coral bleaching (loss of zooxanthellae) on the right, (2) a polyp in the process of dividing (the figure-eight shape on the right), (3) in the center, several new budding polyps between parent polyps, and (4) a pair of serpulid tubes at right center.

Distribution: East Africa; Red Sea to Australia; Philippines; Samoa and Tuamotus (Batangas, Luzon, Philippines).

G.C. Williams
G.C. Williams

82

266. *Favia stelligera* (Dana, 1846)

Identification: Colonies are spherical, flat and lumpy, or columnar with rounded, upright, club-like lobes. The skeletons of individual polyps are small (each about 5 mm in diameter), circular in shape, and crowded. Color is golden brown or greenish brown.

Natural History: This species occurs in a variety of reef habitats. Pyrgomatid barnacles (#694, 695) are visible in the photo.

Distribution: Western Indian Ocean: Red Sea to Australia; Philippines; Coral Sea and Hawai'i (Bohol, Philippines).

G.C. Williams

267. *Favia* sp.

Identification: Colonies form rounded, hemispherical mounds, usually 10-30 cm in diameter. The individual polyp calyces are ovoid. Similar in appearance to *Favia maritima* (Nemenzo, 1971), but the species pictured here has smaller calyces. Color dark brown or greenish brown.

Natural History: A rarely encountered species in various reef habitats including turbid water areas. Intertidal to shallow subtidal.

Distribution: Australia (Darwin, Australia).

G.C. Williams

268. *Favites* cf. *abdita* (Ellis & Solander, 1786)

Identification: Colonies are rounded or encrusting with a bumpy surface appearance. The walls between the polyps are thick with prominent and straight radiating ridges (septa). Color is usually brown with bright green oral discs of the polyps.

Natural History: Abundant in a wide range of reef habitats.

Distribution: Red Sea to Australia; Melanesia to Samoa (Solomon Islands).

Ken Howard
G.C. Williams

269. *Favites* cf. *complanata* (Ehrenberg, 1834)

Identification: Massive to dome-shaped colonies. The individual polyp skeletons are 8-12 mm in diameter. Color is golden brown with green oral discs.

Natural History: Locally common on some reef flats.

Distribution: Red Sea to Society Islands and Tuamotu Archipelago (New Georgia Group, Solomon Islands).

G.C. Williams

270. *Favites flexuosa* (Dana, 1846)

Identification: Flat to hemispherical colonies with individual polyp skeletons, 15-20 mm in diameter. Color is golden brown with green polyp centers.

Natural History: A commonly species in a variety of reef habitats.

Distribution: Red Sea to Solomon Islands and Fiji (Mborokua Island, Solomon Islands).

G.C. Williams

271. *Favites* sp.

Identification: Colonies are usually somewhat rounded or dome-shaped. The crowded polyps usually share walls with adjacent polyps and therefore have little or no space between them. At least 7 species are recognized in the western Pacific; commonly encountered on reef flats and in areas of turbid water. The various species are difficult to distinguish underwater.

Natural History: Polyps are seen only at night.

Distribution: Melanesia. The genus extends from East Africa; Red Sea to Philippines; Melanesia; Line Islands; Tuamotu Archipelago; and southern Japan (New Georgia Group, Solomon Islands).

G.C. Williams
G.C. Williams

272. cf. *Goniastrea australensis* (Milne Edwards & Haime, 1857)
Brain Coral

Identification: Colonies are encrusting, flat and circular in shape, or dome-shaped. The valleys and walls that comprise the surface patterns are sinuous, uniform in depth or height, and usually differently colored. Color variable: mostly cream, brown, grey, or green. This species is virtually impossible to tell apart underwater from *Platygyra daedalea* (Ellis & Solander, 1786).

Natural History: Both species are common in a variety of reef habitats throughout much of the western Pacific.

Distribution: Australia to much of Melanesia; Micronesia and Polynesia; southern Japan (Batangas, Luzon, Philippines).

273. *Platygyra lamellina* (Ehrenberg, 1834)
Brain Coral

Identification: Colonies are flat, hemispherical, or spherical. The sinuous walls comprising the surface ornamentation are uniform in height and width. At least 5 species of *Platygyra* inhabit the western Pacific, all of which are difficult to distinguish.

Natural History: Polyps are usually extended only at night.

Distribution: Red Sea to Solomon Islands; Polynesia (Florida Group, Solomon Islands).

84

274. *Leptoria phrygia* (Ellis & Solander, 1786)
Brain Coral

Identification: Colonies are often mound-like or form upright columns with a very lumpy or lobular surface. The narrow valleys are the same width as the walls. The walls and valleys are very sinuous and uniform in width and height. These are usually colored differently: dark green, cream, or brown.
Natural History: Commonly encountered on reef slopes but generally absent in areas of turbid water; shallow subtidal.
Distribution: Northern Australia and Coral Sea to Solomon Islands (Mborokua, Solomon Islands).

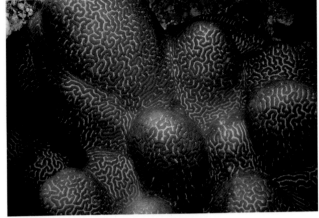
G.C. Williams

275. *Diploastrea heliopora* (Lamarck, 1816)

Identification: Large dome-shaped colonies up to a meter or more high and 4 or more meters in diameter. The individual polyp skeletons are mound-like (1-2 cm in diameter) and very densely situated on the very uniform surface of the colony. Color is grey, cream, or green.
Natural History: The polyps are seen only at night. Depth: 5-15 m.
Distribution: Madagascar; Red Sea to Australia; Philippines; Solomon Islands; Fiji; Samoa and Marshall Islands; southern Japan (Mborokua, Solomon Islands; Batangas, Luzon, Philippines).

G.C. Williams

G.C. Williams

276. *Echinopora lamellosa* (Esper, 1795)

Identification: Colonies sometimes resemble shelf fungus–composed of whorls or overlapping tiers of thin plates. The outer margin of the tiers is often salmon to pinkish, while the interior is mostly grey or brownish.
Natural History: Often encountered on shallow reef flats.
Distribution: Red Sea to Australia; Philippines; Micronesia and Samoa (Batangas, Luzon, Philippines).

G.C. Williams
G.C. Williams

277. *Echinopora mammiformis* (Nemenzo, 1959)

Identification: Colonies are usually plate-like and may have several bent, upright branches arising from the basal portion. The smooth polyp mounds appear breast-shaped. The surface of the colony between the polyps is smooth as spines are absent.
Natural History: Often abundant in quiet turbid-water habitats. In other species of the genus, numerous low spines are present on and between the polyp mounds.
Distribution: Australia to Philippines and Melanesia (New Georgia Group, Solomon Islands).

G.C. Williams

278. *Moseleya latistellata* Quelch, 1884

Identification: Colonies are disc-like, usually 10-20 cm in diameter. The individual polyp skeletons have angular walls which are shared by neighboring polyps. Color is golden-brown to chocolate brown.

Natural History: Polyps are observed only at night; intertidal to shallow subtidal.

Distribution: Australia; Philippines; Indonesia; New Guinea (Darwin, Australia).

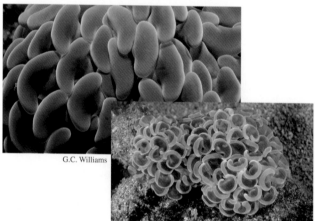

Brian Boer

279. *Euphyllia glabrescens* (Chamisso & Eysenhardt, 1821)

Identification: Colonial corals with distinctive polyps. Unlike other corals the species of *Euphyllia* can be differentiated from each other by their unusual tentacles, which are large and distinctive, and are extended during the day. The tentacles of *E. glabrescens* are tubular with rounded knob-like tips. Color is dark grey or brown with white or cream tips.

Natural History: The tentacles of this species may resemble those of the solitary hard coral *Heliofungia actiniformis* (#247) and some smaller individuals of the sea anemone *Stichodactyla mertensii* (#195).

Distribution: Red Sea to Australia; Malaysia; Philippines; Micronesia and Samoa (Sipadan, Borneo).

G.C. Williams

David K. Mulliner
Bert Hoeksema

280. *Euphyllia ancora* Veron & Pichon, 1980

Identification: Colonial corals in which the tentacles are tubular with tips that are distinctively T-shaped or anchor-shaped. They are often inflated and are either uniformly green or are brown with cream, white, or green margins.

Natural History: This species is sometimes very abundant on deeper reef flats, below 10 m.

Distribution: Australia; Philippines; Indonesia and parts of Melanesia; southern Japan (Batangas, Luzon, Philippines).

281. *Catalaphyllia jardinei* (Saville-Kent, 1893)

Identification: Fan-shaped corals with a large fleshy oral disc and tubular, pink or magenta-tipped tentacles. The greenish disc may have one to several mouths. Polyp color is metallic green with violet-tipped tentacles.

Natural History: Encountered only in areas of turbid water; conspicuous but not common. These corals live semi-buried in soft sediments, at the base of reef slopes, in sandy bottoms, with polyp and tentacles extending over the substratum; generally in depths greater than 20 m.

Distribution: Seychelles to Philippines; New Guinea; Solomon Islands; New Caledonia; Indonesia; southern Japan (Sulawesi, Indonesia).

282. *Plerogyra sinuosa* (Dana, 1846)
Bubble Coral or Grape Coral

Identification: Colonial corals often covered with clusters of bubble-like vesicles, each over 1 cm in length. Color is grey, bluish, cream, green or brown.

Natural History: Polyps and tentacles are extended to feed only at night. However, during the day, clusters of large inflated grape-like or bubble-like vesicles, which contain symbiotic algae, are extended to cover the surface of the coral. This species is capable of stinging the unsuspecting diver.

Distribution: Red Sea to Indonesia; Philippines; Micronesia and Melanesia (Batangas, Luzon, Philippines; Mindanao, Philippines).

T. M. Gosliner   Mike Miller

283. *Physogyra lichtensteini*
Milne Edwards & Haime, 1851
Bubble Coral or Grape Coral

Identification: These corals are similar to *Plerogyra*, but the vesicles are smaller (less than 1 cm in length).

Natural History: The clusters of small vesicles are extended during the day while the polyp tentacles are seen only at night. Like *Plerogyra*, this species can inflict a sting on the unwary diver.

Distribution: Madagascar; Red Sea to Australia; Okinawa; Micronesia and Melanesia (New Georgia Group, Solomon Islands).

G.C. Williams

284. *Flabellum* sp.

Identification: Solitary and free-living (semi-buried) corals that are non-reef-building. The single polyps are usually oval in shape, and the mouth is elongated. The skeleton is cup-like and often has projections at the base that protrude into the sandy or rubbly sea bottom. Color is highly variable–cream, pink, red, or orange are common. Many species have been described.

Natural History: These corals are encountered in deeper areas off or between coral reefs, mostly in regions of sand, gravel, or rubble; usually below 50 m in depth.

Distribution: Southwestern Indian Ocean. The genus has a worldwide distribution in deeper water (Natal, South Africa).

G.C. Williams
Fred McConnaughey

285. *Turbinaria* cf. *reniformis* Bernard, 1896
Plate Coral

Identification: Colonies are plate-like or scroll-like. The polyp mounds are circular in shape, and show conspicuous spaces between adjacent polyps. Color is golden brown or yellowish green.

Natural History: This coral often forms large mono-specific stands that can cover extensive areas such as ridge tops and slopes.

Distribution: Central Indian Ocean to Australia; Melanesia and western Polynesia (Milne Bay, Papua New Guinea).

G.C. Williams

286. *Turbinaria* cf. *frondens* (Dana, 1846)
Plate Coral

Identification: This coral forms frond-like plates. The plates are covered with numerous cylindrical or turret-shaped protuberances formed by individual polyps. Color usually grey to yellowish brown.

Natural History: This coral is common in several different reef habitats.

Distribution: Thailand; Australia; Philippines; Fiji; Samoa; Japan (Batangas, Luzon, Philippines).

G.C. Williams

287. *Turbinaria* sp.
Plate coral

Identification: Colonies form large and robust flat or convoluted plates, rolled scrolls, or upright columns. The polyp mounds are short, conical, or tubular, and usually well-spaced. Color is grey, brown, golden yellow, or green. At least 10 species inhabit the western Pacific; these are common and often abundant members of reef communities. The coral pictured here resembles both *Turbinaria bifrons* and *T. peltata*.

Natural History: Usually encountered on reef slopes.

Distribution: Philippines. The genus ranges from eastern Africa; Red Sea to Australia; Micronesia; Melanesia; Society Islands; southern Japan (Batangas, Luzon, Philippines).

G.C. Williams

288. *Tubastraea diaphana* Dana, 1846

Identification: The polyp tubes of these corals are usually arranged in clumps. Each tube is elongated and usually flares out toward the tip. Polyp tubes are dark chocolate brown as are the large flower-like polyps that extend at night.

Natural History: Often encountered in open areas of reef flats and slopes.

Distribution: Australia and Philippines (Batangas, Luzon, Philippines).

G.C. Williams

289. *Tubastraea* cf. *faulkneri* Wells, 1982

Identification: This flamboyant coral forms clumps of large polyp mounds. The clumps are fleshy and vivid pink in color. The large sunflower-like polyps are bright yellow-orange and emerge at dusk.

Natural History: Common and often very abundant in caves and on vertical walls. The nudibranch *Phestilla melanobrachia* (#624) and the snail *Epitonium billeeanum* (#493) feed on this coral.

Distribution: Australia; Philippines and Melanesia (Batangas, Luzon, Philippines).

88

290. *Tubastraea micrantha* Ehrenberg, 1834

Identification: This distinctive and beautiful coral often forms tall branched colonies (over 1 meter in height) with numerous cylindrical polyp tubes. The tubes are often apricot-colored with a greenish sheen or are uniformly dark green. The extended polyps are green or apricot-colored.

Natural History: Frequently encountered on reef flats or slopes with strong prevailing currents. The nudibranch *Phestilla melanobrachia* (#624) feeds on this coral.

Distribution: Australia; Philippines and Melanesia (Batangas, Luzon, Philippines).

G.C. Williams   G.C. Williams

291. *Dendrophyllia/Tubastraea* spp.

Identification: These two genera are, for the most part, only distinguishable by close examination of the internal skeletal structure of mature polyps. They both have tubular polyp skeletons that often flare toward the tips, and are arranged in clusters or along upright branches. The large brightly-colored polyps are extended mostly at night. At least ten species of both genera inhabit the western Pacific.

Natural History: Common on vertical walls, in caves, and rarely in open areas of reef flats. These corals are not reef builders and do not contain symbiotic algae in their tissues.

Distribution: Melanesia. The genera range in the Atlantic and Mediterranean and throughout the Indo-Pacific from Africa to Polynesia, and eastern Pacific (Madang, Papua New Guinea).

G.C. Williams

292. *Balanophyllia bairdiana*
Milne Edwards & Haime, 1848
Cup Coral

Identification: Solitary cup-like corals that do not have symbiotic algae and are not reef-forming. Brightly colored, often pink, orange, or yellow. More than 10 species of this genus occur in the western Pacific, but the genus has a worldwide distribution, mostly in temperate and tropical regions.

Natural History: Often encountered on the walls and ceilings of caves, overhangs, and vertical surfaces.

Distribution: Australia. The genus is cosmopolitan (New South Wales, Australia).

Carol Buchanan

Ceriantipatharia - Black Corals and Tube Anemones

Antipatharia - Black or Thorny Corals

Black corals are characterized by having an internal axis of a tough dark protein called horn. The axis is covered by minute thorns or prickles. Calcium carbonate is not present in the skeleton. The polyps usually have six, pointed, finger-like tentacles surrounding the mouth. Black corals, although not abundant, are frequently encountered in deeper parts of reefs and on vertical walls.

89

Kathy deWet

293. *Cirripathes* sp. 1

Spiral Wire Coral

Identification: This characteristic black coral has an unbranched spiral or corkscrew-like growth form. The polyps are distributed all around the stem. Color of living colonies is lemon-yellow or dull orange. The axis is dark brown to coal black.

Natural History: Several species of *Cirripathes* inhabit coral reef areas, particularly on barrier reef walls, and deeper areas of reef flats. Depth: 10-50 m.

Distribution: Indonesia. The genus inhabits much of the Indo-Pacific from Africa to Polynesia (Manado, Indonesia).

294. *Cirripathes* sp. 2

Whip Coral or Sea Whip

Identification: These black corals are long, slender and whip-like. The polyps are contained all around the central axis. As in other black corals, the dark internal axis of horn is covered with minute spines. Several species of *Cirripathes* have been named including *C. anguina, C. contorta, C. propinqua,* and *C. spiralis.*

Natural History: Frequently encountered on deeper reefs.

Distribution: Melanesia (Fiji).

Bruce Watkins

G.C. Williams
G.C. Williams

295. *Cirripathes* sp. 3

Wire Coral

Identification: A black coral that is very sinuous and often forms large open spirals. The polyps have six tapering finger-like tentacles, and are present on all sides of the stem. Color of living colonies is usually pale yellow or greenish.

Natural History: Commonly encountered in deeper portions of reefs or along reef margins. The pontoniine shrimp, *Dasycaris zanzibarica* (#728) and *Pontonides unciger* (#747) and a goby are often found living as commensals on this species.

Distribution: Philippines (Batangas, Luzon, Philippines).

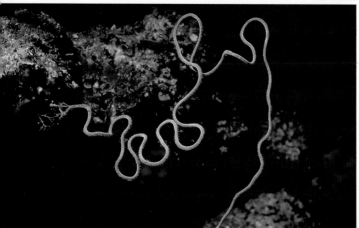

296. *Stichopathes* sp. 1

Wire Coral

Identification: Very similar to *Cirripathes* except that the polyps are contained in one or two longitudinal rows along the stem, leaving a substantial part of the stem smooth and bare. Several species of *Stichopathes* are found on coral reefs. Microscopic examination of the numerous spines that cover the internal axis is necessary to identify the species of *Stichopathes* and *Cirripathes.*

Natural History: Depth: 10-50 m.

Distribution: Melanesia (Madang, Papua New Guinea).

297. *Stichopathes* sp. 2

Whip Coral

Identification: The colonies of this species are distinctive in that they are whip-like. The polyps are contained on one side of the stem only. Color cream or grey.

Natural History: Frequently encountered in deeper portions of reefs or on reef margins.

Distribution: Philippines. The genus is found throughout much of the Indo-Pacific as well as the tropical Atlantic (Batangas, Luzon, Philippines).

G.C. Williams

298. *Stichopathes* sp. 3

Wire Coral

Identification: Colonies are unbranched and very sinuous. Color is a rich chocolate brown with cream-colored polyps. The tentacles are finger-like and distinctly pointed at the tips.

Natural History: Frequently encountered on reef margins.

Distribution: Philippines (Batangas, Luzon, Philippines).

D.W. Behrens

299. *Antipathes abies* Linnaeus, 1758

Bottlebrush Black Coral

Identification: The colonies are elongate and intricately branched around a single central stem, which results in a bottlebrush-like appearance. Several species of bottlebrush-like black corals have been described. This species name refers to the fir tree genus, *Abies*, for the bottlebrush-like arrangement of fir needles around the branches of some species.

Natural History: Small xanthid crabs (*Quadrella maculosa*) often inhabit the branches of these black corals–single pairs are commonly encountered, composed of one male and one female.

Distribution: East Africa; Mauritius; Indonesia; New Guinea; Philippines to Polynesia (Batangas, Luzon, Philippines).

G.C. Williams

Nicholas Galluzzi

300. *Antipathes* sp. 1

Bushy Black Coral

Identification: Finely branched, tree-like and often very bushy. Some colonies may reach 1-2 m in height. Color of living colonies is usually dull orange to tan.

Natural History: This is a luxuriant and beautiful species looking more plant-like than animal-like as it sways in the surge. Depth: 10-50 m.

Distribution: Indonesia. The genus is circumtropical and ranges throughout much of the Indo-Pacific from Africa to Polynesia (Sipadan, Borneo).

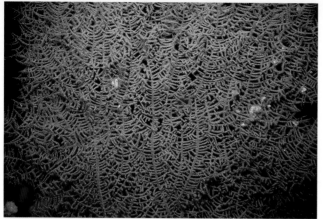

G.C. Williams

301. *Antipathes* sp. 2

Black Coral

Identification: A fan-like, planar species with dense and intricate branching, often up to 50 cm in height. Color is usually rust orange. Identification to species is difficult and depends on branching patterns and features of the minute spines covering the stem..

Natural History: Many species of *Antipathes* inhabit coral reef regions, on reef flats or on wall faces.

Distribution: Melanesia (New Georgia Group, Solomon Islands).

302. *Antipathes* sp. 3

Black Coral

Identification: Colonies are relatively small, rounded and compact, usually 15-20 cm in diameter, with very fine branching. Color is mostly lemon yellow. Over 30 species have of *Antipathes* have been described.

Natural History: Commonly encountered on ledges and vertical surfaces.

Distribution: Malaysia (Sipadan, Borneo).

Nicholas Galluzzi
G.C. Williams

Ceriantharia - Tube Anemones

Tube anemones are bottom-dwelling solitary animals that are actually more closely related to black corals than they are to sea anemones. They are inhabitants of areas of soft sediment. Cerianthids live in elongated tubes made of thread-like specialized cells (ptychocysts) interwoven with mucus. Two rings of tentacles surround the mouth - an inner ring of short tentacles and an outer ring of long slender tentacles.

G.C. Williams

◄ 303. Cerianthids

Tube Anemones

Identification: Mostly nocturnal, anemone-like animals inhabiting tough but flexible tubes. Tube anemones do not have hard skeletons like corals. Two separate rings of tentacles surround the mouth (see inset), unlike the sea anemones and their relatives that have only one ring. Three genera (*Cerianthus*, *Pachycerianthus*, and *Arachnanthus*) with several species each are probably the most commonly encountered. Positive identification (even to genus level) is impossible without examination of internal anatomy.

Natural History: The phoronid worm, *Phoronis australis* (#885) lives attached to the outside of the tube of various cerianthids.

Distribution: Worldwide distribution including the entire Indo-Pacific from Africa to Polynesia (Batangas, Luzon, Philippines).

Scyphozoa - Jellyfish

Mike Miller

304. *Cassiopeia andromeda* (Forskål, 1775)
Upside-Down Jellyfish

Identification: The bell is disc-shaped with eight branching oral arms that are elaborately fringed. Color is brown, grey, or greenish with white triangular markings around the margin of the bell. This species is remarkably similar in appearance to, and may be the same as, *C. xamachana* Bigelow, 1892, from the western Atlantic and Caribbean.

Natural History: It lies upside-down in shallows, usually in calm sandy areas, absorbing sunlight for photosynthesis by the symbiotic algae in its tissues. Intertidal to 10 m in depth.

Distribution: Tropical western Pacific (Mindanao, Philippines).

305. *Mastigias papua* (Lesson, 1837)

Identification: The bell of this species has conspicuous white spots. The margin of the bell is shallowly scalloped or fringed, and does not bear tentacles. This species is superficially very similar to and easily confused with *Phyllorhiza punctata*.

Natural History: This jellyfish is epibenthic, found swimming just off the shallow sea bottom in lagoons or embayments, often on rubble or muddy habitats.

Distribution: Western Pacific (Batangas, Luzon, Philippines).

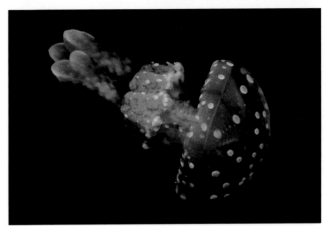

Mike Miller

306. *Nausithoe punctata* Kölliker, 1853
Tubular Sponge Polyp

Identification: The suspected polyp stage of a coronate jellyfish inhabits sponges, which appear as numerous separate tubes with four-part internal partitions and many slender tentacles at the opening of each tube. There is some confusion as to the proper generic name for this species; *Stephanoscyphus* has also been used for these scyphozoan polyps. Mayer (1910) states that this name was applied before it was known that the polyps merely prepresented a different part of the life cycle of *Nausithoe punctata*.

Natural History: Found in *Suberites*, *Desmacella* (# 18), *Mycale*, *Reniera*, *Esperia*, *Myxilla* and other sponges.

Distribution: Circumtropical and subtropical: Atlantic and Indo-Pacific (Indonesia).

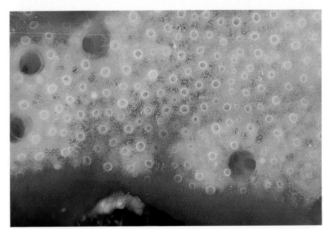

Lovell & Libby Langstroth
Mike Severns

307. cf. *Lipkea* sp.
Attached Jellyfish

Identification: Unlike most jellyfish, the stauromedusans are attached and not free-swimming. The genus *Lipkea* is distinguished by being cup-shaped (mostly less than 1 cm in length), without a basal stalk, and with 8 pointed tentacles that lack terminal knobs. Color is milky and translucent.

Natural History: Infrequently found attached to other bottom-dwelling organisms such as pteroeidid sea pens (pictured here).

Distribution: Indonesia. The genus was originally described from the Mediterranean, but the range is here extended to the western Pacific Ocean. (Manado, Indonesia).

93

Ctenophores - Comb-Jellies

Most comb jellies are transparent gelatinous predators inhabiting open water, and therefore are not covered in this book. However, one aberrant group of comb jellies, the unusual platyctene ctenophores, are flattened, bottom dwelling animals that are often brightly colored. Because of their resemblance to the more common flatworms, they are seldom recognized as ctenophores.

Some platyctenes are attached directly to the sea bottom, while others inhabit the surface of a variety of benthic organisms including soft corals and sea stars. These epizoic platyctenes may occur in aggregations that appear as mottled patterns on the host animals.

Five genera of platyctenes have been described worldwide, and several species are frequently encountered on Indo-Pacific coral reefs. Rankin (1955) gives a detailed description of the structure and biology of all five genera.

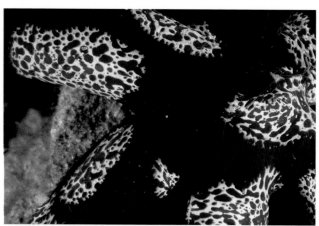

Mike Severns

308. *Coeloplana astericola* Mortensen, 1927
Identification: The body is flattened and oval-shaped. This species is dark reddish brown with a white reticulated pattern, producing large, bold, brown spots. A few short tube-like papillae often occur on the dorsal surface. On close inspection one can see the thread-like tentacles, which are pinnately branched. The tentacle sheaths are inconspicuous.
Natural History: Commonly found living on the sea star, *Echinaster luzonicus* (#968).
Distribution: Indonesia; Philippines; New Guinea; Solomon Islands (Batangas, Luzon, Philippines).

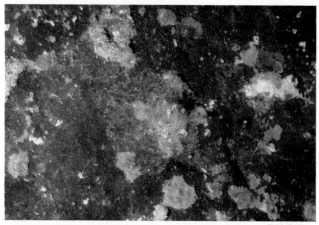

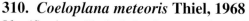

T. M. Gosliner
T. M. Gosliner

309. *Coeloplana* cf. *gonoctena* Krempf, 1921
Identification: A nearly transparent species, its only color is white and brown speckling. This species has two lateral flask-shaped tentacular sheaths and several small papillae on the body surface.
Natural History: This species has been reported to live on the surface of soft corals, including species of the genus *Cladiella*.
Distribution: South Africa; Vietnam; Hawaiian Archipelago (Maui, Hawai'i).

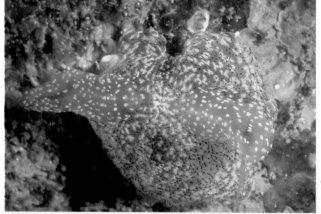

310. *Coeloplana meteoris* Thiel, 1968
Identification: The body is often somewhat globose with ends forming erect and robust chimneys. The thread-like tentacles are long and feather-like. The color is translucent, with brown specks and white spots.
Natural History: Usually encountered free-living on soft sediment. Often very abundant, up to 64 individuals have been observed per square meter.
Distribution: Somalia; Tanzania; Madagascar; Australia (Msimbati, Tanzania).

311. *Coeloplana* sp. 1

Identification: The body is flattened, oval-shaped and transparent with a network-like pattern of thin white lines. The shaped tentacle sheaths appear as low raised ridges, milky grey in color.

Natural History: In the Philippines, it is commonly found living on the soft coral *Sarcophyton*.

Distribution: Philippines (Batangas, Luzon, Philippines).

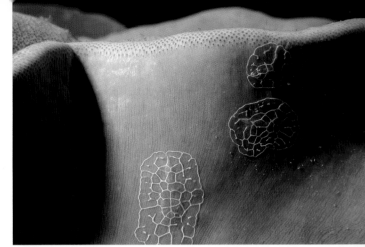

D.W. Behrens

312. *Coeloplana* sp. 2

Identification: The body is squat with ends forming urn-shaped chimneys. The body margin has a white band. Inside of this is a pattern of orange markings, which also surrounds the tentacle sheaths.

Natural History: Usually encountered on the surface of sea cucumbers.

Distribution: Melanesia (Madang, Papua New Guinea).

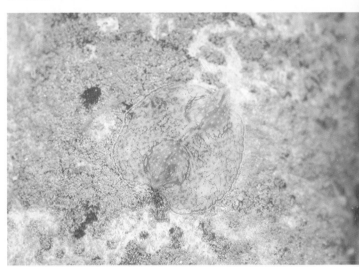

313. *Lyrocteis imperatoris* Komai, 1941

Identification: Two lobes arise from a common attached portion, which contains a broad mouth on the opposite end from the lobes. Two long and finely-branched tentacles emerge from the tips of the lobes, and can be completely retracted into them. A conspicuous furrow runs along each side of the body. Color is variable: red, pink, yellow or cream. Up to 26 cm in length.

Natural History: This unusual animal apparently feeds on soft corals. Depth: 58-73 m.

Distribution: Central Japan to Ryukyu Islands (Okinawa).

Leslie Newman & Andrew Flowers
Robert Bolland

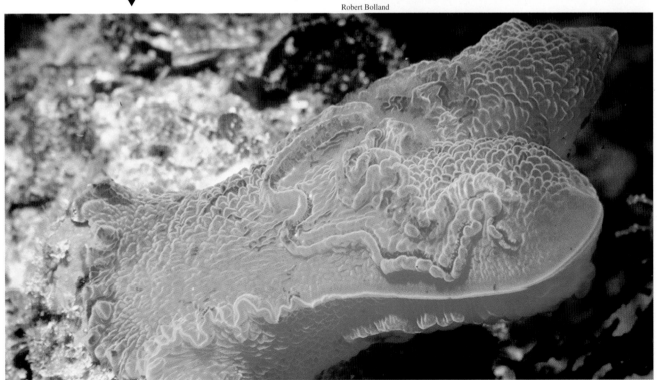

Platyhelminthes - Flatworms

Flatworms are bilaterally symmetrical animals whose soft bodies have become dorso-ventrally flattened. They have an obvious head region which may or may not bear eyes and tentacles. The body is covered with cilia, which are longer on the ventral surface, for motility. They have no special respiratory organs, and no body cavity; a common opening serves as both the mouth and anus. Most flatworms are hermaphroditic.

Acoel Flatworms

Acoel flatworms are very small, and are found on numerous substrates. Winsor (1990) reviews the Acoela of tropical

T. M. Gosliner

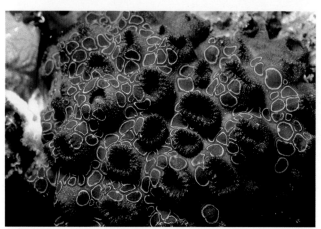

Bruce Watkins
D.W. Behrens

Australia. Some species have eyes or anterior sensory areas, and all have a sensory ganglion. Most species are thought to graze on the surface flora and fauna trapped in the mucus of the host. Gut analysis suggests that their diet consists of tiny crustaceans, mainly copepods, plus detritus and diatoms (L. Winsor, pers. comm.). A number of acoel taxa reproduce asexually, by fragmentation. Species like *Waminoa*, shown below, reproduce this way for much of the year. In a given population there are relatively few mature individuals, most showing evidence of regeneration following fragmentation (L. Winsor, pers. comm.). Some species are known to reproduce sexually. Asexual specimens cannot be identified, as the taxonomy is largely based upon characteristics of the reproductive organs.

314. cf. *Waminoa* sp. 1
Identification: This species of acoel is a flat disc in the shape of a pumpkin silhouette. The olive green color may be due to an algal or diatom symbiont, which is unknown (Trench, 1986). There is a yellow spot at the base of the caudal notch.
Natural History: Found on the bubble coral, *Plerogyra sinuosa* (#282), giving it the appearance of green spots.
Distribution: Australia; New Guinea; Indonesia; Philippines (Batangas, Luzon, Philippines).

315. cf. *Waminoa* sp. 2
Identification: This acoel is also disc-shaped and brown, matching the color of its host. It has a distinctive bluish white ring within its margin. This ring is probably an accumulation of concrement granules which reflect light (L. Winsor, pers. comm.). A yellow spot, which is found on the tail is probably an accumulation of pigment.
Natural History: Found in association with the zoanthid, *Palythoa* (#200), giving it the appearance of having a tunicate overgrowth. Has also been found on the mantle of the nudibranch *Phyllodesmium kabiranum* (#628), on sea urchin spines and the tentacles of several sea anemones. The brown coloration, matching the color of the host, is due to algal symbionts in the worm's tissue.
Distribution: Indonesia and Philippines (Indonesia).

316. cf. *Waminoa* sp. 3
Identification: This acoel is disc-shaped like the two above species, but transparent in body color. It has a distinctive white line down the center of the body, which is thought to be an accumulation of concrement granules (L. Winsor, pers. comm.).
Natural History: Associated with soft corals, and anemones.
Distribution: Philippines (Batangas, Luzon, Philippines).

317. cf. *Amphiscolops* sp.

Identification: This acoel has a long, highly distendable body. When stretched out it is pointed at the head end, and forked at the tail. Grey to tan with a faint white reticulating pattern.
Natural History: Found on hard coral and reef surfaces.
Distribution: Circumtropical: Ryukyu Islands; Japan; Bermuda; South America and Mediterranean (Batangas, Luzon, Philippines).

D.W. Behrens

318. *Convoluta* sp.

Identification: This minute acoel flatworm is usually seen only as a large mass of green on the bottom substratum. The genus is made up of several similar species, some of which may more appropriately be designated to separate genera.
Natural History: Photosynthetic algae live in the tissue of this worm, giving it the green color. Trench and Winsor (1987) review the details of such symbiotic relationships.
Distribution: South Africa (Natal, South Africa).

Polyclad Flatworms

Polyclad flatworms get their name from having "many" branches to their gut. They bear numerous eyes on the head. The presence of head tentacles varies. In those species having them, the tentacles are either formed by a fold in the body margin, or are found centrally on the head. Polyclads feed on a variety of colonial ascidians and smaller marine animals, which may include other flatworm species. This behavior demonstrates non-specificity in their diet. Many of the flamboyantly colored species are believed to display aposematic or warning coloration. Several of the species presented here mimic nudibranch species. Australia is believed to have 270 species; 90% are undescribed. Refer to Newman and Cannon (1994 a, b) for a review of the genera *Pseudoceros* and *Pseudobiceros*.

G.C. Williams

319. *Acanthozoon* sp.

Identification: The ground color of this polyclad is white. The pseudotentacles are formed by two tall folds in the body margin. The dorsal surface of this species is covered with tall white papillae tipped in orange. There is a white line along the margin, within which is an undulating black pattern. There is also an irregular black pattern down the midline of the body. Length: to 100 mm.
Natural History: A very common, yet undescribed species. This cryptic species is found under boulders and rubble.
Distribution: Indonesia; New Guinea (Papua New Guinea).

320. *Bulaceros porcellanus*
Newman & Cannon, 1996

Identification: Transparent cream, the midline is a broken white and brown line. There are random white dashes dispersed over the body. The marginal band is orange-brown with transparent rim. Length: 20 mm. Pseudotentacles bear knobs at their tips.
Natural History: A small cryptic species found under rubble.
Distribution: Australia; New Guinea (Madang, Papua New Guinea).

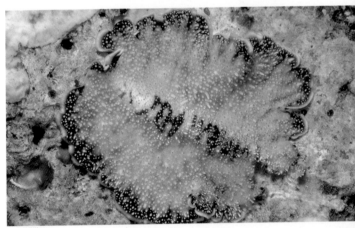

G.C. Williams

Leslie Newman & Andrew Flowers

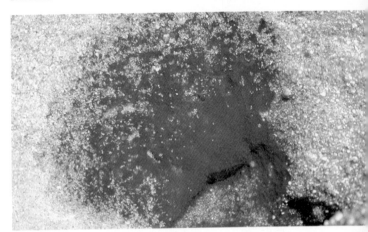

97

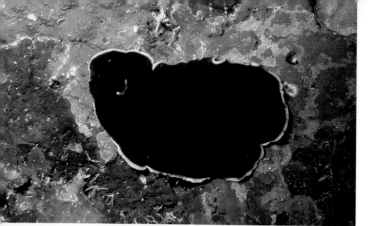

Leslie Newman & Andrew Flowers

321. *Callioplana marginata* Stimpson, 1857

Identification: This species is distinguished by the nuchal tentacles which arise a distance back from the margin of the head and appear as two orange conical horns. The tentacles are yellow, with a white area between them. The body is dusky with a dark brown midline and marginal bands consisting of white and orange. Length: 70 mm.

Natural History: A fast moving worm, common on reef crests.

Distribution: Mozambique; Sri Lanka; Micronesia; New Guinea; Japan; Australia (Madang, Papua New Guinea).

Lovell & Libby Langstroth

322. *Eurylepta fuscopunctatus* (Prudhoe, 1977)

Identification: A highly undulating white species with tightly spaced black lines set along the margin, above and below. The center white clearing has diffuse brown patches or a white on dull white, reticulated pattern. The margin of the body may be adorned with varying sized black spots. This species has narrow, erect pseudotentacles.

Natural History: Common from reef crest to slope, often on sponges.

Distribution: Maldives; Australia and Micronesia (Gosei Maru, Truk Lagoon).

Carol Buchanan
T. M. Gosliner

323. *Eurylepta* sp. 1

Identification: Dull white to tan with large deep, white, blotches and stripes. The entire body is covered with black spots, some of which become bars at the margin. There is a reddish region behind the head. The pseudotentacles are long and black.

Natural History: Nothing is known at present.

Distribution: Australia (Nelson Bay, New South Wales, Australia).

324. *Eurylepta* sp. 2

Identification: This long slender species has two reddish brown pseudotentacles. The body is grey with a white line around the margin and two thick black longitudinal stripes. A thinner broken stripe occurs along the midline.

Natural History: Nothing is known at present.

Distribution: Philippines (Batangas, Luzon, Philippines).

98

325. *Eurylepta* sp. 3

Identification: A white species with irregular brown spots down the center of the body. The margin has a series of black bars.
Natural History: Nothing is known at present.
Distribution: Tanzania and Philippines (Hamilo, Luzon, Philippines).

T. M. Gosliner

326. *Eurylepta* sp. 4

Identification: White with dramatic thin, black slash marks. The head tentacles are black with some red pigmentation. There is a yellow midline with black transverse streaks.
Natural History: Nothing is known at present.
Distribution: East Africa (Msimbati, Tanzania).

T. M. Gosliner

327. *Maiazoon orsaki* Newman & Cannon, 1996

Identification: A dull cream-tan polyclad has a highly ruffled margin. It has a thin white line mid-dorsally, and a black line along the edge of the body, inside of which is a wide fading burgundy marginal band. This species has 3-5 female and two male pores. The pseudotentacles are square and ruffled. Length: 35 mm.
Natural History: Found on coral reef faces and on sand bottoms, active at night.
Distribution: Maldives; Philippines; Melanesia; Micronesia (Madang, Papua New Guinea).

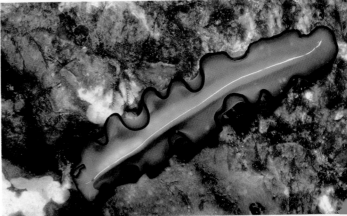

Leslie Newman & Andrew Flowers

Leslie Newman & Andrew Flowers

328. *Paraplanaria oligoglena* (Schmarda, 1859)

Identification: Nearly transparent, this species has a finely ruffled margin. There is a pattern of tan and orange with white and brown specks. Nuchal tentacles are striped and very small. Margin with orange and white spots. Length: 80 mm.
Natural History: Moves swiftly. Probably feeds on crustaceans and other polyclads.
Distribution: Australia; New Guinea; Marshall Islands; Hawai'i (Heron Island, Australia).

Leslie Newman & Andrew Flowers

329. *Phrikoceros* sp. 1

Identification: In this recently described genus, individuals have one male and female gonopore. The pseudotentacles are square and ruffled. Dark brown with tan specks, which form bars at the margin. The margin has a series of small blue specks. The margin is only shallowly ruffled.
Natural History: Nothing is known at present.
Distribution: Australia (Great Barrier Reef, Australia).

Mike Miller

330. *Phrikoceros* sp. 2

Identification: A bright, orangish red polyclad with densely set white specks. There is a wide clear area, free of these specks along the margin. The pseudotentacles are square and ruffled. The margin is highly undulate.
Natural History: Nothing is known at present.
Distribution: Philippines (Philippines).

Mike Miller

331. *Phrikoceros* sp. 3

Identification: A dull white species covered with thin, densely set white lines. Brown specks of varying sizes are sprinkled over the body surface. The pseudotentacles are square and ruffled.
Natural History: Nothing is known at present.
Distribution: Melanesia (Fiji).

T. M. Gosliner

332. *Phrikoceros/Pseudobiceros* sp.

Identification: The pseudotentacles are square and ruffled. The body is deep brown in color. The body is covered with large irregular white spots which vary greatly in size. The spots near the margin are smaller, closer together and yellowish in color.
Natural History: Nothing is known at present.
Distribution: Hawaiian Archipelago (Midway Atoll, Hawaiian Islands).

333. *Prosthiostomum trilineatum*
Yeri & Kaburaki, 1918

Identification: A long thin, yellowish tan species, up to 50 mm long. It has a white medial stripe which is edged with red-orange. Anteriorly there is a tranverse orange band.
Natural History: Found under rocks to depths of 5 m.
Distribution: Singapore; New Guinea; Belau; Philippines; Hawai'i; Japan (Batangas, Luzon, Philippines).

D.W. Behrens

334. *Pseudobiceros bedfordi* (Laidlaw, 1903)

Identification: One of the most well-known and frequently seen Indo-Pacific flatworms. Brown to black with yellow transverse stripes which are pink at the center and brown-black at their edges. These stripes may vary in width depending on location. Margin black with white dots. The remainder of the body is covered with yellow dots. The pseudotentacles are ruffled. Members of the genus are characterized by having two penises. Length: 80 mm.
Natural History: It is reported to be a swift crawler, and feeds on tunicates and crustaceans, which it engulfs.
Distribution: Mozambique; Singapore; Belau; Vietnam; Indonesia; New Guinea; Philippines; Australia; Federated States of Micronesia (Madang, Papua New Guinea).

T. M. Gosliner

335. *Pseudobiceros fulgor*
Newman & Cannon, 1994

Identification: A striking species, the pseudotentacles are formed by folds in the margin. Color is brown or pink, with densely set, thin white reticulating lines or stripes. It has a black marginal band with numerous yellow dots and dashes, or bright yellow lines.
Natural History: This species secretes large amounts of clear mucus and disintegrates when handled.
Distribution: Indonesia; Micronesia; Philippines; Australia; Marshall Islands (Queensland, Australia).

Carol Buchanan
Mike Miller

336. *Pseudobiceros gloriosus*
Newman & Cannon, 1994

Identification: Body velvety black with three marginal bands. The inner band is wide and orange, the middle band is narrow and pink. The outer band is narrow and dark burgundy. Large specimens, 90 mm long, have a thin pink midline.
Natural History: Under ledges on reef slopes.
Distribution: Melanesia and eastern Australia (Fiji).

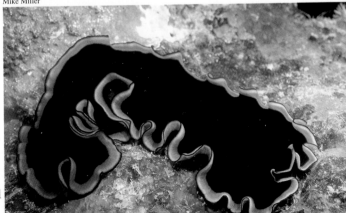

101

Leslie Newman & Andrew Flowers

337. *Pseudobiceros gratus* (Kato, 1937)

Identification: Translucent white, the margin is highly undulating. There is a thin black line along the margin and three or four wide black stripes down the body. The medial stripe terminates prior to meeting the lateral stripes. In some specimens this stripe may have a light center. Length: to 50 mm.

Natural History: Extremely fragile and common on reef crest.

Distribution: Mozambique; Sri Lanka; Australia; Indonesia; New Guinea; Philippines; Micronesia; Japan and Hawai'i (Great Barrier Reef, Australia).

Jack Randall

338. *Pseudobiceros hancockanus* (Collingwood, 1876)

Identification: The margin of the body of this species is highly undulating; the pseudotentacles square or ruffled. Its body is black with a wide orange band and white rim. Length: to 140 mm.

Natural History: Common on reef slope, usually associated with sponges.

Distribution: Maldives; Lakshadweep; Malaysia; Philippines; Singapore; eastern Australia; Fiji; Japan and Marshall Islands (Fiji).

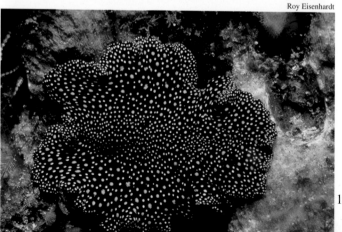

T. M. Gosliner
Roy Eisenhardt

339. *Pseudobiceros* sp. 1

Identification: This species is bright lime green, with a white and black marginal stripe, and white mid-dorsal stripe. The pseudotentacles are ruffled and the margins are highly undulate.

Natural History: On sand bottoms, active nocturnally.

Distribution: Philippines (Batangas, Luzon, Philippines).

340. *Pseudobiceros* sp. 2

Identification: Similar in coloration to *Thysanozoon flavomaculatum*, except it lacks papillae and has no white line along the margin, and the yellow spots are interspersed with white spots. At the margin of the body, the spots are exclusively white.

Natural History: Little is known about this species.

Distribution: Melanesia (Solomon Islands).

341. *Pseudobiceros* sp. 3

Identification: This green polyclad has an extremely undulating body margin. The entire body is covered with white specks.

Natural History: Nothing is known of its natural history.

Distribution: New Guinea; Indonesia; Philippines (Batangas, Luzon, Philippines).

Mike Miller

342. *Pseudobiceros* sp. 4

Identification: This polyclad is white in color. The entire body is covered with yellow spots of varying sizes, which are haloed in black.

Natural History: Nothing is known of its natural history.

Distribution: Philippines (Batangas, Luzon, Philippines).

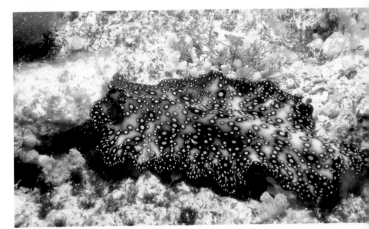

Mike Miller

343. *Pseudobiceros* sp. 5

Identification: A dull brown polyclad, it has a complicated marginal stripe. The outer most edge is a thin white line. Inside this is brown, similar to the body color. This brown turns to orange, which is followed by a black line.

Natural History: Found on sand bottoms, active at night.

Distribution: Philippines (Batangas, Luzon, Philippines).

Mike Miller
T. M. Gosliner

344. *Pseudobiceros* sp. 6

Identification: This species has a highly ruffled margin. The body is black with an orange and white marginal band. Three white stripes follow the middle of the body.

Natural History: Mimics the chromodorid nudibranch *Chromodoris magnifica* (#576).

Distribution: Philippines (Puerto Galera, Mindoro, Philippines).

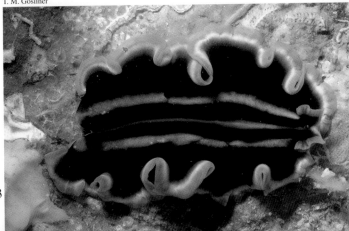

D.W. Behrens

345. *Pseudobiceros* sp. 7

Identification: The body is dark brown with white specks forming a barred pattern. It has an orange clearing along the midline. The margin is black.

Natural History: Shallow water silty habitats.

Distribution: Philippines (Batangas, Luzon, Philippines).

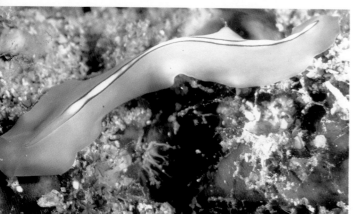

Terry Schuller

346. *Pseudoceros bifurcus* Prudhoe, 1989

Identification: In this genus the pseudotentacles are simple folds. Cream to bluish lavender in color, this species has a distinctive white stripe down its midline. The stripe is orange at the anterior end, and posteriorly it is bordered with a thin, deep purple edge. Eyes are apparent on the head just anterior to this line. Length: 60 mm.

Natural History: Found only on reef slopes. Feeds on colonial tunicates.

Distribution: Madagascar; Comoro Islands; Indonesia; eastern Australia; Philippines (Indonesia).

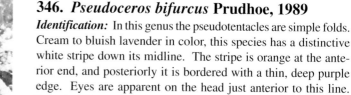

T. M. Gosliner
Leslie Newman & Andrew Flowers

347. *Pseudoceros bimarginatus* Meixner, 1907

Identification: A easily identified species, it is dull white. The head has long tentacles, which are folds of the body margin. The anterior end may be white or purple. There is a thin bright white mid-dorsal line. The margin is striped, from outer edge inward: narrow yellow, black and wide orange. Length: 30 mm.

Natural History: This species' bright coloration suggests it is displaying aposematic coloration. Observed day and night on sand and boulders on the reef flat.

Distribution: Somalia; Djibouti; Philippines; eastern Australia; Marshall Islands (Batangas, Luzon, Philippines).

348. *Pseudoceros confusus*
Newman & Cannon, 1995

Identification: The body is cream, with a narrow white center line, which does not connect to the marginal band. The margin is made up of five lines: a thin yellow, a thin black, a wide orange, a wide black and a thin white inside. Similar to *Pseudoceros bimarginatus*, except the orange marginal line is bordered by two black lines. Length: 30 mm.

Natural History: Cryptic on colonial ascidians, upon which it feeds.

Distribution: Australia (Great Barrier Reef, Australia).

349. *Pseudoceros contrarius*
Newman & Cannon, 1995

Identification: White or cream with a thin yellow midline. The marginal band has three colors: a thin yellow inner band, a thin black middle band and a wider, orange outer band. It is similar to *Pseudoceros bimarginatus*, except the three marginal color bands are reversed. Anteriorly the black band forms a deep V with a white spot, connecting to the yellow midline.
Natural History: Found on reef crests.
Distribution: New Guinea (Madang, Papua New Guinea).

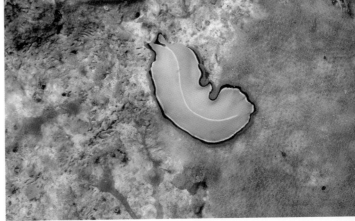

Leslie Newman & Andrew Flowers

350. *Pseudoceros depiliktabub*
Newman & Cannon, 1994

Identification: Velvety black. The marginal band is yellow cream with a distinct bright orange rim. A hint of green may be found between the black body and marginal band. Length: 25 mm.
Natural History: Found under coral rubble. Depth: 4 m.
Distribution: New Guinea (Madang, Papua New Guinea).

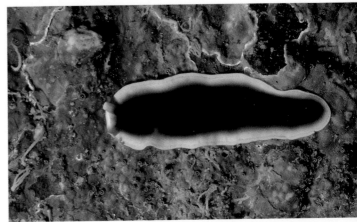

Leslie Newman & Andrew Flowers

351. *Pseudoceros dimidiatus*
(Graff in Saville-Kent, 1893)

Identification: Highly variable in color pattern. Body black with an orange margin and wide yellow lateral stripes or zebra-like bars. The barred variation was misidentified by George and George (1979) as *P. zebra*. It may have three wide black and two wide yellow longitudinal stripes. In some specimens the yellow stripes may be thinner than the black ones. Length: 80 mm. Four variations are shown here.
Natural History: Displays aposematic coloration.
Distribution: South Africa; Red Sea; Mozambique; Tanzania; New Caledonia; New Guinea; Solomon Islands; Australia; Philippines; Indonesia; Marshall Islands; Hawai'i (Indonesia; Solomon Islands).

Terry Schuller
Roy Eisenhardt

105

T. M. Gosliner

352. *Pseudoceros ferrugineus* Hyman, 1959

Identification: The body is deep red in this striking flatworm. It also has a yellow marginal band. The center of the body contains densely set white dots. The area between the yellow marginal band and the speckled area may appear as a wide, deep purple or burgundy band.

Natural History: An aposematic species, it feeds on colonial ascidians.

Distribution: New Guinea; Belau; eastern Australia; Philippines; Hawai'i (Dakak, Mindanao, Philippines).

Leslie Newman & Andrew Flowers

353. *Pseudoceros goslineri* Newman & Cannon, 1994

Identification: Variably colored, cream mottled with orange, pink and red dots. The marginal band is composed of irregular pink and purple dots and spots. Named in recognition of Terry Gosliner. Length: 70 mm.

Natural History: Cryptic; found on reef slopes.

Distribution: Tanzania; Maldives; New Guinea to eastern Australia (Great Barrier Reef, Australia).

Leslie Newman & Andrew Flowers
Leslie Newman & Andrew Flowers

354. *Pseudoceros imitatus* Newman, Cannon & Brunckhorst, 1994

Identification: The surface of this polyclad is bumpy, forming pustules. The body color is creamy grey with yellow to white spots. There is a black reticulate pattern encircling groups of spots. The pseudotentacles are black. Length: 20 mm.

Natural History: Resembles the phyllidiid nudibranch, *Phyllidiella pustulosa* (#597), both in color and texture (Newman, Cannon & Brunckhorst, 1994). The flatworm may derive protection by resembling the nudibranch which secretes a noxious chemical to ward off predators. Depth: 12 m.

Distribution: New Guinea and Australia (Heron Island, Australia).

355. *Pseudoceros jebborum* Newman & Cannon, 1994

Identification: The body is cream orange fading to grey black near the margin. The marginal band is cream with a bright yellow edge. Similar to *Pseudoceros paralaticlavus* (#359), but *P. jebborum* has a black band, not a black background. Length: 70 mm.

Natural History: Found under reef rubble.

Distribution: Eastern Australia; New Guinea; Hawai'i (Madang, Papua New Guinea).

356. *Pseudoceros leptostichus* Bock, 1913

Identification: The body color is cream near the center, fading to white. The margin is black, interrupted with yellow at rim. There are large dark brown blotches along the midline. The body is covered with brown spots becoming orange near the margin. Length: to 30 mm..

Natural History: Found on colonial ascidians under rubble.

Distribution: New Guinea and Australia (Heron Island, Australia).

Leslie Newman & Andrew Flowers

357. *Pseudoceros lindae*
Newman & Cannon, 1994

Identification: This species is somewhat variable in color pattern. It may be brown to purple in body color. Some specimens have a series of large creamy yellow spots along the outer edge of the body, and numerous golden yellow oval spots over the center of the body. Margin is blue. The pseudotentacles have a hint of blue. Length: to 50 mm.

Natural History: Under ledges on reef slopes. Active at night.

Distribution: South Africa; Indonesia; New Guinea; eastern Australia (Heron Island, Australia).

Leslie Newman & Andrew Flowers

358. *Pseudoceros monostichos*
Newman & Cannon, 1994

Identification: Cream with a distinct narrow black-brown median line, which bisects the eyespots anteriorly. This line is bordered with a white line, then light brown. The narrow margin has four indistinct bands, fading from yellow to green, to blue, to purple at the rim.

Natural History: Under rubble and ledges.

Distribution: New Guinea and Australia (Madang, Papua New Guinea).

Leslie Newman & Andrew Flowers
Leslie Newman & Andrew Flowers

359. *Pseudoceros paralaticlavus*
Newman & Cannon, 1994

Identification: Velvety black with white-grey midline. The marginal band is yellow then white-grey. The pseudotentacles are black with a yellow rim. Length: 50 mm.

Natural History: Found on a yellow colonial ascidian under boulders.

Distribution: New Guinea and eastern Australia (Heron Island, Australia).

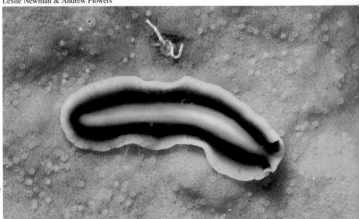

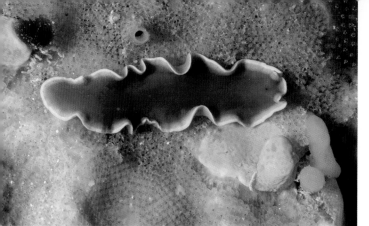

360. *Pseudoceros prudhoei* Newman & Cannon, 1994

Identification: Brown-orange with two marginal bands, the inner sky blue to mauve, the outer band is distinct yellow. Ventrally the same color. Length: 30 mm.

Natural History: Under boulders and rubble at the reef crest. Depth: 7 m.

Distribution: New Guinea and Australia (Heron Island, Australia).

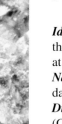

361. *Pseudoceros sapphirinus* Newman & Cannon, 1994

Identification: A jet black species, it has a thin white line around the edge of the body, and light blue to purple stripe near but not at the margin.

Natural History: Observed crawling across live *Acropora* both day and night. Secretes dark red mucus when disturbed.

Distribution: Eastern Australia; Philippines; Marshall Islands (Great Barrier Reef, Australia).

362. *Pseudoceros scintillatus* Newman & Cannon, 1994

Identification: A black flatworm with large irregular yellow-green spots, the margin of which are encircled with white. There is a distinct orange marginal band. Length: to 10 mm.

Natural History: Found associated with tunicates and under boulders.

Distribution: Previously known only from Australia; Philippines (Batangas, Luzon, Philippines).

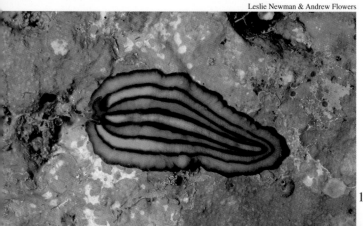

363. *Pseudoceros tristriatus* Hyman, 1959

Identification: The body color is brilliant blue with three wide orange longitudinal stripes, edged in black. The two lateral stripes are connected posteriorly. Length: to 50 mm.

Natural History: Found in silty areas under boulders at low water mark.

Distribution: Mozambique; Australia; Indonesia; New Guinea; Federated States of Micronesia (Madang, Papua New Guinea).

364. *Pseudoceros zebra* (Leuckart, 1828)

Identification: The body is white and the marginal stripe golden brown. There are a few black lateral marks which are very irregular in shape.

Natural History: Found under coral heads on shallow patch reefs.

Distribution: South Africa; Tanzania; Red Sea (Msimbati, Tanzania).

T. M. Gosliner

365. *Pseudoceros* sp. 1

Identification: In the photograph this flatworm is the animal to the left. Pure white in color, the margin of this species is made up of three thin lines, tan at the outer edge followed by orange and black, inside.

Natural History: This species closely resembles the chromodorid nudibranch *Chromodoris preciosa,* also shown on the right.

Distribution: New Guinea and Australia (Madang, Papua New Guinea).

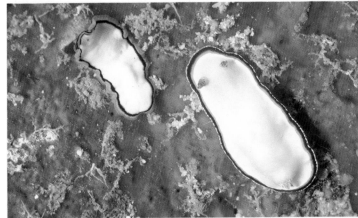

T. M. Gosliner

366. *Pseudoceros* sp. 2

Identification: The basic body color is dusky purple centrally, fading to white at the margins. There is a purple line around the edge of the body. The center of the body contains a broken striped pattern of orangish yellow oblong spots and stripes.

Natural History: Little is known about this species.

Distribution: Philippines (Batangas, Luzon, Philippines).

T. M. Gosliner
Mike Miller

367. *Pseudoceros* sp. 3

Identification: An interesting white polyclad, its head fades from brown to purple. It has a blue margin, and three long narrow longitudinal marks down the center of the body. These markings are composed of purple to maroon specks which are most dense at their margin.

Natural History: Little is known about this flatworm.

Distribution: Philippines (Philippines).

368. *Pseudoceros* sp. 4

Identification: The ground color is a granular white. This species has a wide golden brown band around the margin of the body, and three black longitudinal stripes.

Natural History: This species bears a striking resemblance to the chromodorid nudibranch *Chromodoris elizabethina* (#572).

Distribution: New Guinea (Madang, Papua New Guinea).

369. *Pseudoceros* sp. 5

Identification: Light purple in body color, with a dark purple margin. There are lighter blotches scattered over the surface of the body, and a yellow mid-dorsal stripe.

Natural History: Found under coral rubble in reef passes.

Distribution: Seychelles (Aldabra Atoll).

370. *Pseudoceros* sp. 6

Identification: Creamy white in color, it has a bright yellow marginal stripe. There are several irregularly-shaped, black blotches on the dorsal surface.

Natural History: Little is known about this species.

Distribution: Philippines (Philippines).

371. *Pseudoceros* sp. 7

Identification: White in body color, the margin bears a thin red line followed by blue stripe. There are double orange stripes along the midline. Mid-laterally, there are wider yellow-orange stripes along each side.

Natural History: Virtually nothing is known about this animal.

Distribution: Micronesia (Belau).

372. *Pseudoceros* sp. 8

Identification: A beautiful species, it is purple along the margin and has bright orange pseudotentacles. The white body is covered with large and small diffuse violet spots, surrounded with varying degrees of orange. Length: 30 mm.
Natural History: Virtually nothing is known about this flatworm.
Distribution: Australia (Marsh Shoal, Australia).

Carol Buchanan

373. *Pseudoceros* sp. 9

Identification: This polyclad bears a striking resemblance to the chromodorid nudibranch *Chromodoris geometrica* (#573). As in the nudibranch, the grey body has opaque white tubercles and a network of black lines. The pseudotentacles vary from brown to greenish, similar to the rhinophores of the nudibranch.
Natural History: Found in the same habitat as *C. geometrica*, on patch reefs and under coral rubble.
Distribution: Philippines and New Guinea (Papua New Guinea).

T. M. Gosliner

374. *Pseudoceros* sp. 10

Identification: The body is whitish grey and is broken by black areas which form white spots and marks. It has an orange marginal band.
Natural History: Found in shallow water silty habitats.
Distribution: Philippines (Batangas, Luzon, Philippines).

D.W. Behrens
Roy Eisenhardt

375. *Thysanozoon nigropapillosum* (Hyman, 1959)

Identification: Members of this genus have two male pores. The dorsal surface is black, and is covered with short papillae, which are tipped in yellow. There is a bold white line around the edge of the body.
Natural History: Found on reef slopes to 20 m in depth.
Distribution: Maldives; Sri Lanka; Indonesia; New Guinea; Solomon Islands (Solomon Islands).

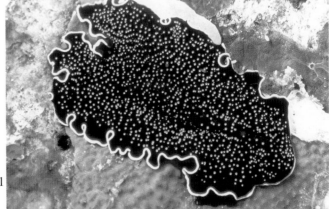

T. M. Gosliner

Leslie Newman & Andrew Flowers

Leslie Newman & Andrew Flowers

Laura Losito

Nemertea - Ribbon Worms

Unlike flatworms, ribbon worms are narrow and elongate and have a complete digestive tract with a separate mouth and anus. All species are predatory and utilize an eversible proboscis to capture and secure prey. Often, the proboscis is armed with a sharp stylet.

◀ 376. *Notospermus tricuspidatus* (Quoy & Gaimard, 1833)

Identification: This nemertean can be recognized by its uniform greenish black coloration with a prominent opaque white zigzag marking immediately behind the head.

Natural History: Found underneath coral rubble in relatively shallow water on the outer face of barrier reefs.

Distribution: Kenya; Australia; New Guinea; Indonesia; Loyalty Islands; and Ryukyu Islands (Madang, Papua New Guinea).

377. *Lineus* sp.

Identification: This species is uniformly green in color and has lateral slits on the side of the head.

Natural History: Little is known about the natural history of this species.

Distribution: Australia (Heron Island, Great Barrier Reef, Australia).

378. *Baseodiscus delineatus* (Delle Chiaje, 1825)

Identification: This moderately-sized nemertean has a greenish body color with many irregular cream to yellowish longitudinal lines.

Natural History: This species is found in the open on shallow living reefs.

Distribution: Cosmopolitan in tropical seas: Indo-Pacific, Atlantic and Mediterranean (Madang, Papua New Guinea).

379. *Baseodiscus hemprichii* (Ehrenberg, 1831)

Identification: This is an extremely large nemertean which may exceed one meter in length when fully extended. The two black bands on the head and the single, mid-dorsal black longitudinal line are distinctive. This and the following species have been reviewed by Gibson (1979).

Natural History: This species is found in shallow water on sandy bottoms, where it is nocturnally active.

Distribution: Mozambique; Mauritius; Red Sea; to Australia; Philippines; Japan; Samoa and Hawaiian Islands (Batangas, Luzon, Philippines).

112

380. *Baseodiscus quinquelineatus* (Quoy & Gaimard, 1833)

Identification: This species is similar to the preceding species in that it is large in size (more than a meter in length) and is white with black lines. It has five longitudinal black lines on the dorsal surface, one on each side, and two on the ventral surface.
Natural History: It is nocturnal, inhabiting shallow, sandy habitats.
Distribution: Australia; New Caledonia; New Guinea, Indonesia; Singapore; Philippines and Japan (Batangas, Luzon, Philippines).

D.W. Behrens

381. *Emplectonema* sp.

Identification: This species can be recognized by the relatively uniform greenish speckling on the body. The head has a prominent longitudinal crest on its dorsal surface.
Natural History: Found under coral rubble in shallow water.
Distribution: Philippines (Batangas, Luzon, Philippines).

382. Hoplonemertean sp.

Identification: The uniform bright red-orange color distinguishes this moderately small species. The yellow viscera are visible in the posterior third of the semi-transparent body.
Natural History: Found in shallow water on living reefs.
Distribution: Australia (Great Barrier Reef, Australia).

T. M. Gosliner

Annelida - Segmented Worms
Polychaeta - Polychaetes

In annelids each segment usually bears two pairs of parapodia and stiff chitinous bristles called setae (Fig. 3F). Many polychaetes have modified head and anterior segments with the development of antennae, tentacles, eyes and jaws. The setae and segments facilitate movement in both soft and hard bottom habitats. The setae also have a defensive function in many polychaetes. The polychaetes are the most diverse group of annelids, the dominant group in marine environments. Relatives of the earthworms (oligochaetes) and parasitic leeches are also present in the marine environment, but are not treated here. These groups have reduced body segments, parapodia and setae.

Leslie Newman & Andrew Flowers
T. M. Gosliner

383. *Laetmonice* cf. *moluccana* Horst, 1916
Sea Mouse

Identification: Members of the Aphroditidae are recognized by their fuzzy appearance with bundles of elongate golden setae. In *Laetmonice*, the setae are tipped with harpoon-like barbs (Fig. 3F). The present species is about 30 mm in length.
Natural History: Most members of the Aphroditidae are found in silty or muddy habitats, but this species is found under coral heads and coral rubble on shallow patch reefs.
Distribution: Australia; Indonesia and Philippines (Batangas, Luzon, Philippines).

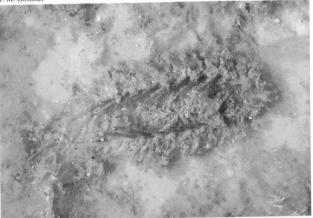

113

Carol Buchanan

Carol Buchanan

Marc Chamberlain

D.W. Behrens
T. M. Gosliner

Polynoid Worms

384. *Asterophilia carlae* Hanley, 1989

Identification: The orange coloration with numerous opaque white tubercles is distinctive. This species is closely related to *Gastrolepidia clavigera* (below), but lacks ventral lamellae and has 3-4 opaque mounds on the posterior margin of each scale.

Natural History: This species is an obligate commensal on asteroids, especially *Linckia levigata* (#954). The opaque white mounds appear to mimic the tube feet of the sea star.

Distribution: Australia; Indonesia; Vietnam; Fiji (Solitary Islands, New South Wales, Australia).

385. *Asterophilia* sp.

Identification: Similar to *A. carlae*, but lack the mounds on the posterior end of the scales and has different setae.

Natural History: This species is also commensal on asteroids and has been found on several variously colored species.

Distribution: Australia: Western Australia, Northern Territory, New South Wales (Solitary Islands, New South Wales, Australia).

386. *Gastrolepidia clavigera* Schmarda, 1861

Identification: This scale worm has at least 21 pairs of large, soft scales and the underside has large lamellae at the base of each parapodium. The color pattern of the worm closely mimics that of the host.

Natural History: This species is an obligate commensal on sea cucumbers, especially species of *Bohadschia* (such as *B. argus* #1024, shown here), *Thelenota*, *Stichopus* and *Holothuria*.

Distribution: South Africa; Madagascar and Mozambique to Australia; Malaysia; Philippines and Marshall Islands (Sipadan, Borneo; Batangas, Luzon, Philippines).

387. *Iphione muricata* (Savigny, 1818) and *Lepidonotus cristatus* (Grube, 1875)

Identification: In this photo two scale worms are shown. *Lepidonotus cristatus* on the left has paired bulbous appendages on the elytra or scales, and is brightly colored. The distinct color patterns may represent distinct species. *Iphione muricata* has 12 pairs of simple elytra and two conspicuous antennae on the head. It can curl into a ball when disturbed.

Natural History: Both species are found on the undersurface of dead coral heads on shallow patch reefs.

Distribution: Both species are widely distributed from the Western Indian Ocean of South Africa to the Western Pacific. *Iphione* is found as far east as the Marshall Islands (Sabang, Mindoro, Philippines).

114

388. *Paralepidonotus indicus* (Kinberg, 1856)

Identification: This scale worm has pairs of large brown scales and two elongate antennae projecting anteriorly. The underside has small lamellae at the base of each parapodium.

Natural History: Like the preceding one, this species is found on the undersurface of dead coral heads in shallow water. Gut contents reveal pieces of amphipods and other polycheate worms.

Distribution: Mozambique; Maldives; Australia; Philippines (Batangas, Luzon, Philippines).

T. M. Gosliner

389. *Loimia medusa* (Savigny, 1818)
Spaghetti Worm

Identification: The elongate, white tentacles of this species are readily visible and distinctive, while the bulk of the body is hidden within rocky crevices and under boulders. May represent a complex of several species.

Natural History: This species, like other terebellid polychaetes, uses its mucus-covered tentacles to pick up particulate organic material from the substrate. Found from shallow water to more than 30 m depth.

Distribution: Circumtropical: Western Indian Ocean to Hawaiian Islands and the tropical Atlantic (Maui, Hawaiian Islands).

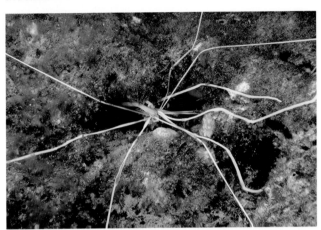
Mike Severns

390. cf. *Amphitrite* sp.

Identification: This is another terebellid polychaete. The elongate feeding tentacles, as well as the branched respiratory branchiae, are visible on the head.

Natural History: Found on the underside of stones and in rocky crevices in shallow water.

Distribution: Australia (Solitary Islands, New South Wales, Australia).

Carol Buchanan
G.C. Williams

391. *Chaetopterus variopedatus* Renier, 1804

Identification: This bizarre polychaete has its body divided into three distinct regions, which produce water currents and filter particulate matter through its tube.

Natural History: This species lives in a parchment tube usually covered with sand grains. The tubes are often cemented to the undersurface of coral heads on shallow reefs.

Distribution: Cosmopolitan (Batangas, Luzon, Philippines).

Amphinomidae - Fire Worms

392. *Chloeia flava* (Pallas, 1766)

Identification: Fire worms (Amphinomidae) are a group of polychaetes with needle-like setae that can inflict a painful sting. Species of *Chloeia* are shorter than other amphinomids and can further be disinguished by different colored markings on the dorsal surface. *Chloeia flava* has a distinct pattern of brownish markings on each segment, seen in the individual shown here.

Natural History: Species of *Chloeia* are found on sand and mud bottoms, actively crawling through the surface in search of prey or carrion, often attracted to lights at night.

Distribution: South Africa; Red Sea to Australia; Indonesia and Philippines (Batangas, Luzon, Philippines).

393. *Chloeia fusca* McIntosh, 1885

Identification: This species can be recognized by the two parallel longitudinal dark lines found on the dorsal surface.

Natural History: Found in the same habitat as the preceding species.

Distribution: Southern Africa; Amirante Islands, Seychelles; Red Sea; India to Indonesia; New Caledonia; Philippines and Japan (Batangas, Luzon, Philippines).

394. *Chloeia* cf. *parva* Baird, 1870

Identification: This species has a unique color pattern and could not be identified positively. It appears most similar to *Chloeia parva*.

Natural History: Found in the same habitat as the previous two species, it is also nocturnally active.

Distribution: Indonesia and Philippines (Batangas, Luzon, Philippines).

395. *Pherecardia striata* (Kinberg, 1857)

Identification: In contrast to species of *Chloeia*, this more typical fire worm has an elongate body shape. There are bundles of stiff white setae associated with each segment. This species gets its name "*striata*" from the series of pinkish brown and white longitudinal stripes on the back.

Natural History: Like most amphinomids, *Pherecardia striata* is predatory, and a scavenger. Here it is seen feeding on a dead smelt.

Distribution: Mozambique to the Hawaiian Islands (Batangas, Luzon, Philippines).

396. *Pherecardia* sp.

Identification: This species may differ from the preceding one in having smaller bundles of setae and larger, prominent branchiae above each setal bundle.

Natural History: Found on shallow water patch reefs at 1-2 m depth.

Distribution: Tanzania (Msimbati, Tanzania).

397. *Eunice* cf. *australis* Quatrefages, 1865

Identification: This species is darkly colored with a white bar on the head and white bar on segment four. The cirri have white and dark bands.

Natural History: Found under rocks in intertidal and subtidal.

Distribution: South Africa; Madagascar; Red Sea to Australia and New Zealand (Langebaan Lagoon, South Africa).

398. *Eunice* sp.

Bobbit worm

Identification: This undescribed species, with its 5 pairs of massive, spring-loaded jaws, appears like a frightening apparition from a science fiction movie. Its five tentacles make it a member of the Eunicidae. Members of this family are large polychaetes. Based on the twenty-five mm width of the body, this species may reach 5-9 m in length.

Natural History: This species is a voracious predator. It was observed to feed on a file fish of more than 150 mm in length. When the file fish ventured too close to the worm it emerged slightly from its burrow and seized the fish in its jaws with lightning speed. In an instant, the worm had pulled the fish beneath the sand surface and had begun to consume it. The worm is found in 1-2 m of water in coarse sand.

Distribution: New Guinea and Philippines (Batangas, Luzon Philippines).

399. *Palola* sp.

Palolo worm

Identification: This is the spawning stage of a normally bottom dwelling species of *Palola*. It is the gamete containing portion of the body, and lacks a head.

Natural History: At the new moon, these worms emerge for a mass spawning and portions of their body break away from the head region. These parts, or epitokes, contain eggs or sperm and swim rapidly through the water column spreading gametes to be fertilized. In Samoa, this spawning takes place in October or November, while in the Philippines spawning occurs in late February or early March.

Distribution: Widely distributed in the western Pacific as far east as Samoa (Puerto Galera, Mindoro, Philippines).

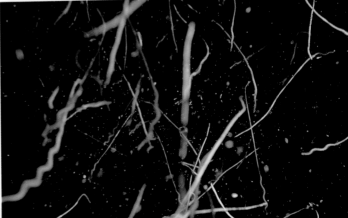

117

400. *Hesione splendida* Savigny, 1818

Identification: This polychaete has eight pairs of tentacular cirri. Its striking coloration is distinctive, with brown striations and spots, and yellow pigment on the head and at the base of the notopodial setae.

Natural History: A carnivorous species that is found on the undersurface of coral rubble on shallow reefs.

Distribution: Circumtropical: South and East Africa to Australia and Hawai'i; Atlantic Ocean (Heron Island, Great Barrier Reef, Australia).

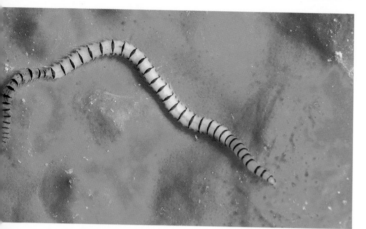

401. *Odontosyllis* sp.

Identification: This species of polychaete can be distinguished by its white body with black lines between segments.

Natural History: This species appears to be associated with sponges, under rubble on shallow reefs.

Distribution: Australia (Heron Island, Great Barrier Reef, Australia).

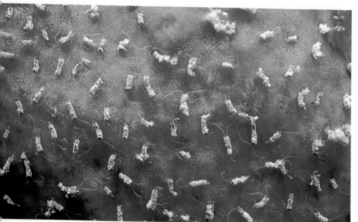

402. *Polydorella prolifera* Augener, 1914

Identification: This spionid lives in a neatly constructed particulate tube.

Natural History: Individuals apparently distribute themselves evenly over the surface of their substratum, in this case a red encrusting sponge, at a distance equalling the length of their tentacles.

Distribution: India; Indonesia; Australia (Indonesia).

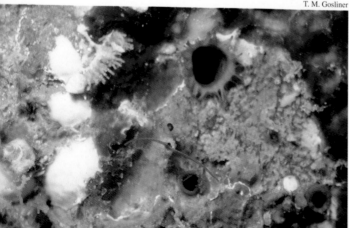

403. Spionid sp.

Identification: Spionid polychaetes are a group of tube building annelids which are abundant in a variety of habitats. They have a pair of elongate tentacles emerging from the head, which may be extended as shown in the photo, or coiled like ram's horns.

Natural History: Spionids are suspension feeders and use mucus stretched between the tentacles to capture plankton and other small prey.

Distribution: New Guinea (Madang, Papua New Guinea).

Serpulidae - Feather Duster Worms and Sabellidae - Fan Worms

In two families of polychaetes, the Serpulidae and Sabellidae, the head is modified into series of pinnate tentacles or branchiae that are arranged in a circular, spiral fashion. These tentacles are utilized for filter feeding and respiration. In serpulids the worm lives in a calcareous tube which it secretes, while in sabellids the tube is variable in construction, but never calcareous. In most serpulids one of the tentacles has been modified to form an operculum, which can seal off the tube, once the animal withdraws.

Marc Chamberlain

404. *Filograna implexa* Berkeley, 1828
Identification: This colonial species builds irregular masses of fine calcareous tubes. The small tentacular crown consists of about a dozen simple, capitate tentacles.
Natural History: It is frequently found in harbors and on pier pilings in disturbed settings; intertidal to considerable depths.
Distribution: Cosmopolitan: In many places it has been accidentally introduced through discharge of ballast water by ships. Malaysia (Sipadan Island, Borneo).

405. *Filogranella elatensis* Ben Eliahu & Dafni, 1980
Identification: This serpulid is similar to the preceding one, but has larger individuals with pinnate, reddish tentacles.
Natural History: This species is a filter feeder and is found on the edges of reefs, in coral rubble and on sandy substrata.
Distribution: Originally described from the Red Sea; also found from Australia; New Guinea; Indonesia; Philippines and southern Japan (Sulu Sea, Indonesia).

Fred McConnaughey

406. *Protula magnifica* Straughan, 1967
Identification: The calcareous tube of this species is about a centimeter in diameter. There is no operculum to close off the tube. The feeding structures consist of two spirally coiled tufts of pinnate tentacles.
Natural History: The calcareous tube is cemented to coral reefs and the tentacular crown often emerges between living coral heads. Russell Hanley (pers. comm.) has observed these worms to periodically shed and regenerate the tentacular crown.
Distribution: Red Sea; India; Sri Lanka; Australia; New Guinea; Indonesia and Philippines (Manado, Sulawesi, Indonesia).

Terry Schuller
Marc Chamberlain

407. *Protula* sp.
Identification: This species is characterized by its twin spiral tentacular crowns, each consisting of more than ten whorls. The color of the pinnate tentacles is orange with white patches. Like other members of the genus, it lacks an operculum.
Natural History: Found in coarse sand. Depth: 15-30 m.
Distribution: Luzon and Mindoro Islands, Philippines (Batangas, Luzon, Philippines).

119

Roy Eisenhardt

408. *Pomatostegus stellatus* (Abildgaard, 1789)

Identification: This small species of serpulid has tubes that emerge from living coral heads. The pinnate tentacles are arranged in a double horseshoe. The operculum has several distinct tiers.

Natural History: Found on living reefs, in shallow water.

Distribution: Cosmopolitan: western Indian Ocean to Australia; Caribbean; tropical Atlantic and European waters (Solomon Islands).

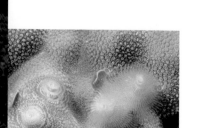

Roy Eisenhardt

D.W. Behrens

409. *Spirobranchus giganteus* (Pallas, 1766)
Christmas Tree Worm

Identification: The twin spirals of evenly spaced tentacles characterize the common Christmas tree worm. The tentacles are extremely variable in color: yellow, blue, purple, red, orange, or brown. There is a round, yellowish operculum.

Natural History: This species lives in calcareous tubes which penetrate living coral heads, from the intertidal zone to more than 30 m depth.

Distribution: Circumtropical: western Indian Ocean and western Pacific to the Caribbean and eastern Atlantic (Solomon Islands; Tuamotus).

410. *Bispira* sp.

Identification: The brownish parchment tube of this sabellid is clearly visible in the photo. This species could not be positiviely identified.

Natural History: The tube of this species is largely embedded in sand at the edge of a shallow patch reef.

Distribution: Philippines (Philippines).

James Hargrove
Jack Randall

411. cf. *Myxicola* sp. 1

Identification: The uniformly colored yellowish tentacular crown is deeply divided at one end. The tentacles are finely pinnate.

Natural History: Found on shallow patch reefs where the tubes are largely embedded in crevices.

Distribution: Seychelles (Praslin, Seychelles).

120

412. *Sabellastarte indica* (Savigny, 1818)

Identification: The large tentacular crown is horse-shoe shaped. The pinnate tentacles are variously banded.
This species is similar to *S. magnifica* from the Caribbean.
Natural History: The tubes are generally buried in sandy substrata or found in cracks in shallow reefs.
Distribution: Circumtropical: East Africa; Red Sea; Australia; New Guinea; and Philippines and from Senegal in the eastern Atlantic (Batangas, Luzon, Philippines; Indonesia).

Bruce Watkins

413. *Sabellastarte sanctijosephi* (Gravier, 1908)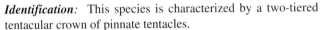

Identification: This species is characterized by a two-tiered tentacular crown of pinnate tentacles.
Natural History: The tube is embedded in substrate between coral heads of *Millepora* sp. on shallow patch reefs.
Distribution: South Africa; Philippines; Solomon Islands; Australia; Cook Islands; Tonga (Cebu, Philippines)

Sipuncula - Peanut Worms

Sipunculans are a group of unsegmented worms that are thought to be related to annelids and the echiurans. The body is divided into a trunk and a portion that can be everted, called the introvert. The anterior end of the introvert contains a complete or incomplete ring of tentacles around the mouth, which is used for deposit feeding. The Sipuncula has been reviewed by Stephen and Edmonds (1972) and Cutler (1994), and contains 320 species worldwide.

Jack Randall

414. *Phascolosoma nigrescens* Keferestein, 1865

Identification: This species is tan with black mottling over the surface and a series of distinct pigment rings on the introvert. It has a complete ring of short, unbranched tentacles around the mouth.
Natural History: This species inhabits dead and living reefs and forms burrows in blocks of coral.
Distribution: Circumtropical: East Africa; Madagascar; Red Sea, throughout the western Pacific to Baja California and into the Caribbean and eastern Atlantic (Batangas, Luzon, Philippines).

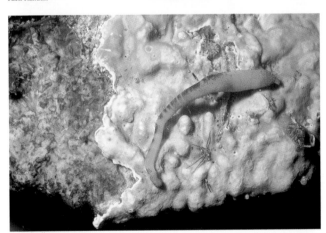

T. M. Gosliner

415. *Themiste lageniformis* Baird, 1868

Identification: The bulk of the body is whitish with a black ring below the end of the introvert. The introvert bears five or six tentacles which are highly branched.
Natural History: This species is found on the underside of coral rubble that is partially buried in coarse coral gravel on shallow patch reefs.
Distribution: South Africa; Tanzania; Madagascar; Red Sea to New Zealand; Australia; Indonesia; Philippines and Japan; also known from the eastern Atlantic of West Africa and the Caribbean (Msimbati, Tanzania).

T. M. Gosliner

Echiura - Tongue Worms

The echiurans are a small group of unsegmented worms that are closely allied to the annelids and sipunculans. Some species have chitinous setae as in the annelids. The anterior end contains a proboscis that may be bilobed or undivided. The 130 species worldwide were reviewed by Stephen and Edmonds (1972). Echiurans are infaunal burrowing organisms that feed by ingesting sediments.

416. *Bonellia sp.*

Identification: The species differences within this group are poorly known. This animal is frequently seen with only its elongate, strongly bilobed proboscis protruding from holes in coral. The bulk of the body is contained within the coral.

Natural History: The proboscis of *Bonellia* is covered with mucus and is protruded from coral to remove nutrients from sandy portions of reef communities. The female is large and conspicuous. The males are dwarfed and live in the same burrow as the female. The males rarely feed and function largely to fertilize the females.

Distribution: Widely distributed in the Indo-Pacific from the western Indian Ocean to New Guinea; Indonesia and Philippines (Madang, Papua New Guinea).

417. cf. *Archibonellia*

Identification: This species, with a slightly bilobed proboscis, appears to belong to this poorly known genus. The trunk of the body is longer and narrower than in the preceding species. Obviously more taxonomic and biogeographical study of these organisms is required.

Natural History: Found underneath coral rubble that is partially embedded in sandy substrate; on shallow patch reefs.

Distribution: Both described members of *Archibonellia* are known only from the tropical portions of western Australia and New Guinea (Madang, Papua New Guinea).

Leslie Newman & Andrew Flowers

T. M. Gosliner

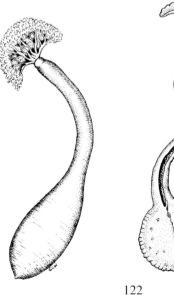

Mollusca: Solenogasters, Chitons, Snails, Nudibranchs, Bivalves, Tusk Shells, Octopods and their relatives

Aplacophora - Solenogasters

Solenogasters are worm-like mollusks that lack a shell, but have well-developed calcium carbonate spicules and a radula.

418. *Epimenia australis* (Thiele, 1897)

Identification: This species is recognizable by its worm-like appearance and mottled grey-brown body color, prominent bumps on the body, and large body size. The animals are generally about 100 mm in length, but may reach 300 mm, making it the largest known species of this group of mollusks.

Natural History: Most neomenoid solenogasters feed upon coelenterates. *Epimenia australis* has been found in association with alcyonaceans and pennatulaceans, most recently in association with *Scleronepthyea* sp. (#110) (Scheltema & Jebb, 1994).

Distribution: Indonesia; New Guinea; Japan (Madang, Papua New Guinea).

T. M. Gosliner

419. Aplacophoran sp.

Identification: This small white aplacophoran is probably a new species. It cannot be placed in a genus at the present.

Natural History: This species is found on the undersurface of dead coral rubble on patch reefs. Depth: 1-10 m.

Distribution: New Guinea (Madang, Papua New Guinea).

T. M. Gosliner

Polyplacophora - Chitons

Chitons are primitive, flattened mollusks with eight shell valves. They have a broad foot which enables them to withstand considerable wave shock. When removed from their rock, they tend to roll up into a ball for protection, like a pill bug.

420. *Acanthopleura spinosa* (Brugière, 1792)

Identification: Although most chitons are relatively small, *A. spinosa* may exceed 100 mm in length. It is easily recognized by the rough texture of the shell plates (valves) and the series of spiny appendages that protrude from the girdle.

Natural History: This species is found grazing on algae on the surface of elevated reefs just above the low tide line. It is usually present in open areas exposed to some wave action.

Distribution: Northern Australia; Indonesia and Philippines (Batangas, Luzon, Philippines).

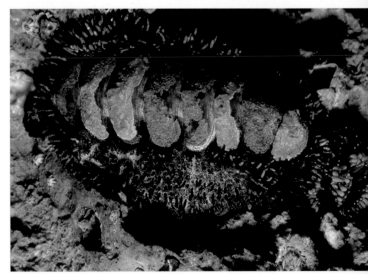
G.C. Williams

421. *Tonicia lamellosa* (Quoy & Gaimard, 1835)

Identification: This species is recognizable by the pattern of numerous eye-like sensory structures, called esthetes, on the surface of the valves.

Natural History: *Tonicia lamellosa* is common under rocks, just below the low tide mark. Most chitons are grazers and many species feed upon a mixture of algal and animal material.

Distribution: Tonga; Samoa; Fiji; northeastern Australia; New Caledonia; Indonesia; Philippines; Okinawa; Thailand (Batangas, Luzon, Philippines).

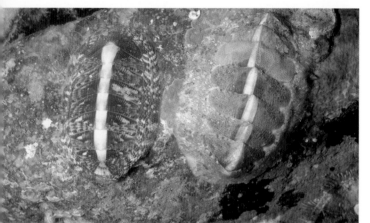

422. *Chiton discolor* Souverbie, 1866

Identification: Similar in appearance to *Tonicia lamellosa*, but lacks numerous sensory eyes on the surface of the valves. The soft tissue surrounding the shell plates (girdle) contains several rows of fine plates, while the girdle of *Tonicia* lacks these.

Natural History: *Chiton discolor* is shown here on the left, together with *Tonicia lamellosa* under rocks. It grazes on algae just below the low tide mark in water no deeper than 1 m.

Distribution: New Caledonia and Philippines (Batangas, Luzon, Philippines).

423. *Cryptoplax larvaeformis* (Quoy & Gaimard, 1835)

Identification: Owing to the reduction of the shell plates and increased girdle, this species looks more like a worm than a mollusk. It is far more flexible than other chitons and may reach a length of 150 mm.

Natural History: Nocturnal, seen actively crawling over exposed rocks at low tide. During the day, they are found underneath rocks. When exposed during the day, members of this species move out of the sunlight, rapidly seeking dark crevices and holes.

Distribution: Thailand; northern Australia; New Caledonia; New Guinea; Tonga; Loyalty Islands; Philippines (Batangas, Luzon, Philippines).

Gastropoda - Snails and Slugs

424. *Cellana sandwicensis* (Pease, 1861)
Opihi limpet

Identification: This limpet is readily distinguishable by its dark color and prominently raised ribs on the shell. It is one of four species of *Cellana* found in the Hawaiian Islands. Opihi have been an important food source for native Hawaiians.

Natural History: *Cellana sandwicensis* is found right at the low tide mark and slightly below. It inhabits basaltic rocks on exposed shores. Like many other limpets, this species defends a home range and produces an eroded home scar in the rock where it resides. It is an herbivore, feeding on several green algae.

Distribution: Endemic to the Hawaiian Islands (Napili, Maui, Hawaiian Islands).

425. *Emarginula* sp.

Identification: Species of *Emarginula* can be recognized by the slit on the front part of the shell. Species are differentiated by the elaboration of the sculpture on the shell. The species inhabiting the Indo-Pacific tropics are poorly known.

Natural History: Many members of this family of limpets are predators on colonial tunicates. Though this species has not been observed feeding, it was found in the vicinity of tunicates and is probably a predator on them. Depth: 10 m.

Distribution: Philippines (Batangas, Luzon, Philippines).

T. M. Gosliner

426. *Scutus unguis* (Linnaeus, 1758)
Shield Limpet

Identification: This species is sometimes thought to be an opisthobranch gastropod or a lamellarid, since the mantle usually completely envelops the shell. The mantle and foot are uniformly black and a bit of the shell is visible in this photo.

Natural History: Species of *Scutus* are commonly found under rocks in moderately shallow depths. They appear to feed upon colonial tunicates.

Distribution: South Africa; Red Sea; Australia; Philippines; New Guinea (Madang, Papua New Guinea).

T. M. Gosliner

427. *Haliotis asinina* Linnaeus, 1758
Ass's Ear Abalone

Identification: *Haliotis asinina* gets its name from the elongate shell, shaped like a donkey's ear. The shell is relatively smooth and the mantle and epipodial tentacles almost completely cover it.

Natural History: *Haliotis asinina* is a common inhabitant of shallow water coral rubble on the edges of reefs; common from the intertidal zone to a few meters depth. It is generally found under the surface of coral during the day and appears to be nocturnal. When disturbed it crawls rapidly to seek darker areas.

Distribution: Thailand; Malaysia; Australia; Indonesia; Borneo; Philippines; Japan; Okinawa; Taiwan; New Caledonia; Belau; Solomon Islands (Batangas, Luzon, Philippines).

T. M. Gosliner
T. M. Gosliner

428. *Haliotis clathrata* Reeve, 1846
Abalone

Identification: There are many species of abalones inhabiting the Indo-Pacific tropics. *Haliotis clathrata* is a small (mostly less than 20 mm in length), uncommon species that has a prominently raised whorl near the posterior end of the shell. It has elongate epipodial tentacles that extend from the sides of the foot and are readily visible in the photo. Shorter, highly branched tentacles are also present.

Natural History: *Haliotis clathrata* is found underneath rocks in areas with coralline algae. Like all abalones, it is an herbivore. Depth: 10-20 m.

Distribution: Madagascar; New Guinea; Borneo; Philippines (Batangas, Luzon, Philippines).

125

T. M. Gosliner

429. *Tectus pyramis* (Born, 1778)
Pyramid Top Snail

Identification: This species of top snail may reach a height of 70 mm. The surface of the shell is often encrusted with algae or other marine life, making the appearance of this species relatively cryptic.

Natural History: Tectus pyramis is a common inhabitant of shallow water reefs and areas of coral rubble. It is an herbivore and scrapes microscopic algae off rocks.

Distribution: Reunion Island and Mauritius to Australia; Samoa; Indonesia; Philippines; Chuuk; Guam; Okinawa; Japan; Marshall Islands (Batangas, Luzon, Philippines).

430. *Ethalia* cf. *nucleus* (Phillippi, 1849)

Identification: A small turban-shaped snail, the whorls of the shell have several brown lines broken with white. Epipodial tentacles are apparent extending from the sides of the foot.

Natural History: This species lives in sandy habitats. When disturbed, it is capable of swimming by thrashing its foot back and forth, laterally. Depth: 3-10 m.

Distribution: Philippines (Batangas, Luzon, Philippines).

T. M. Gosliner

431. *Turbo petholatus* Linnaeus, 1758
Turban Snail

Identification: *Turbo petholatus* is easily recognized by its smooth shell with well-rounded whorls and elaborate banding pattern.

Natural History: Like most species of turbans, *Turbo petholatus* is found in shallow rubble areas, where it feeds upon microscopic algae. Depth: intertidal to 10 m.

Distribution: Western Indian Ocean to Tahiti, though absent from the Hawaiian Islands (Manado, Sulawesi, Indonesia)

Marc Chamberlain
T. M. Gosliner

432. *Phasianella solida* (Born, 1778)
Solid Pheasant Shell

Identification: Pheasant shells, of which there are many Indo-Pacific species, are high-spired with glossy, brightly patterned shells. The name pheasant shell comes from the bright colors of the shell, similar to the plumage of a pheasant. They have a calcareous operculum.

Natural History: This species, like other pheasant shells, is a common herbivore which inhabits sea grass beds. The snails crawl along seagrass blades and feed upon epiphytic diatoms and algae.

Distribution: Western Indian Ocean (Zanzibar, Tanzania).

126

433. *Nerita polita* Linnaeus, 1758
Polished Nerite

Identification: There are many species of *Nerita* inhabiting the Indo-Pacific tropics. *Nerita polita* can be readily distinguished by its smooth banding pattern and shiny surface.

Natural History: Most nerites are found on rocks above the low tide mark, and *Nerita polita* is no exception. This species, like other members of the family, is an herbivore and is well adapted for living in dry conditions. Nerites are most active at night.

Distribution: Western Indian Ocean to Hawaiian Islands (Batangas, Luzon, Philippines).

T. M. Gosliner

434. *Titiscania limacina* Bergh, 1875

Identification: At first appearance, *Titiscania limacina* is easily mistaken for an opisthobranch slug. This species is closely related to the nerites, but entirely lacks a shell as an adult. A series of white defensive glands is situated on either side of the body. The animals reach 20 mm maximum length .

Natural History: Found below rocks in shallow water where the rocks are embedded in oxygen-poor sediments; intertidal zone to 10 m depth.

Distribution: Western Indian Ocean to Panama and Baja California (Batangas, Luzon, Philippines).

T. M. Gosliner

435. *Dendropoma maxima* (Sowerby, 1825)
Worm Snail

Identification: Vermetid gastropods are unusual snails in that they secrete the calcareous tube in which they live. This is one of the largest vermetids, which can be recognized by the presence of a large operculum at the aperture.

Natural History: This species secretes mucous strands which help to ensnare prey. The plankton is ingested by creating ciliary currents with the gill and directing prey into the mantle cavity. Found imbedded in massive coral heads.

Distribution: Australia to Philippines. Its range in the Indian Ocean is poorly known. This species has been found in the Hawaiian Islands as a fossil, but is now absent in Hawai'i (Kay, 1979) (Bohol, Philippines; Sapwauhfik, Pohnpei).

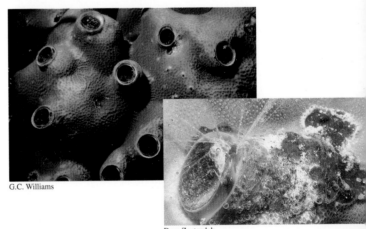
G.C. Williams

Dave Zoutendyk

436. *Serpulorbis grandis* Gray, 1850
Worm Snail

Identification: This species has often been misidentified as *Dendropoma maxima*, but is readily distinguishable by the elaborately mottled animal and by the fact that an operculum covering the shell opening is entirely absent.

Natural History: Feeds much like a bivalve, by capturing plankton with ciliary currents produced by its gill. It is associated with living reefs and builds its spirally coiled tube on dead coral. It may be partially overgrown with living coral or other growth.

Distribution: Australia; New Guinea; and Philippines (Batangas, Luzon, Philippines).

Lynn Funkhouser

127

Marc Chamberlain

Roy Eisenhardt

T. M. Gosliner
T. M. Gosliner

Marc Chamberlain

437. *Sabia conica* (Schumacher, 1817)

Identification: These cap-shaped shells are frequently found attached to other snails such as this *Fusinus*. The specimens illustrated here are covered with pink coralline algae.

Natural History: Horse-hoof shells, like vermetids, are unusual gastropods in that they are filter feeders, or may feed on fecal material from the snails to which they are attached. They also begin life as males and switch sex to females during their life.

Distribution: Indian Ocean; Hawaiian Islands and eastern Pacific (Kona, Hawai'i).

438. *Cassis cornuta* (Linnaeus, 1758)
Horned Helmet

Identification: C. cornuta is the largest species of helmet snail and can be readily distinguished from all other species by the prominent projections from the back of the shell. A common species of helmet, *Cypraecassis rufa*, can be distinguished by its short knobs and bright orange opening of the shell.

Natural History: Common on sandy banks on the inside of barrier and fringing reefs in depths of 10-20 m. Members of the Cassidae feed upon echinoderms, especially sea urchins and sand dollars. Feeds primarily upon irregular urchins, but is also a predator of the crown-of-thorns *Acanthaster planci* (#964).

Distribution: Indian Ocean to Hawaiian Islands and southern Polynesia (Kona, Hawai'i).

439. *Semicassis bisulcata*
(Schubert & Wagner, 1829)
Bonnet Snail

Identification: This bonnet snail has spiral lines on the upper whorls, but a smooth body whorl. It is usually marked with a series of brown checks and may reach 70 mm in length.

Natural History: This species lives in shallow sandy habitats where it presumably feeds upon heart urchins including *Maretia planulata* (#1017).

Distribution: Reunion Island and Mauritius; Australia; Indonesia; Philippines; Japan; Fiji and Marshall Islands (Batangas, Luzon, Philippines).

440. *Neverita didyma* (Röding, 1798)
Moon Snail

Identification: This relatively large Indo-Pacific moon snail can be distinguished by its relatively flat, brown shell and by the white animal with black spotting.

Natural History: This species, like other moon snails, is found in sandy and muddy substrates where it feeds upon other mollusks, primarily bivalves, but also other gastropods. It feeds on them by drilling a circular hole through the shell. Also illustrated here is the egg case of an unidentified naticid. Almost all naticids produce a sand collar of eggs. This species is nocturnally active. Low tide margin to 70 m depth.

Distribution: South Africa; Persian Gulf; Australia; Philippines and Japan (Batangas, Luzon, Philippines).

441. *Naticarius orientalis* (Gmelin, 1791)
Moon Snail

Identification: The shell of this species is drab brown to cream. Living specimens frequently have a thick brownish coat or periostracum on the shell. The animal is a beautiful red with white stripes.

Natural History: This species is found in clean sandy areas in shallow water and is active at night. During the day the animal is inactive, generally buried under sand.

Distribution: Singapore; Japan; Philippines and Australia (Batangas, Luzon, Philippines).

T. M. Gosliner

442. *Polinices mammilla* (Linnaeus, 1758)
Moon Snail

Identification: This species is readily distinguishable by its high-spired, glossy white shell, with an all white animal.

Natural History: Like the preceding species, *Polinices mammilla* feeds upon other mollusks. This species is also found from intertidal sand and mudflats to the shallow subtidal zone. It is active during the day; intertidal to 80 m.

Distribution: South Africa and Red Sea to Hawaiian Islands (Zanzibar, Tanzania).

T. M. Gosliner

443. *Tanea undulata* (Röding, 1798)
Moon Snail

Identification: The shell is white with brown undulating lines. The tan animal has broad black bands on the head and foot.

Natural History: This species is found in silty sand and is active at night.

Distribution: Japan; Philippines and Indonesia (Batangas, Luzon, Philippines).

T. M. Gosliner
T. M. Gosliner

444. *Lambis scorpius* (Linnaeus, 1758)
Spider Conch

Identification: This species of spider conch can be recognized by the alternating black and white radial lines on either side of the shell aperture. Spider conchs are distinguished from other conchs by the elongate spines extending from the aperture of the shell.

Natural History: This species inhabits patch reefs and boundaries between rocky and sandy areas. Like other conchs, it probably feeds on microscopic algae. It is able to move using its muscular foot and claw-like operculum.

Distribution: Tanzania; Kenya and Madagascar; Australia; New Guinea; Fiji; Samoa; Indonesia; Philippines and Marshall Islands (Batangas, Luzon, Philippines).

129

Mike Miller

T. M. Gosliner

T. M. Gosliner
G.C. Williams

445. *Strombus bulla* (Röding, 1798)

Identification: A medium-sized conch. It can be distinguished by the elongate posterior extension of the orange shell.

Natural History: Most members of this family are herbivorous and live in habitats with soft substrate. Like other conchs, *Strombus bulla* has well developed eyes and can use its long, pointed operculum to move about or bury into mud or sand. This specimen has commensal anemones on the shell.

Distribution: Western Pacific: Indonesia; Philippines; Japan; New Guinea; Samoa (Batangas, Luzon, Philippines).

446. *Strombus minimus* Linnaeus, 1771

Identification: This is one of the smallest species of conchs, reaching a maximum length of 25 mm. It has a smooth shell with mottled brown pigment.

Natural History: Most conchs feed on microscopic algae in relatively shallow water. It is found in shallow sandy habitats.

Distribution: Malay Peninsula; New Guinea; Indonesia; Philippines; Okinawa; Solomon Islands; New Caledonia and Fiji (Batangas, Luzon, Philippines).

447. *Terebellum terebellum* (Linnaeus, 1758)

Identification: The elongate conical shell of this close relative of the conchs is distinctive. There is only one species in this genus, though the color pattern on the shell varies considerably.

Natural History: Little is known about the feeding of this species, though it is thought to be herbivorous, as are other members of the Strombidae. It is found in coarse sandy habitats.

Distribution: Reunion Island; Madagascar; Mauritius; Red Sea; Philippines; Samoa; Tonga and Marshall Islands (Batangas, Luzon, Philippines).

Cypraeidae - Cowries

The true cowries are well represented in the Indo-Pacific tropics. Of the approximately 200 worldwide species monographed by Burgess (1985), 75% are known from the Indo-Pacific tropics.

448. *Cypraea annulus* Linnaeus, 1758
Gold Ringed Cowrie

Identification: The shell of *Cypraea annulus* can be readily identified by its smooth, flattened shape, greenish color with bright yellow ring. It is similar to the money cowrie, *Cypraea moneta*, which differs by its yellow color and more irregular shape. The mantle of *C. annulus* contains a gray "finger print" pattern and elongate, simple and compound papillae.

Natural History: A common inhabitant under stones in shallow water, from the low tide mark, to 5 m. Hundreds of individuals may be found under a single rock in certain habitats.

Distribution: Indian Ocean, Red Sea to the Central Pacific as far east as Kure Island in the leeward Hawaiian Islands and the Cook Islands. It has not been found alive in the Society Islands or the high Hawaiian Islands (Batangas, Luzon, Philippines).

449. *Cypraea aurantium* Gmelin, 1791
Golden Cowrie

Identification: This species reaches 120 mm in length and can be distinguished by the golden orange color and globose shape. The mottled orange or brown mantle is ornamented with numerous compound papillae.

Natural History: Cypraea aurantium is found in deep holes in reefs, usually on the exposed sides of barrier reefs. This species used to be considered one of the rarest cowries and its shell was priced accordingly. The species is fairly common in areas with appropriate habitat, but is not often seen due to its nocturnal behavior and cryptic habitat. Depth: 8-40 m.

Distribution: Fringes of the western Pacific; Solomon Islands; Belau; Philippines; Guam; Marshall Islands; Fiji and Tahiti (Kwajalein Atoll, Marshall Islands).

Jack Randall

450. *Cypraea chinensis* Gmelin, 1791

Identification: The shell reaches a length of 50-60 mm and is characterized by reddish orange mottling and purple marginal spots. The mantle is red with a few white spots. The mantle has thin nodular papillae.

Natural History: Cypraea chinensis is found on shallow reefs under coral heads and coral rubble. It presumably feeds on sponges. Depth: 1-40 m.

Distribution: Western Indian Ocean and Red Sea to Hawaiian Islands and Tahiti (Batangas, Luzon, Philippines).

Mike Miller

451. *Cypraea cribraria* Linnaeus, 1758

Identification: Cypraea cribraria is a member of a complex of six species with a reddish orange shell with white circles. It is the most common and widespread member of this complex. The shell may reach 50 mm in length and lacks basal spots which are present in closely allied species. The mantle is bright red with simple papillae.

Natural History: This species feeds on unidentified reddish colored sponges and is found under rocks on shallow reefs. Depth: 10 m.

Distribution: South Africa; Madagascar; Red Sea; Australia; New Guinea; Indonesia; Philippines; Japan; Fiji; Samoa; Guam and Marshall Islands. Absent in Hawai'i and Society Islands (Madang, Papua New Guinea).

T. M. Gosliner

452. *Cypraea helvola* Linnaeus, 1758
Honey Cowrie

G.C. Williams

Identification: This is one of the most common and widespread species of Indo-Pacific cowries. It can be recognized by its orange-brown shell with numerous small white spots. Purple pigment is often present on the anterior and posterior portions of the shell. The mantle is ornamented with numerous compound papillae.

Natural History: Frequently found under stones on intertidal reef platforms and in shallow water. This species probably feeds largely upon sponges.

Distribution: Western margin of the Indian Ocean to Hawai'i and Society Islands (Sodwana Bay, South Africa).

T. M. Gosliner

453. *Cypraea limacina* Lamarck, 1810

Identification: This cowrie is characterized by the presence of white nodules on the surface of the shell. The mantle is orange-brown with elongate unbranched and branched papillae.

Natural History: *Cypraea limacina* is found on the under-surface of coral rubble from the intertidal zone to the shallow subtidal.

Distribution: South Africa; Madagascar; Red Sea to Fiji and Marshall Islands (Msimbati, Tanzania).

Mike Miller

454. *Cypraea mappa* Linnaeus, 1758
Map Cowrie

Identification: The shell of this large cowrie is characterized by the presence of an irregular undulating pattern where the two mantle margins meet. The mantle contains numerous simple, conical papillae and a few thicker compound ones.

Natural History: Usually found on the undersurface of coral slabs near the outer margin of fringing and barrier reefs. Depth: 1-16 m.

Distribution: Madagascar and Tanzania to Society Islands (Batangas, Luzon, Philippines).

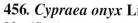
Mike Miller
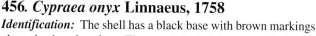
Mike Miller

455. *Cypraea miliaris* Gmelin, 1791

Identification: The yellowish orange shell is ornamented with white spots of variable size. The mantle is brownish to black with very elongate, branched papillae.

Natural History: This species is found in moderate depths crawling on sand. Like many other cowries, it is nocturnally active. Depth: 3-45 m.

Distribution: Malaysia; Myanmar; Thailand; Australia; Philippines and Japan (Batangas, Luzon, Philippines).

456. *Cypraea onyx* Linnaeus, 1758

Identification: The shell has a black base with brown markings along the dorsal surface. The mantle is dark brown to black with short, scattered compound papillae.

Natural History: This species is most commonly found in silty and sandy habitats, where it is nocturnal.

Distribution: South Africa and Tanzania; Japan; Philippines and Marshall Islands (Batangas, Luzon, Philippines).

457. *Cypraea stolida* Linnaeus, 1758

Identification: This is a small species of cowrie less than 50 mm in length. It has a characteristic dark brown patch on the shell. The mantle is brown, yellow or white, and bears short, compound papillae.

Natural History: This species is generally found in shallow water under dead coral rubble. Depth: 10-40 m.

Distribution: South Africa; Madagascar; Reunion Island; Australia; Fiji; Solomon Islands; Philippines; Okinawa (Heron Island, Australia).

Leslie Newman & Andrew Flowers

458. *Cypraea talpa* Linnaeus, 1758
Mole Cowrie

Identification: This moderately-sized cowrie is more cylindrical in shape than most cowries, and black with gold and brown banding. The mantle is black with numerous minute white spots. Large simple papillae are scattered over the surface of the mantle.

Natural History: *Cypraea talpa* is usually found under large dead or living coral heads in moderate depths. Subtidal to 20 m depth.

Distribution: One of the most widespread Indo-Pacific cowries. South Africa; Red Sea; Hawai'i and Cocos Island, off the coast of Costa Rica (Molokini, Hawaiian Islands).

Mike Severns

459. *Cypraea teres* Gmelin, 1791

Identification: The shell is variously mottled with reddish brown bands and pigment spots. The mantle is bright reddish orange with opaque white on the simple to compound papillae. *Cypraea teres* is part of a complex involving several distinct species.

Natural History: In the photo, the large foot of this species is evident. The posterior portion of the foot is autotomized or shed when the animal is disturbed. This species feeds upon red sponges with which it is often associated.

Distribution: Western Indian Ocean to Hawai'i, but some records are doubtful, owing to confusion with other members of the species complex (Bali, Indonesia).

460. *Cypraea tigris* Linnaeus, 1758
Tiger Cowrie

Identification: This is probably the best known of the large cowries. The white shell with black spots of variable size and density is distinctive. The mantle contains a "finger print pattern" of grey and black. The papillae are undivided and elongate with an opaque white tip.

Natural History: This species is commonly found on elevated intertidal reef platforms, in seagrass beds, in areas of sand and rubble, and on the outside of barrier and fringing reefs. It is generally under coral rubble or in crevices in living reef, but may be found crawling in the open.

Distribution: South Africa; Red Sea to Hawai'i; Tuamotus and Society Islands (Manado, Indonesia).

Dave Zoutendyk
Terry Schuller

133

Mike Miller

461. *Cypraea vitellus* Linnaeus, 1758
Pacific Deer Cowrie

Identification: A medium-sized cowrie with a brown shell covered with opaque white spots and fine darker brown lines along the margin. The mantle is variously colored with brown, black and gray. The large papillae are distinctive.

Natural History: This species is frequently found under rocks and coral heads from the lower intertidal zone to a few meters depth. It is especially common in the Western Indian Ocean.

Distribution: South Africa; Red Sea; to Hawai'i; Tuamotus and Society Islands (Batangas, Luzon, Philippines).

Ovulidae - Ovulids, Egg Cowries

Ovulids are related to cowries and are predators on anthozoan coelenterates. More than 200 species have been described, most of which occur in the Indo-Pacific tropics. Cate (1973) monographed and revised the family based on shell morphology. Liltved (1989) added much to our understanding of the living animals and their natural history.

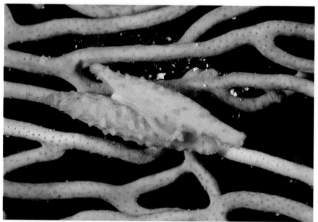

T. M. Gosliner

◄ 462. *Aclyvolva lanceolata* (Sowerby, 1848)

Identification: The reddish shell is distinguishable by its thin shape with a narrow opening. The mantle is red with opaque white, compound papillae. Length: to 25 mm.

Natural History: *Aclyvolva lanceolata* feeds on an unidentified red orange gorgonian, as depicted in the photo. The papillae ornamenting the mantle of this ovulid remarkably mimic the polyps of the host gorgonian. Depth: 1-20 m.

Distribution: Indian Ocean; New Guinea; Indonesia and Philippines (Dakak, Mindanao, Philippines).

◄ 463. *Aclyvolva* sp.

Identification: This species could not be positively identified. Its shell is not as elongate as that of *Aclyvolva lanceolata* and may be *A. haynesi*, which is known only from Western Australia.

Natural History: This species has been found on a gorgonian that appears to be *Subergorgia mollis* (#141). The mantle of the ovulid closely mimics that of the host gorgonian.

Distribution: Indonesia (Kurkap Island, Indonesia).

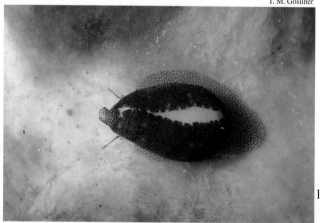

Mike Severns
T. M. Gosliner

464. *Calpurnus lacteus* (Lamarck, 1810)

Identification: This is a small ovoid ovulid that reaches a maximum length of about 20 mm. The shell is uniformly white, with a series of denticulate teeth along the margin of the lip. The mantle is brownish with small white spots and larger black spots.

Natural History: This species commonly preys upon colonies of soft corals of the genera *Sarcophyton* (#81-85) and *Sinularia* (#86-90). It has been found on the underside of the colony as well as on the top. The pattern on the mantle closely resembles that of its host. Depth: low tide mark to 80 m.

Distribution: South Africa; Red Sea; Indian Ocean to Society Islands (Madang, Papua New Guinea).

465. *Calpurnus verrucosus* (Linnaeus, 1758)

Identification: This species is larger than *Calpurnus lacteus*, reaching 35-40 mm in length. The most obvious difference in the shell is that *C. verrucosus* has a circular depression at either end of the dorsal surface of the shell that contains an orange or brown line. The tips of both ends of the shell are purplish. The foot and mantle are ornamented with black or brown spots. Specimens from the western Indian Ocean have larger, lighter colored spots than Pacific specimens.

Natural History: Feeds upon *Sarcophyton* spp. It is usually found right in the open, and is not at all camouflaged. It may be exhibiting warning coloration and may be distasteful to predators. Depth: 3-20 m.

Distribution: South Africa; Mozambique; Madagascar; Indian Ocean; Fiji; Australia; Singapore; Indonesia; Philippines; Japan; Marshall Islands (Batangas, Luzon, Philippines).

Mike Miller

466. *Crenavolva rosewateri* Cate, 1973

Identification: The shell is rose pink with a medial band of white. The ends of the shell are ornamented with an orange spot. The mantle of this beautiful species is white with bright red spots and yellow papillae.

Natural History: *Crenavolva rosewateri* is known to feed upon melathaeid gorgonians. Depth: 30 m.

Distribution: South Africa; Australia; New Caledonia; Philippines and Japan (Solitary Islands, New South Wales, Australia).

Carol Buchanan

467. *Crenavolva tigris* Yamamoto, 1971

Identification: This is one of the most striking and recognizable ovulids with a bright orange mantle with black stripes and spots. The shell is uniformly yellowish orange and reaches a length of 15 mm.

Natural History: *Crenavolva tigris* is known to feed upon gorgonians of the genus *Euplexaura* and is shown here on a member of this genus. Depth: 18-73 m.

Distribution: Originally described from Japan, it has also been found in Australia and Indonesia (Solitary Islands, New South Wales, Australia).

Carol Buchanan
T. M. Gosliner

468. *Cymbovula deflexa* (Sowerby, 1848)

Identification: The shell of *Cymbovula deflexa* is spindle shaped and up to 20 mm in length. It is pinkish grey with a bright orange band around the margin. The animal is translucent white, except for a red line that runs from either eye to the tip of the tentacle.

Natural History: The gorgonian, *Rumphella* sp., is the only known food of this species. Depth: 12 m.

Distribution: South Africa; Australia; Indonesia; Solomon Islands; Philippines; Singapore and Japan (Sodwana Bay, Natal, South Africa).

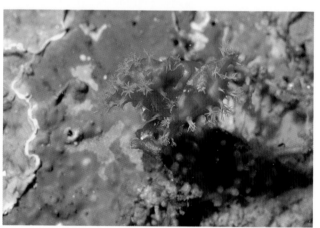

469. *Delonovolva formosa* (Adams & Reeve, 1848)

Identification: This species is characterized by its spindle shape with an intense violet color and yellow markings on the tips of the shell. The mantle is black with white spots and lacks any tubercles.

Natural History: Found at moderate depths on an unidentified red gorgonian.

Distribution: Borneo and now also known from Australia (Solitary Islands, New South Wales, Australia).

Carol Buchanan

470. *Dentiovula dorsuosa* (Hinds, 1844)

Identification: Species of *Dentiovula* are characterized by denticles that continue from the lip to the anterior and posterior extensions of the shell. *Dentiovula dorsuosa* has pink markings on the anterior and posterior ends of the shell, as well as a yellow stripe. The mantle has a red "finger print" pattern and small, short white papillae.

Natural History: This species of *Dentiovula* feeds on an unidentified red gorgonian.

Distribution: Western Pacific: Indonesia; Singapore; Malaysia; New Guinea; Philippines; Japan (Papua New Guinea).

Marc Chamberlain

471. *Dentiovula eizoi* Cate & Azuma in Cate, 1973

Identification: *Dentiovula eizoi* has an irregularly spindle-shaped shell and a series of teeth on the posterior end. The animal is bright reddish purple throughout.

Natural History: This species has been found on the reddish purple gorgonian, *Acalycigorgia* sp. The mantle is covered with compound papillae that mimic the polyps of the gorgonian. Depth: 25-91 m.

Distribution: South Africa; New Guinea and Japan (Madang, Papua New Guinea).

T. M. Gosliner
T. M. Gosliner

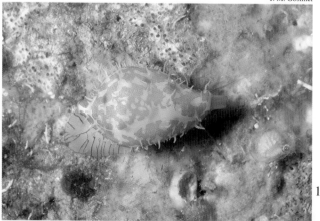

472. *Diminovula alabaster* (Reeve, 1865)

Identification: The shell of this species reaches about 10 mm in length. It is characterized by having a series of orange spots on the dorsal surface and a bright yellowish orange line on the margin. The mantle is white with a red "finger print" pattern with simple white papillae. The foot is ornamented with red lines.

Natural History: Found on and feeds upon soft corals of the genus *Dendronepthya*. Depth: 10 m.

Distribution: Indian Ocean; Australia; New Guinea; Indonesia; Singapore; Malaysia; Taiwan; Okinawa and Japan (Madang, Papua New Guinea).

136

473. *Hiatovula brunneiterma* (Cate, 1969)

Identification: Species of *Hiatovula* can be recognized their thin spindle shape with indented apices. *Hiatovula brunneiterma* is readily indentified by the brown markings across the apices. It has a spotted mantle and black lines on the head, foot and tentacles.

Natural History: This species is shown on an unidentified gorgonian.

Distribution: Australia; New Guinea; Solomon Islands and Philippines (Madang, Papua New Guinea).

T. M. Gosliner

474. *Hiatovula depressa* (Sowerby, 1875)

Identification: This species is similar to the preceding one, except that there are yellow markings at the apices rather than brown ones. The animal has yellow tentacles rather than black ones.

Natural History: *Hiatovula depressa* was found on the gorgonian *Rumphella* sp. (#153). Depth: 10 m.

Distribution: Western Pacific: Australia; New Guinea; Indonesia and Philippines (Batangas, Luzon, Philippines).

T. M. Gosliner

475. *Ovula costellata* Lamarck, 1810

Identification: The ovoid shell is thickly calcified for an ovulid, and numerous teeth are present along the margin of the lip. The shells reach approximately 40 mm in length. The mantle is white with numerous yellow-tipped papillae. The edge of the foot has a narrow black line.

Natural History: This species is known to feed on octocorals of the genus *Sarcophyton*. Depth: 12-30 m.

Distribution: South and East Africa; Fiji; Australia; Tonga and Japan (Solitary Islands, New South Wales, Australia).

Carol Buchanan

476. *Ovula ovum* (Linnaeus, 1758)

Egg Cowrie

Identification: *Ovula ovum* is larger than *O. costellata*, reaching a length of 100 mm. It is readily distinguished by its black mantle.

Natural History: The soft corals, *Sarcophyton* and *Sinularia* spp. are the primary food sources. Juvenile specimens have a few large tubercles and appear to mimic the toxic nudibranch *Phyllidia madangensis*. When specimens grow larger than the maximum size of the nudibranch they change their appearance. Depth: intertidal zone to 20 m.

Distribution: Western Indian Ocean to the Tuamotu Archipelago (Manado, Indonesia; Madang, Papua New Guinea).

T. M. Gosliner

Laura Losito

137

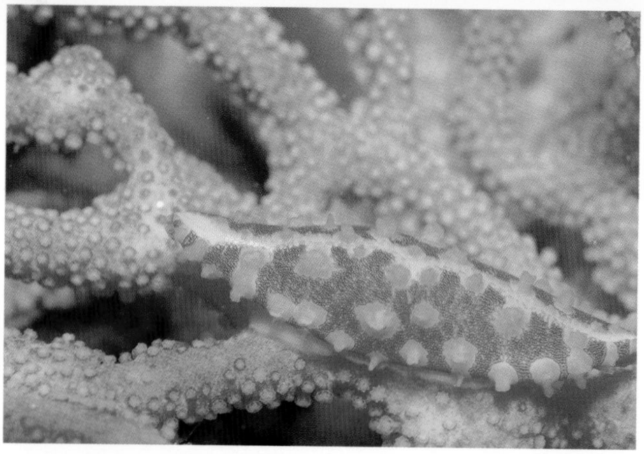

Carol Buchanan

477. *Phenacovolva angasi* (Reeve, 1865)

Identification: Species of *Phenacovolva* are elongate and spindle shaped. There are approximately 30 described species. *Phenacovolva angasi* has shorter apices to the shell than other species treated here. The shell is uniformly yellowish white, about 32 mm in length. The mantle has numerous small red circles and white or yellow compound papillae.

Natural History: It is found in association with an unidentified red gorgonian, which it closely resembles. Depth: 30 m.

Distribution: Western Pacific: Australia; Philippines and Japan (Solitary Islands, New South Wales, Australia).

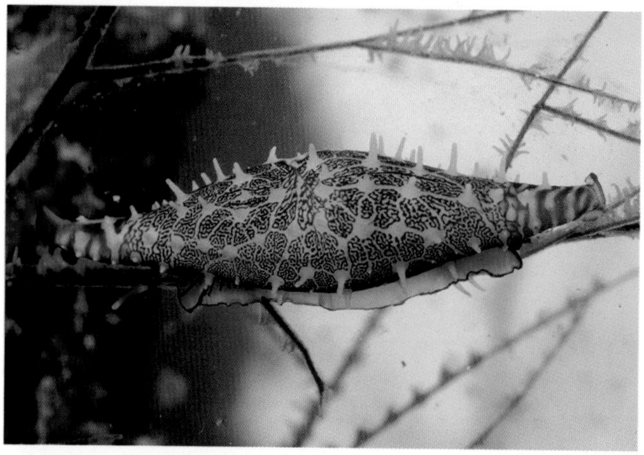

T. M. Gosliner

478. *Phenacovolva gracilis* (Adams & Reeve, 1848)

Identification: The present species can be distinguished by the reddish markings on the shell (about 30 mm in length) with a bright orange marginal line. The animal has a red "finger print" pattern on the mantle with white papillae.

Natural History: This species feeds upon the black coral, *Antipathes* sp. The specimen photographed here was found on a single isolated black coral colony situated on a steep sandy slope. Depth: 10 m.

Distribution: Western Pacific: New Guinea; Borneo; Philippines and Japan (Madang, Papua New Guinea).

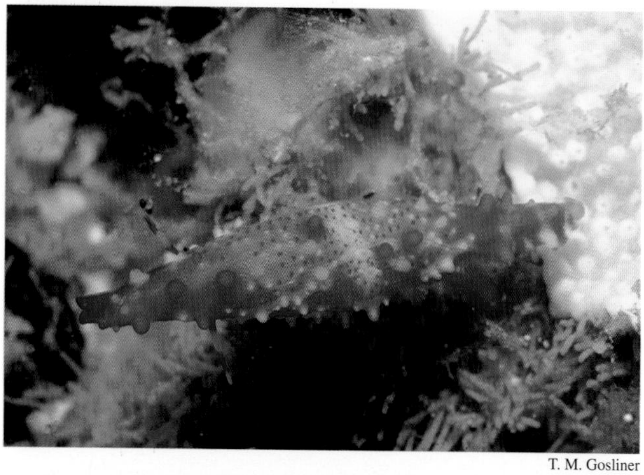

T. M. Gosliner

479. *Phenacovolva rosea* (A. Adams, 1854)

Identification: The shell of this species reaches a length of more than 50 mm and is reddish in color with a medial white band. The animal is pinkish with darker rounded papillae and small brownish or black spots.

Natural History: In South Africa this species is known to feed upon the gorgonians, *Leptogorgia palma* and *Homophyton verrucosum*. In Japan and the present specimens, it feeds upon plexaurid gorgonians. Depth: 15-73 m.

Distribution: South Africa to Australia; New Guinea; Philippines; Taiwan; China and Japan (Madang, Papua New Guinea).

T. M. Gosliner

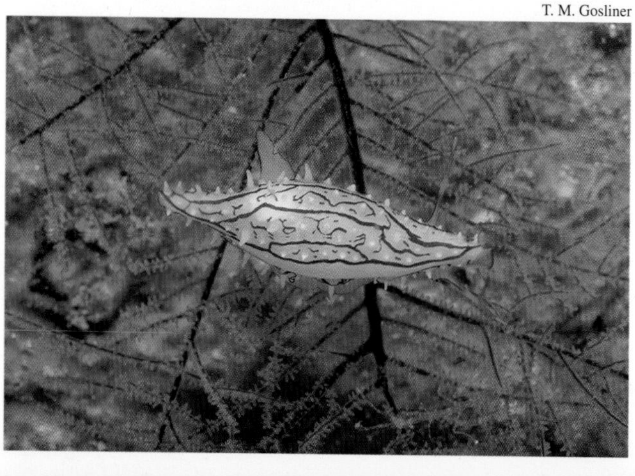

480. *Phenacovolva weaveri* Cate, 1973

Identification: The shell has a yellow line above its lip, but lacks the reddish markings found in *P. gracilis*. The shell reaches a length of 38 mm. The mantle is white with red brown vermiform markings and white-tipped papillae.

Natural History: This species feeds on black corals of the genus *Antipathes*. Depth: 25-115 m.

Distribution: South Africa to Australia; Philippines; Japan and Hawaiian Islands (Madang, Papua New Guinea).

481. *Primovula* sp. 1

Identification: This species cannot be identified with certainty. It is recognized by its white shell and mantle with numerous small red circles. The compound papillae resemble the polyps of its host gorgonian.

Natural History: Found on an unidentified species of red gorgonian, probably *Acabaria* sp. Depth: 20 m.

Distribution: Australia (Solitary Islands, New South Wales, Australia).

Carol Buchanan

482. *Primovula* sp. 2

Identification: The small shell of this species is about 10 mm in length and is bright yellow. The mantle, head and foot are uniformly translucent white. It is not readily identifiable with any described species.

Natural History: This species is found on an unknown yellow gorgonian found on a patch reef inside a lagoon formed by a large barrier reef. Depth: 20-30 m.

Distribution: New Guinea (Madang, Papua New Guinea).

T. M. Gosliner

483. *Primovula* sp. 3

Identification: The shell of this species reaches a maximum of 8 mm in length. The foot and mantle have reddish spots scattered over their surfaces. A red line extends from the eye along most of the length of each tentacle. This species was not readily identifiable with any known species.

Natural History: Specimens were associated with the soft coral, *Carijoa* sp. 2 (#70) on shallow patch reefs. Depth: 2-3 m.

Distribution: Madagascar; Tanzania (Tulear, Madagascar).

T. M. Gosliner

T. M. Gosliner

484. *Prosimnia semperi* (Weinkauff, 1881)

Identification: The strong striations on the shell give this species a rough, uneven appearance. The shell is widest near the posterior end and tapers to a funnel-like posterior opening. The mantle is variously colored depending on the host, but the body color is usually yellow or red. The specimen photographed here has white compound papillae. Cate (1973) divided this species into three discrete subspecies, based on minor shell differences.

Natural History: *Prosimnia semperi* feeds on several species of melathaeid gorgonians on patch reefs. Depth: 10-30 m.

Distribution: South Africa; Madagascar; Australia; Fiji; New Guinea; Philippines; Japan (Madang, Papua New Guinea).

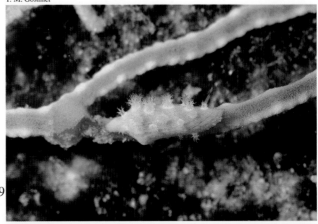

139

G.C. Williams

T. M. Gosliner

Charlie Arneson
G.C. Williams

485. *Volva volva* (Linnaeus, 1758)

Identification: This large (up to 120 mm in length), spindle-shaped species was subdivided into four subspecies by Cate (1973). One of these is isolated in the Caribbean, while the others are found in the Indo-Pacific. The mantle is translucent white with irregular brown patches which extend onto the surface of the long simple papillae. The apex of each papilla has a dark brown spot.

Natural History: *Volva volva* is found on open sandy substrates where it feeds on sea pens, *Actinoptilum molle,* in South Africa. Depth: 30-100 m.

Distribution: Circumtropical: South Africa; Fiji; Indonesia; Australia; Philippines; Japan; Caribbean (Kosi Bay, Natal, South Africa).

486. *Trivia oryza* (Lamarck, 1810)

Identification: Triviids look like cowries, but are more closely related to velutinids than to cypraeids and ovulids. There are more than 150 species of *Trivia*. The number of Indo-Pacific species is difficult to determine since most species have not been studied alive and are separated by minor differences in the shell. The group was monographed by Cate (1979), and Liltved (1989) summarized much of our current knowledge. It is readily distinguished by the opaque white lines on the foot.

Natural History: This species, like all other known members of the family, feeds on compound tunicates. Specimens are most commonly found on the undersurface of rocks and coral rubble.

Distribution: South Africa; Gulf of Oman to Tuamotus (Madang, Papua New Guinea).

487. *Coriocella nigra* Blainville, 1824

Identification: This species, like other velutinids is often mistaken for an opisthobranch. We consider *Chelynotus semperi* a synonym. The internal shell is completely covered by the tissue of the mantle. This species is characterized by its brown to black color with a reticulate pattern and elongate finger-like appendages extending from the mantle. The living animal may exceed 100 mm in length.

Natural History: Observed crawling on the surface of shallow reef and rubble, especially at night.

Distribution: South Africa; Madagascar; Philippines; New Guinea; Indonesia; Solomon Islands (Batangas, Luzon, Philippines).

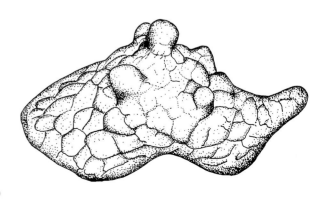

488. *Lamellaria* spp.

Identification: At present, it is virtually impossible to identify Indo-Pacific lamellarids to species. The taxonomic distinctions between species are poorly known and little is known about their anatomy. Three different species are depicted here to show some of the variability within the genus. All have internal shells, and the mantle forms a characteristic siphon at the head.

Natural History: All Indo-Pacific lamellarids have been found in association with compound tunicates, and bear an uncanny resemblance to their colonial prey.

Distribution: Owing to taxonomic confusion, it is not possible to ascertain the distribution of species of *Lamellaria* (Ramsgate, Natal, South Africa; Batangas, Philippines; Msimbati, Tanzania).

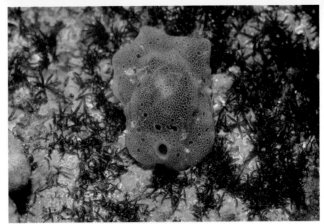
T. M. Gosliner

T. M. Gosliner

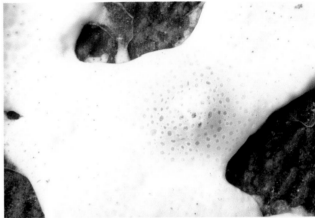
T. M. Gosliner

489. *Thyca crystallina* (Gould, 1846)

Identification: Eulimid gastropods are variable in form. Species of *Thyca* have a cap-shaped shell with numerous raised ridges. Color varies from tan to blue. Their color does not always match that of their host.

Natural History: Eulimid gastropods are parasitic upon echinoderms. *Thyca crystallina* is most commonly found on the sea stars *Linckia multifora* (#955) and *L. laevigata* (#954, shown here). It inserts its proboscis into the groove from which the tube feet insert and sucks coelomic fluid and tissue from its host. Frequently, a large female snail and a smaller male are found together on the same host. The male may be present under the shell of the female.

Distribution: This species has been found from Madagascar to the Hawaiian Islands (Batangas, Luzon, Philippines).

D.W. Behrens
Mike Miller

490. *Luetzenia asthenosomae* Warén, 1980

Identification: This species has a uniformly white shell with several distinct whorls.

Natural History: This species is found parasitizing the fire urchin *Asthenosoma varium* (#997).

Distribution: Australia and Philippines (Batangas, Luzon, Philippines).

141

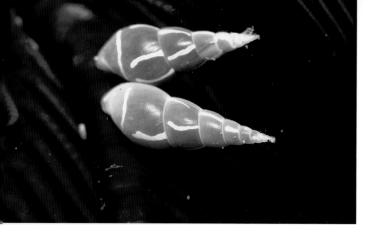

491. Eulimid sp.

Identification: The shell, as in the case with most eulimids, is glossy white, with a high spire. The dark black and yellow pigment seen in the photo is that of the mantle shining through the shell.

Natural History: This species is parasitic upon crinoids and blends in well with the black and yellow pigment of its host.

Distribution: Australia (Solitary Islands, New South Wales, Australia).

492. *Euthymella elegans* (Hinds, 1843)

Identification: *Euthymella elegans* is one of at least 1000 species of Triphoridae that likely inhabit the Indo-Pacific tropics (Marshall, 1983). Most species are small and have a high spire and are brightly colored with elaborate ornamentation.

Natural History: Members of the Triphoridae are known to feed on sponges. They are generally found on sponges on the undersides of rocks and coral rubble. Intertidal zone to 37 m.

Distribution: Western Pacific: Straits of Malacca; Philippines; Australia; Japan; Solomon Islands; New Caledonia; Tahiti (Batangas, Luzon, Philippines).

493. *Epitonium billeeanum*
(DuShane & Bratcher, 1965)

Identification: This species is recognizable by its yellow or orange body and shell color.

Natural History: Feeds on dendrophyllid corals of the genus *Tubastraea* (#289). Its body color closely resembles that of its coral prey, as do the irregularly shaped egg capsules of the snail, which are also evident in this photo.

Distribution: South Africa and Tanzania to the tropical Pacific coast of the Americas (Heron Island, Great Barrier Reef, Australia).

494. *Epitonium* sp.

Wentletrap

Identification: Like most typical epitoniid snails, this species has a shiny transparent shell with elevated axial ridges. The pigmented tissues of the digestive gland, gonad and mantle are visible through the shell. This species cannot be identified with certainty, but is similar to *Epitonium pallisi*.

Natural History: Virtually all epitoniids are predators on anthozoan coelenterates, and Indo-Pacific species most commonly feed upon sea anemones and stony corals.

Distribution: Indonesia (Manado, Sulawesi, Indonesia).

495. *Cymatium intermedius* (Pease, 1869)

Identification: This species is similar to the following species, but differs largely in the shell color and sculpture.

Natural History: *Cymatium intermedius* feeds on bivalves, especially oysters, and inhabits tide pools and fringing reefs. Intertidal zone to 30 m depth.

Distribution: Australia; Philippines; Guam; Marshall, Hawaiian and Marquesas Islands (Molokini, Hawai'i).

Mike Severns

496. *Cymatium nicobaricum* (Röding, 1798)

Identification: The prominent lines on the shell and the brown periostracum covering the shell are distinctive. Most species of *Cymatium* have brightly colored spotting on the body.

Natural History: Found in sand and rubble habitats, this species is known to feed on a wide variety of other gastropods. Depth: intertidal zone to 20 m.

Distribution: Originally described from the Nicobar Islands in the Indian Ocean, but known also from Seychelles to Hawaiian Islands and the Caribbean (Mahé, Seychelles).

T. M. Gosliner

497. *Charonia tritonis* (Linnaeus, 1767)
Triton's Trumpet ▶

Identification: This relative of *Cymatium* is one of the largest and best known tropical gastropods. It may exceed 500 mm in length.

Natural History: This species feeds on many different echinoderms, including the sea urchin *Heterocentrotus mamillatus* (#1014), the cushion star *Culcita novaeguinae* (#933) and most notably the crown-of-thorns, *Acanthaster planci* (#964). Because it is predatory upon the crown-of-thorns, this gastropod is protected by law in many parts of the world. Owing to the ecological importance of this species in controlling an important coral predator, divers should resist the temptation to collect this species. Depth: 3-40 m.

Distribution: South Africa; Tanzania; Red Sea; Hawaiian Islands; also found in the Caribbean (Kona, Hawai'i).

Nicholas Galluzzi
Kathy deWet

Alex Kerstitch

498. *Cymbiola nobilis* (Lightfoot, 1786)
Noble Volute

Identification: This is one of many species of *Cymbiola* inhabiting the waters of the western Pacific. The greatest diversity is in northern and western Australia. Species are distinguished by unique shell and color patterns of the living animal. *Cymbiola nobilis* can be distinguished by its dark smooth protoconch and a dark animal with yellow orange spots.

Natural History: Little is known about the biology of these predatory gastropods. Depth: intertidal zone to 80 m.

Distribution: China Sea; Singapore; southern Philippines and New Guinea (Coral Sea).

Mike Miller

499. *Cymbiola vespertilio* (Linnaeus, 1758)
Bat Volute

Identification: This species can be readily identified by the distinct axial sculpture of the protoconch, and a dark brown and white animal.

Natural History: *Cymbiola vespertilio* has been found in relatively sandy and silty habitats, where it is likely predatory upon other gastropods.

Distribution: Philippines; Indonesia; New Guinea and northern Australia (Batangas, Luzon, Philppines)

Mike Miller
Lynn Funkhouser

500. *Harpa articularis* Lamarck, 1822
Harp Snail

Identification: This is one of the larger of the nine Indo-Pacific species of *Harpa*. It can be distinguished by its relatively narrow ribs with chestnut brown markings.

Natural History: Harps are known to feed on crustaceans, especially crabs. The foot is wrapped around the crab and a mucous secretion combined with sand is used to asphyxiate the prey. Harps may autotomize the posterior end of the foot when disturbed.

Distribution: Eastern Indian Ocean to western Pacific: Malaysia; Indonesia; Philippines; Okinawa; Australia and Fiji (Batangas, Luzon, Philippines).

501. *Harpa harpa* (Linnaeus, 1758)
Harp Snail

Identification: Can be distinguished from the preceding species by the more numerous and finer brown lines on the shell ribs.

Natural History: Presumably feeds upon crustaceans. Like other harps, found on sandy substrates in shallow waters.

Distribution: Reunion Island and Mauritius to Marshall Islands; Philippines; Tonga and Samoa (Batangas, Luzon, Philippines).

502. *Nassarius albescens* (Dunker, 1846)

Identification: This is a relatively small species (less than 20 mm) with white and brown markings on the shell and an all white animal.

Natural History: Found on clean coralline sand flats, this species is also a scavenger.

Distribution: South Africa; Tanzania; Red Sea to Philippines and Australia (Msimbati, Tanzania).

T. M. Gosliner

503. *Nassarius glans* (Linnaeus, 1758)

Identification: This species is recognized by its fairly large shell (up to 45 mm in length). The newest whorls are smooth with thin brown spiral lines.

Natural History: Most species of *Nassarius* are scavengers. This species is found on sandy substrates in relatively shallow water.

Distribution: Philippines; Australia; Marshall Islands (Batangas, Luzon, Philippines).

Mike Miller

504. *Nassarius papillosus* (Linnaeus, 1758)

Identification: Recognized by its large size (up to 45 mm), honey color and beaded texture.

Natural History: As in the preceding species, this is likely a scavenger on dead animal material. It inhabits shallow sand habitats and is most active nocturnally.

Distribution: Reunion Island and Mauritius to Hawaiian Islands (Kona, Hawai'i).

Marc Chamberlain
Mike Miller

505. *Malea pomum* (Linnaeus, 1758)
Apple Snail

Identification: Similar to the following species but with a smaller, thicker shell with a lighter color. The spire is also shorter.

Natural History: Inhabits shallow water sandy and silty habitats and is nocturnally active.

Distribution: Reunion Island and Mauritius to Hawaiian Islands (Batangas, Luzon, Philippines).

506. *Tonna perdix* (Linnaeus, 1758)
Partridge Tun

Identification: Tuns are recognizable by their large animal with a thin brown shell with white markings.

Natural History: Tuns are predators upon bivalves and sea cucumbers. Active at night, they inhabit clean sandy areas in moderate depths. Depth: intertidal zone to 20 m.

Distribution: South Africa; Tanzania to Hawaiian and Society Islands (Manado, Indonesia).

Kathy deWet

507. *Chicoreus brunneus* (Link, 1807)

Identification: This species may reach 100 mm in length and is identifiable by its chocolate brown color with darker brown lines and moderately elevated spire. *Chicoreus* has been recently reviewed by Houart (1992).

Natural History: Species of *Chicoreus* are predators upon other gastropods. Most species inhabit coral rubble and living coral heads.

Distribution: South Africa; Australia; Philippines; Marshall Islands (Msimbati, Tanzania).

T. M. Gosliner

508. *Murex ternispina* Lamarck, 1822

Identification: Species of the genus *Murex* can be identified by their shell with an elongate siphonal canal, usually with a variable number of elongate spines along the body of the shell as well as the siphonal canal. This species usually has black or purple tips to its spines.

Natural History: Species of *Murex* are known to be scavengers and to prey upon bivalves. Most species inhabit soft substrate communities. Depth: 0-62 m.

Distribution: Sri Lanka; Andaman Islands; Singapore; Indonesia; Hong Kong; Philippines; Okinawa; Japan; New Guinea and Australia (Batangas, Luzon, Philippines).

T. M. Gosliner
Leslie Newman & Andrew Flowers

509. *Coralliophila neritoidea* (Lamarck, 1816)
Coral Snail

Identification: Species of *Coralliophila* usually have a bright purple aperture to the shell. It has a relatively low spire compared to other members of the genus.

Natural History: Species of *Coralliophila* are predators on scleractinean corals. In this photo you can see the tissue damage to a *Porites* coral. Unlike other muricids, *Coralliophila* species lack a radula and feed with their muscular proboscis.

Distribution: Mauritius and Reunion Island; New Guinea; Hawaiian Islands, and to Clipperton Island and the Galápagos in the eastern Pacific (Madang, Papua New Guinea).

146

510. *Granulina* sp.

Identification: This minute marginellid is only 2-4 mm in length. It can be distinguished by its fused tentacles and orange head.
Natural History: This species is found on shallow reefs in association with calcareous bryozoans and presumably feeds upon them.
Distribution: South Africa (Sodwana Bay, Natal South Africa).

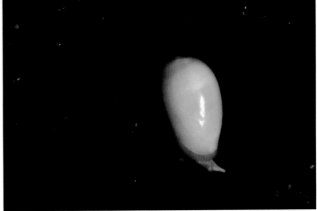

T. M. Gosliner

511. *Marginella elegans* (Gmelin, 1791)

Identification: Marginellids have smooth, shiny shells with a variable number of ridges on the inner side of the shell opening. This species is one of the larger and more attractive species found in the Indo-Pacific.
Natural History: This species is found in sandy habitats where it is presumably predatory upon other gastropods.
Distribution: Restricted to the eastern Indian Ocean: Myanmar; Thailand and Malaysia (Langkawi Island, Malaysia).

T. M. Gosliner

512. *Oliva reticulata* (Röding, 1798)
Netted Olive

Identification: Olive snails are notoriously variable in shell coloration within a single species, even from a single locality. *Oliva reticulata*, including its many synonyms, is one of the larger Indo-Pacific olives and can be recognized by the orange pigment present on the anterior end of the shell aperture.
Natural History: Olives are predators upon other gastropods and are commonly found in well-aerated, clean sand. They burrow just below the sand surface, often with their siphon exposed.
Distribution: Seychelles and Mauritius; Australia; Philippines; Solomon Islands; New Caledonia and Samoa (Batangas, Luzon, Philippines).

G.C. Williams

G.C. Williams

513. *Oliva sidelia* Duclos, 1835

Identification: This is a fairly small (20 mm in length) olive with a distinct suture line between the body whorl and the spire.
Natural History: This species is found burrowing through silty sand in shallow waters.
Distribution: Seychelles and Mauritius; Philippines and Solomon Islands (Batangas, Luzon, Philippines).

514. *Mitra papalis* Linnaeus, 1758
Pope's Miter

Identification: This is one of the largest miters, reaching a length of 165 mm. It can be distinguished by its pattern of orange-brown spots and elevated nodules near the sutures.

Natural History: Most species of miters are predatory and *M. papalis* is no exception, though its prey is not well known. Together with the other large species of miter, *Mitra mitra*, *M. papalis* is unusual in that it lives in sandy rather than rocky habitats.

Distribution: One of the most widespread species. Tanzania; Kenya; Reunion Island and Mauritius; Hawaiian and Society Islands, and Clipperton Island in the eastern Pacific (Batangas, Luzon, Philippines).

515. *Lienardia purpurata* (Souverbie, 1860)

Identification: This beautiful little (3-6mm) purplish snail, is a member of the Turridae, which contains a great diversity of species and genera, distinguished by differences in their shell shape and radular teeth.

Natural History: Turrids are predators on other small gastropods and other small animals. This species is commonly found underneath small pieces of coral rubble in relatively shallow water.

Distribution: Philippines (Batangas, Luzon, Philippines).

516. *Turris babylonia* (Linnaeus, 1758)

Identification: This species can be recognized by its white shell with dark brown to black blotches.

Natural History: *Turris babylonia* lives in shallow waters in sandy habitats. Like other turrids, it is a predatory carnivore, though its specific prey remains unknown.

Distribution: Mauritius and Reunion Island to Philippines; Indonesia; Solomon Islands; New Guinea (Batangas, Luzon, Philippines).

517. *Conus geographus* Linnaeus, 1758
Geographic Cone

Identification: This extremely toxic cone is responsible for several human fatalities each year. This species, with its wide aperture and tent-like markings on the shell, should be treated with respect and never handled from the front end of the shell.

Natural History: This species feeds primarily upon fish and uses its potent toxin to rapidly subdue and kill its prey. It is commonly found in shallow coral rubble communities.

Distribution: Tanzania; Reunion Island; Mauritius and Seychelles to Philippines and Marshall Islands (Philippines).

148

518. *Conus marmoreus* Linnaeus, 1758
Marble Cone

Identification: This species also has a series of tent-like markings but has a darker, more solid shell than other species.
Natural History: Found in shallow water to depths of 90 m, this species is predatory upon other cones, including *Conus flavidus*, upon which it is feeding in this photo.
Distribution: South Africa; Red Sea; Hawai'i and Society Islands (Maui, Hawaiian Islands).

Mike Severns

519. *Conus milneedwardsi* Jousseaume, 1894
Glory-of-India

Identification: The highly elevated spire with tent-like markings distinguish this rare, deep water species of cone.
Natural History: Little is known about the feeding of this species since it is generally dredged or trawled from relatively deep water.
Distribution: South Africa; India; China; Taiwan; and Ryukyu Islands (Natal, South Africa).

G.C. Williams

520. *Conus textile* Linnaeus, 1758
Textile Cone

Identification: The oval shell with a pattern of wide tent-like markings is distinctive. This species is also highly venomous and has been responsible for human fatalities. The tip of the siphon is always a reddish orange with a black band more basally.
Natural History: This species feeds upon other gastropods, usually other species of *Conus*. It is found on the underside of coral rubble, in relatively shallow water.
Distribution: South Africa; Red Sea to Hawai'i and Society Islands (Sapwauhfik, Pohnpei).

Dave Zoutendyk
T. M. Gosliner

521. *Architectonica perspectiva* (Linnaeus, 1758)
Sundial

Identification: Sundials are members of a group of unusual gastropods called heterobranchs. Heterobranchs also include the pulmonate and opisthobranch gastropods. The beautiful markings and texture of this discoidally-shelled snail make it readily recognizable.
Natural History: Found in clean sandy habitats in relatively shallow water. These feed on burrowing coelenterates, probably sea pens and burrowing anemones.
Distribution: Tanzania to New Guinea and Hawaiian Islands (Madang, Papua New Guinea).

149

522. *Heliacus variegatus* (Gmelin, 1791)

Identification: This sundial can be distinguished by its black and white checked shell and its conical operculum.

Natural History: Always found in close association with zoanthids, it is shown here feeding on the zoanthid, *Palythoa nelliae*. It inhabits shallow rocky habitats, intertidal zone to 10 m depth, where zoanthids abound.

Distribution: South Africa to Hawaiian Islands (Ramsgate, Natal, South Africa).

Opisthobranchia - Sea Slugs

The opisthobranchs are a group of marine gastropods which have an external shell, internal shell or entirely lack a shell as adults. They include the bubble snails, sea hares, side-gilled slugs, sacoglossans and nudibranchs. Field guides to Indo-Pacific opisthobranchs include the works of Bertsch and Johnson (1981), Willan and Coleman (1984), Gosliner (1987), Coleman (1989) and Wells and Bryce (1993).

Acteonoidea

The Acteonoidea have been considered traditionally as cephalaspideans, but current studies show them to be an early divergence from the opisthobranch line.

523. *Hydatina physis* (Linnaeus, 1758)

Identification: The bright reddish pink animal with a white shell with numerous black spiral lines is distinctive.

Natural History: *Hydatina physis* is found in shallow water. It feeds on cirratulid polychaetes and incorporates toxins from its prey for its own defense.

Distribution: Circumtropical: South Africa; Red Sea to Hawaiian Islands. Also in the Caribbean (Ft. Dauphin, Madagascar).

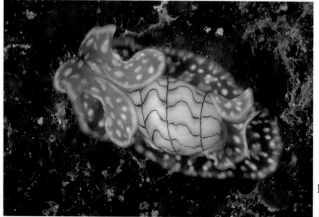

524. *Micromelo undata* (Brugière, 1792)

Identification: Distinguished by its bluish green animal with opaque white spots and shell with dark brown undulating bands.

Natural History: Like *Hydatina physis*, this species feeds upon cirratulid polychaetes and incorporates their toxins. It is found in shallow water rocky habitats. Intertidal to 5 m.

Distribution: Circumtropical: South Africa; Red Sea to Hawaiian Islands. Also in the Caribbean (Kauai, Hawai'i).

Cephalaspidea

Cephalaspideans are a group of opisthobranchs which usually have a bubble shell, an internal shell or may entirely lack a shell.

525. *Bulla vernicosa* Gould, 1859
Bubble Snail

Identification: The brown mottled shell is characteristic of species of *Bulla*. Each species has its own color pattern of spots.
Natural History: *Bulla* are herbivorous and have been observed eating filamentous algae. Animal material has also been found in the stomachs of some species. Species burrow in soft muddy sand, but sometimes emerge from the substrate at night where they may form breeding aggregations.
Distribution: Fiji; New Guinea; Philippines; Okinawa; Japan; Hawaiian Islands (Batangas, Luzon, Philippines).

526. *Haminoea cymbalum*
(Quoy & Gaimard, 1835)

Identification: There are 65 species of *Haminoea* known from the Indo-Pacific, including many undescribed species. Most are cryptically colored on algae, but a few, like *H. cymbalum*, are brightly colored. Several species have yellow and purple spotting, but differ from *H. cymbalum* in the pattern of spots.
Natural History: Species of *Haminoea* are herbivores, feeding almost exclusively on filamentous algae. Many species are found in intertidal rock pools or in shallow sandy habitats.
Distribution: Mozambique; Madagascar to Hawai'i (Madang, Papua New Guinea).

527. *Chelidonura electra* Rudman, 1970
Electric Swallowtail

Identification: This species is larger and paler than *C. pallida*. It also lacks any black pigment. It is large, reaching 70 mm.
Natural History: Frequently found in silty or sandy habitats inside bays. Feeds on acoel flatworms.
Distribution: Originally described from the Solomon Islands; also in Madagascar and New Guinea (Madang, Papua New Guinea).

528. *Chelidonura hirundinina*
(Quoy & Gaimard, 1824)

Identification: One of 13 described and several undescribed species of *Chelidonura* known from the Indo-Pacific. It reaches about 30 mm in length and can be distinguished from all other species by blue, white or yellow T-shaped markings on the head.
Natural History: Common on rocky or sandy substrates in shallow water. Like most other members of the genus, it feeds upon small acoel flatworms like *Convoluta* (#318) and swallows them whole with its muscular pharynx.
Distribution: Circumtropical: South Africa; Aldabra Atoll; Madagascar; to Hawaiian Islands. Also known from the Caribbean (Madang, Papua New Guinea).

T. M. Gosliner

G.C. Williams

T. M. Gosliner
T. M. Gosliner

151

529. *Chelidonura inornata* Baba, 1949

Identification: This species can be recognized by its black pigment with small white spots and broad white band on the head. There are also red-orange spots on the head.

Natural History: This species is often found in large breeding associations on shallow living reefs. It also feeds upon small flatworms.

Distribution: Western Pacific: Australia; Fiji; New Guinea; Indonesia; Philippines; and Japan; Marshall Islands (Milne Bay, Papua New Guinea).

530. *Chelidonura pallida* Risbec, 1951

Identification: This species can be readily identified by its white body with yellow and black lines around the edges.

Natural History: Found on shallow patch reefs, where it presumably feeds on small flatworms.

Distribution: Malaysia; western Australia; Fiji; New Guinea and New Caledonia (Madang, Papua New Guinea).

531. *Philinopsis cyanea* (Martens, 1879)

Identification: This species can be recognized by its truncate body with extended posterior portion of the head shield. It is extremely variable in color and may be whitish, brown or black, with or without yellow and blue markings.

Natural History: It is a predator upon other opisthobranchs, especially bubble snails of the genera *Haminoea* and *Bulla*. As in *Chelidonura*, it lacks a radula and ingests its prey whole by suction. Buried in sand or mud during the day, it emerges at night to track prey by chemical means.

Distribution: South Africa; Red Sea to Hawai'i (Batangas, Luzon, Philippines).

532. *Philinopsis gardineri* (Eliot, 1903)

Identification: This species is similar in color to some specimens of *P. cyanea* but has a round rather than truncate anterior end of the body. Also, it has a bubble-shaped head like a Boeing 747.

Natural History: This is a member of a species complex whose members probably feed upon polychaete worms rather than opisthobranchs. They are generally found crawling on the surface of soft substrate.

Distribution: Originally described from Tanzania. Madagascar; Fiji; New Guinea; Indonesia; Guam and Philippines; Marshall Islands (Madang, Papua New Guinea).

152

533. *Sagaminopteron nigropunctatum* Carlson & Hoff, 1973

Identification: The Gastropteridae is a family of small opistho-branchs that is well represented in the tropical Indo-Pacific. They have been recently reviewed by Gosliner (1989). This greyish species is cryptic on the sponge where it is found.

Natural History: Found in subtidal reef habitats on its prey, the gray sponge *Dysidea* sp. Depth: 10-40 m.

Distribution: Tanzania; Fiji; Philippines; New Guinea and Guam (Msimbati, Tanzania).

T. M. Gosliner

534. *Sagaminopteron psychedelicum* Carlson & Hoff, 1974

Identification: The distinctive brilliant color pattern of this species distinguishes it from the other three described species.

Natural History: This species feeds upon sponges. In shallow water (intertidal zone to 3m depth) it feeds upon the rubbery green sponge, *Dysidea* cf. *herbacea* (#42). In deeper water it is found upon the gray sponge, *Dysidea* sp., together with *S. nigropunctatum*. Depth: 1-25 m.

Distribution: Tanzania and Reunion Island to Fiji; New Guinea; Philippines; Okinawa and Guam (Madang, Papua New Guinea).

T. M. Gosliner

535. *Siphopteron tigrinum* Gosliner, 1989

Identification: With its orange and bluish purple markings, it is one of 12 described species of this exclusively Indo-Pacific genus. Most species are minute, only 5 mm when mature.

Natural History: Species of *Siphopteron* are generally found on the undersurface of coral rubble in shallow water. Nothing is known about their diet. Depth: 1-10 m.

Distribution: Madagascar; Australia; New Guinea; Belau; Philippines and Okinawa (Madang, Papua New Guinea).

Anaspidea - Sea Hares

Sea hares are a group of herbivorous opisthobranchs that often have widespread distributions. Most species have an internal shell and secrete purple ink as a defense mechanism.

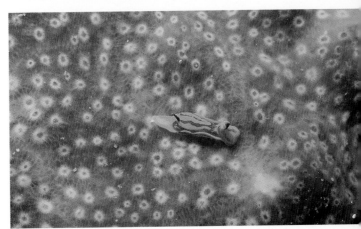

T. M. Gosliner
T. M. Gosliner

536. *Aplysia dactylomela* Rang, 1828

Identification: *Aplysia dactylomela* is one of the more common species in the Indo-Pacific tropics and can be distinguished by the large black rings present on its mantle.

Natural History: Sea hares feed primarily on red algae, but will feed on a variety of algal species when their preferred food is not present. Most species live in shallow waters, where algal diversity is highest. *Aplysia dactylomela* is often more common in subtropical areas where algae are more abundant.

Distribution: Circumtropical: South Africa; Red Sea to Hawaiian Islands and Panama. Also present in the Caribbean (Umgazana, Transkei, South Africa).

153

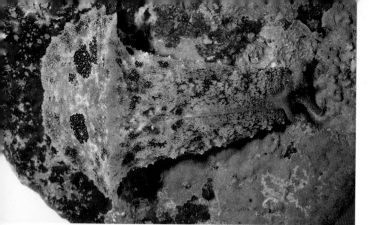

T. M. Gosliner

Marc Chamberlain

T. M. Gosliner
T. M. Gosliner

537. *Dolabella auricularia* (Lightfoot, 1786)

Identification: This is the largest species of sea hare present in the Indo-Pacific tropics and may reach one-half meter in length. It can be immediately distinguished by its discoidal hind end that makes it look like the posterior of the animal was chopped off.

Natural History: During the day this species is found under coral heads or stones, and emerges at night to forage; to 15 m depth.

Distribution: South Africa; Red Sea to Hawaiian and Galapagos Islands and the tropical East Pacific (Madang, Papua New Guinea).

538. *Paraplysia geographica* Adams & Reeve, 1850

Identification: This sea hare is beautifully colored with elaborate patterns of green and white. It is the only member of this genus and is distinguished by its high parapodia.

Natural History: This species is nocturnal and lives in sandy habitats where it feeds on brown algae. It is also capable of swimming by flapping its parapodia. Depth: 1 m.

Distribution: South Africa to Australia; New Guinea; Indonesia; Philippines (Batangas, Luzon, Philippines).

539. *Stylocheilus longicauda* (Quoy and Gaimard, 1825)

Identification: This common, but often cryptic, sea hare can be distinguished by its brownish body with longitudinal lines and blue spots.

Natural History: This species is often found feeding on filamentous cyanobacteria (blue green algae) that are locally abundant colonizers of shallow sandy and rocky habitats.

Distribution: Circumtropical: South Africa; Red Sea; Hawai'i; Galapagos; Baja California; Caribbean (Mbudya Island, Tanzania).

Sacoglossa - Sacoglossans

Sacoglossans are a highly specialized group of herbivorous opisthobranchs. They have a well-developed radula and mouth parts to pierce algal cells and suck out the contents. Representatives of most groups of sacoglossans are associated with species of the green algal genus *Caulerpa*. It has therefore been suggested that much of the initial evolution occurred in association with this alga. Primitive members have a shell, while more derived ones entirely lack a shell as adults.

540. *Volvatella* sp.

Identification: This is one of the more primitive sacoglossans, with a well-developed bubble shell. Several species are described.

Natural History: Species of *Volvatella* are found in association with species of *Caulerpa*. This species is found only on the grape-like *Caulerpa racemosa*. Depth: 10 m.

Distribution: Philippines (Bohol, Philippines).

154

541. *Berthelinia schlumbergeri* Dautzenberg, 1895

Identification: Species of *Berthelinia* and their close relatives, *Julia* spp., are unusual "bivalved gastropods." In contrast to true bivalves (clams and their relatives), the shell in these sacoglossans is formed by the splitting of a typical, coiled larval shell. Several species are present in the Indo-Pacific tropics, although the taxonomy has not been well studied.

Natural History: Species of *Berthelinia* feed upon *Caulerpa* spp., most commonly *Caulerpa racemosa*.

Distribution: South Africa and Aldabra Atoll to Philippines and New Guinea (Madang, Papua New Guinea).

T. M. Gosliner

542. *Lobiger souverbii* Fischer, 1856

Identification: This species has a reduced internal shell, usually with bright blue lines on the tissue below the shell. It can be distinguished by the four elongate lobes which extend from the body.

Natural History: Found on *Caulerpa racemosa*, upon which it feeds. When disturbed, this species elevates its parapodia to startle its prey. If disturbed further it can shed its four parapodial lobes.

Distribution: Circumtropical: South Africa and Tanzania to Hawai'i and the Society Islands; Galápagos. Also known from the Caribbean (Madang, Papua New Guinea).

T. M. Gosliner

543. *Oxynoe viridis* (Pease, 1861)

Identification: This species can be recognized by the parapodia which cover the shell and by its elongate tail. The animal is generally green with scattered blue spots. About five species of *Oxynoe* have been described from the Indo-Pacific tropics.

Natural History: Found in association with *Caulerpa racemosa*. Species of *Oxynoe* autotomize the hind end of their foot when disturbed.

Distribution: South Africa; Tanzania and Madagascar to the Hawaiian and Society Islands (Madang, Papua New Guinea).

T. M. Gosliner
Lynn Funkhouser

544. *Cyerce nigricans* (Pease, 1866)

Identification: This strikingly beautiful species can be distinguished by its black cerata and body with orange and bluish pigment. The cerata are respiratory and defensive in nature.

Natural History: *Cyerce nigricans* is found on the filamentous green alga, *Chlorodesmis* sp., commonly known as turtle weed. When disturbed, this sacoglossan can shed its cerata, which contain a noxious, viscous mucus.

Distribution: Tanzania; Madagascar and Aldabra Atoll to Australia; New Guinea; Guam and Okinawa (Australia).

T. M. Gosliner

545. *Cyerce* sp.

Identification: This is an undescribed species of *Cyerce* which has large, transparent cerata with black lines and yellow pigment.

Natural History: This species is found on shallow patch reefs, but has not been found in direct association with any algal food.

Distribution: Indonesia; New Guinea and Guam (Madang, Papua New Guinea).

T. M. Gosliner

546. *Elysia ornata* (Swainson, 1840)

Identification: Elysia ornata can be distinguished by its green body color, black spots and orange and black parapodial margins.

Natural History: This species is found in association with the feathery green algal genus *Bryopsis*, in shallow water.

Distribution: Circumtropical: South Africa; Madagascar; Reunion Island and Aldabra Atoll to Hawaiian Islands and the Caribbean (Kapas Island, Kuala Terrengenu, Malaysia).

T. M. Gosliner

547. *Elysia* sp.

Identification: This large species is undescribed and differs from *E. ornata* in having a white rather than orange line along the margin of the parapodia. It also has larger parapodia.

Natural History: This species has been found on the blades of the large green seagrass *Enhalis* sp. in shallow water lagoons.

Distribution: Thus far, this species is known only from New Guinea (Madang, Papua New Guinea).

T. M. Gosliner

548. *Thuridilla albopustulosa* Gosliner, 1995

Identification: This recently described species has a bluish body with white pustules and black and orange pigment.

Natural History: Like the following species, *T. albopustulosa* is found on shallow reefs usually under coral rubble.

Distribution: South Africa and Aldabra Atoll to New Guinea; Indonesia; Philippines (Madang, Papua New Guinea)

549. *Thuridilla bayeri* Marcus, 1965

Identification: In contrast to the cryptic coloration of most *Elysia*, *Thuridilla* are brightly colored. *Thuridilla bayeri* has a dark, greenish black body color with white or cream longitudinal lines. Blue patches and red pigment may also be present.

Natural History: Species of *Thuridilla* are generally found crawling in the open or under coral rubble in shallow water and are not found in association with specific algae.

Distribution: Madagascar and Maldives to Australia; New Guinea; Philippines; Fiji; Guam and Marshall Islands (Madang, Papua New Guinea).

T. M. Gosliner

550. *Thuridilla undula* Gosliner, 1995

Identification: This is another beautiful and strikingly colored species that has been recently described. It can be recognized by its blue body color with black and orange pigment.

Natural History: Found openly on shallow reefs.

Distribution: Maldives; New Guinea; Philippines; Guam and Belau (Madang, Papua New Guinea).

Notaspidea - Side-gilled Slugs

Members of this group have the gill situated on the right side of the body. A few representatives, such as *Umbraculum* and *Tylodina*, have an external shell. Most species have a reduced, internal shell, while species of *Pleurobranchaea* and *Euselenops* lack shells as adults. Notaspideans are carnivores and are closely related to nudibranchs.

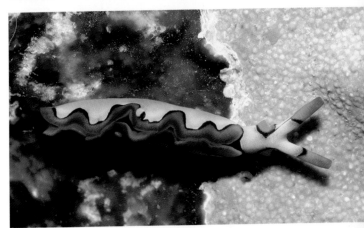

T. M. Gosliner

551. *Umbraculum umbraculum* Lightfoot, 1786

Identification: This is a very unusual opisthobranch since it has a limpet-like shell and the gill is situated on the right side of the body. The mantle is covered by numerous rounded papillae. There appears to be only a single species of *Umbraculum*.

Natural History: This species seems to feed on sponges and is frequently found in shallow pools or on subtidal reefs to 120 m.

Distribution: Circumtropical: South Africa; Tanzania; Madagascar and Aldabra Atoll to Hawaiian Islands; Pacific Central America; Caribbean and Mediterranean (Ft. Dauphin, Madagascar).

T. M. Gosliner
T. M. Gosliner

552. *Berthella martensi* Pilsbry, 1896

Identification: This species is extremely variable in color and ranges from white, yellow, brown or black, with or without mottlings of a different color.

Natural History: This species feeds upon compound tunicates. When disturbed, the animal sheds the mantle along three distinct fracture zones.

Distribution: Indo-Pacific: Reunion Island and Tanzania to Okinawa; Hawai'i and the Pacific coast of Mexico (Madang, Papua New Guinea).

157

553. *Berthellina citrina* (Rüppell & Leuckart, 1828)

Identification: Distinguished by the smooth orange colored mantle. Opaque white spots may be present on the notum, especially in specimens from the western Indian Ocean.

Natural History: Living under coral heads during the day and emerges at night. It appears to feed upon colonial tunicates.

Distribution: South Africa; Red Sea to Hawai'i and Society Islands (Madang, Papua New Guinea).

T. M. Gosliner

554. *Euselenops luniceps* (Cuvier, 1817)

Identification: This is a bizarre looking opisthobranch that intially looks more like a cephalopod than a side-gilled slug. The animal is brownish with white spots. The front of the animal has a large fringed oral veil. A shell is absent in adults.

Natural History: This species is found in sandy and silty habitats and is capable of swimming. It is predatory, but its food is poorly known. Close relatives feed upon sea anemones and other opisthobranchs.

Distribution: South Africa and Tanzania to Hawaiian Islands (Batangas, Luzon, Philippines).

Marc Chamberlain

555. *Pleurobranchus forskalii* (Rüppell & Leuckart, 1828)

Identification: The body color of this species is dark maroon to almost black. Opaque white or black circles or semicircles are frequently present on the dorsal surface. To 200 mm in length.

Natural History: This species is nocturnal and frequently observed crawling over sandy and rubble bottoms.

Distribution: Tanzania; Red Sea; Australia; Fiji; New Guinea; Indonesia; Philippines; Guam; Japan (Manado, Sulawesi, Indonesia).

Jack Randall
Terry Schuller

556. *Pleurobranchus grandis* Pease, 1868

Identification: This large (in excess of 100 mm in length) species of side-gilled slug is variable in its coloration, but usually has different proportions of white, yellow and black pigment. There are several other described species of *Pleurobranchus* from the Indo-Pacific.

Natural History: This species is nocturnal and is seen crawling over the surface of living reefs at night. It presumably feeds on tunicates, but has not been observed feeding.

Distribution: New Guinea; Philippines; Fiji; Society Islands; though the precise distribution is poorly known (Indonesia).

Nudibranchia - Nudibranchs

Nudibranchs are the most diverse group of opisthobranchs, comprising about two-thirds of the known opisthobranch fauna of the Indo-Pacific. Almost 2000 species of nudibranchs are known to occur in the Indo-Pacific. Dorids are the largest group of nudibranchs and are characterized by a circle of gills situated on the hind end of the back, surrounding the anus. The second largest group, the aeolids, have a series of finger like appendages (cerata) located on the back, each of which contains an extension of the digestive tract. At the tip of the cerata are sacs that can store nematocysts, which are obtained from their coelenterate prey and can be used by the aeolid for its own defense.

Doridacea - Dorids

Most dorid nudibranchs feed upon sponges and use their broad radula to scrape tissue from the sponge. Other groups of dorids feed upon tunicates, bryozoans or other opisthobranchs.

557. *Discodoris boholensis* Bergh, 1877

Identification: This species is recognizable by its brown color with black and white spots and white lines on the body. It has a fairly flat appearance with a central hump in the middle of the back. Length: to 70 mm.

Natural History: Feeds on sponges, but is also seen crawling about in the open or underneath rocks in relatively shallow water.

Distribution: Tanzania and Madagascar to Australia; Fiji; New Guinea; Indonesia; Philippines and Okinawa (Batangas, Luzon, Philippines).

T. M. Gosliner

558. *Discodoris* sp.

Identification: The pattern of tubercles that resemble fried eggs distinguish this undescribed species of *Discodoris*.

Natural History: Found in shallow water in the open.

Distribution: Indonesia; Philippines and Marshall Islands (Batangas, Luzon, Philippines).

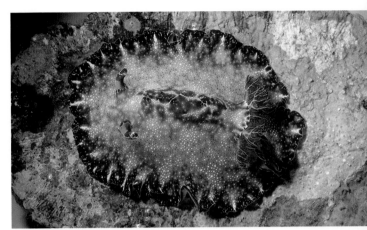

T. M. Gosliner
T. M. Gosliner

559. *Asteronotus cespitosus* (Hasselt, 1824)

Identification: This is a large (70-200 mm in length) and conspicuous dorid characterized by a brownish body with large tubercles which are arranged in concentric rings. The color varies in specimens from different localities.

Natural History: Asteronotus cespitosus is found under large rocks or coral heads during the day and crawls on sand or reef flats during the night. It presumably feeds upon sponges.

Distribution: Tanzania; Madagascar; Seychelles and Mauritius to New Guinea to Hawaiian Islands (Madang, Papua New Guinea).

159

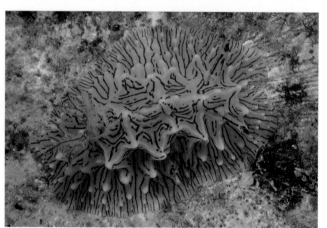

T. M. Gosliner

560. *Halgerda carlsoni* **Rudman, 1978**

Identification: This species has a series of orange-tipped tubercles forming interrupted ridges. Minute, scattered orange spots are present between the tubercles. This species has large gills.

Natural History: This species is never common, but is found occasionally under coral heads and rubble. It presumably feeds on sponges.

Distribution: Solomon Islands; Fiji; New Guinea; Tonga and Okinawa (Madang, Papua New Guinea).

T. M. Gosliner

561. *Halgerda tessellata* **(Bergh, 1880)**

Identification: Most species of *Halgerda* have a series of brightly colored ridges on the back of the animal. *Halgerda tessellata* is characterized by a series of triangular orange ridges with brown pigment and orange markings and white spots.

Natural History: Found on shallow patch reefs among coral rubble.

Distribution: Tanzania; Aldabra Atoll; Seychelles; Red Sea to Australia; New Guinea; Belau and Okinawa (Madang, Papua New Guinea).

562. *Halgerda willeyi* **Eliot, 1903**

Identification: This species has a several yellow ridges with a series of yellow and black lines between them. It is one of the larger members of the genus and may reach 70-80 mm in length.

Natural History: This species is found on living reefs crawling in the open or under rocks. It also feeds on sponges.

Distribution: Tanzania; Red Sea to Australia; New Guinea; Philippines; Okinawa; Loyalty and Marshall Islands (Batangas, Luzon, Philippines).

T. M. Gosliner

Mike Severns

563. *Halgerda sp.*

Identification: This species is undescribed and can be recognized by its large, yellow-orange tubercles and gills. Black pigment is present between the yellow-orange ridges.

Natural History: Little is known about the natural history of this species.

Distribution: Indonesia (Lembeh Strait, Sulawesi, Indonesia).

564. *Jorunna funebris* (Kelaart, 1858)

Identification: The white body with black rings, gills and rhinophores is distinctive. The body has a fuzzy appearance owing to the presence of dense compound papillae called caryophyllidia. The living animal may reach 100 mm in length.

Natural History: Found in relatively shallow water, it is usually associated with the bright blue sponge, *Haliclona* sp. 2 (#28).

Distribution: South Africa; Tanzania; Madagascar to Australia; Fiji; New Guinea; Indonesia; Philippines; Guam; Belau; Okinawa; Marshall Islands (Madang, Papua New Guinea).

T. M. Gosliner

565. *Platydoris scabra* (Cuvier, 1804)

Identification: Platydorids, as the name implies, are a group of flat-bodied dorids with a granular texture and a relatively stiff body. There are approximately 25 described species in this genus, each characterized by a different color pattern. *Platydoris scabra* has a grey blotched body with an orange marginal ring and rhinophores.

Natural History: Found in shallow water under coral heads. Often crawling in the open during the night.

Distribution: Madagascar to Malaysia; Australia; Fiji; Guam; Okinawa to Marshall Islands (Isle St. Marie, Madagascar).

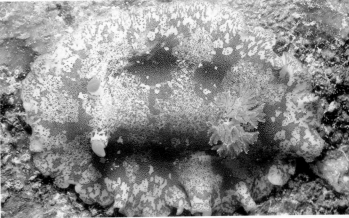

T. M. Gosliner

566. *Hexabranchus sanguineus*
(Rüppell & Leuckart, 1828)
Spanish Dancer

Identification: This is the largest species of nudibranch and may reach 600 mm in length. It is variable in color, but always contains shades of red, pink and orange. It can be readily distinguished from all other dorids by the fact that its six separate gills can be retracted within distinct pockets. The juvenile, shown below, is red and white with a blue margin.

Natural History: This species is commonly seen swimming through the water by undulating its body. When disturbed, it opens its parapodia, revealing intense red pigment. Depth: intertidal to 50 m. May have a commensal shrimp *Periclimenes imperator* (#735).

Distribution: South Africa; Red Sea to Hawai'i and Society Islands (Reunion Island, Mascarene Archipelago; Papua New Guinea)

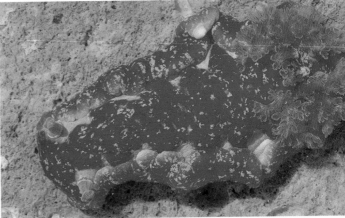

G.C. Williams
T. M. Gosliner

Adult
Juvenile

Chromodorididae - Chromodorids

This family contains the largest number of Indo-Pacific species with more than 360 species, many of which are still undescribed. The family name means colorful dorids, and most species are brightly colored and elaborately patterned. The genera can be distinguished by some external features, but are primarily separated by differences in their radular teeth.

The largest genus, *Chromodoris*, has approximately 165 species and usually has a simple ovoid body shape and is flatter than members of most other genera.

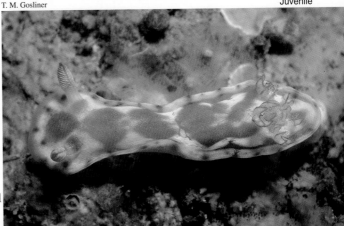

567. *Ardeadoris egretta* Rudman, 1984

Identification: This is a relatively large chromodorid, reaching more than 100 mm in length. Its white body, gills and rhinophores with an orange marginal band are distinctive. The genus and species were named for the common heron, which is abundant on Heron Island, Australia, the type locality for this species.

Natural History: This species is found on relatively shallow reef flats. It is often seen crawling in the open over the surface of living reefs.

Distribution: Australia; New Guinea; Philippines and Okinawa (Madang, Papua New Guinea).

T. M. Gosliner

568. *Ceratosoma alleni* Gosliner, 1996

Identification: This remarkable species of the genus *Ceratosoma* has just recently been described. It is readily distinguished by a series of elongate nodular appendages.

Natural History: This species was found on a shallow reef at 10 m, dominated by a xeniid octocoral, which it closely mimics. It is unlikely that it feeds on octocorals, but rather on sponges.

Distribution: Mindanao and Luzon, Philippines (Dakak, Mindanao, Philippines).

Mike Miller

569. *Chromodoris annae* (Bergh, 1877)

Identification: This species is part of a complex involving species that were once considered to represent a single species. Five of the more common members of this group are depicted here. Most of these species have a combination of blue, black, orange, white and blue rings or lines. *Chromodoris annae* has a blue mantle with darker markings and lacks a mid-dorsal longitudinal line.

Natural History: Found on open rock walls and reef faces where it feeds on aplysilid sponges. Depth: 15-30 m.

Distribution: Christmas Island; Malaysia; New Guinea; Indonesia; Guam; Philippines; Okinawa and Marshall Islands (Batangas, Luzon, Philippines).

T. M. Gosliner
Marc Chamberlain

570. *Chromodoris bullocki* Collingwood, 1881

Identification: Several different color variants have been called *Chromodoris bullocki*. These probably represent a complex of species that can be distinguished by differences in color. The original description of *C. bullocki* is identical to the specimen shown here. It is light pink with a thin white marginal line. The gills and rhinophores are white basally with reddish tips.

Natural History: Found on shallow water reefs where it is observed crawling about in the open.

Distribution: Taiwan; Indonesia and Philippines (Batangas, Luzon, Philippines).

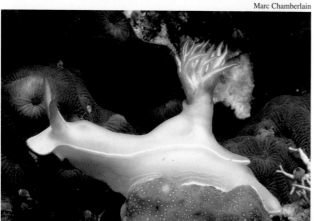

162

571. *Chromodoris coi* **Risbec, 1956**

Identification: *Chromodoris coi* is another strikingly beautiful species with an elaborate color pattern. It differs from *C. kuniei* in having white and black undulating lines surrounding a central area of rose pink.

Natural History: As in *C. kuniei*, the mantle of *C. coi* undulates as the animal crawls. It is also found in sheltered habitats on the inside of barrier reefs.

Distribution: Vietnam to Australia; Fiji; New Guinea; Indonesia; Philippines; Guam; Okinawa and Marshall Islands (Madang, Papua New Guinea).

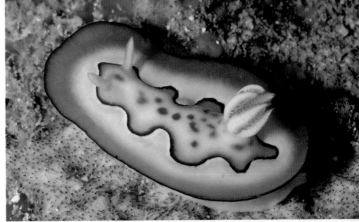

T. M. Gosliner

572. *Chromodoris elizabethina* **Bergh, 1877**

Identification: This species is similar to *C. annae*, but lacks darker punctations on the blue and has mid-dorsal longitudinal black line.

Natural History: This species is found in the same habitat as *C. annae*. Depth: 5-30 m.

Distribution: Aldabra Atoll; Maldives; Christmas Island; Australia; Fiji; Vanuatu; New Guinea; Indonesia; Philippines; Guam; Okinawa and Marshall Islands (Madang, Papua New Guinea).

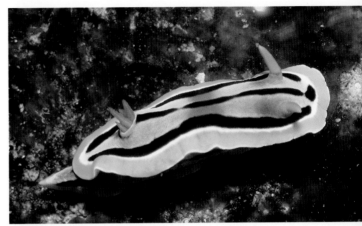

T. M. Gosliner

573. *Chromodoris geometrica* **(Risbec, 1928)**

Identification: This species has a purplish body color with opaque white tubercles and a network of black lines. The gills and rhinophores are tipped with bright green.

Natural History: *Chromodoris geometrica* is found on shallow patch reefs, usually under stones or coral rubble, but may be found in the open. The animal vibrates its gills and the front of the head is alternately raised and lowered when the animal is actively crawling.

Distribution: Tanzania; Maldives; Australia; Fiji; New Guinea; Indonesia; Philippines; Okinawa; Guam; Belau and Marshall Islands (Batangas, Luzon Philippines).

T. M. Gosliner
T. M. Gosliner

574. *Chromodoris kuniei* **Pruvot-Fol, 1930**

Identification: This is a strikingly beautiful species with a yellowish body, purple and blue markings and black spots surrounded by purple.

Natural History: Generally found in moderate depths in fairly sheltered habitats. *Chromodoris kuniei* undulates the sides of its mantle as it crawls along the substrate.

Distribution: Christmas Island to Australia; Tonga; New Caledonia; New Guinea; Indonesia; Philippines; Guam and Marshall Islands (Madang, Papua New Guinea).

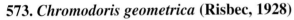

575. *Chromodoris lochi* Rudman, 1982

Identification: This species is pale blue with a dark blue submarginal band and mid-dorsal line.

Natural History: This species is found on the walls on the outside of fringing and barrier reefs where it feeds on sponges.

Distribution: Singapore; Australia; Vanuatu; New Caledonia; Fiji; Solomon Islands; Tonga; New Guinea; southern Philippines and Society Islands (Solomon Islands).

Roy Eisenhardt

576. *Chromodoris magnifica* (Quoy & Gaimard, 1832)

Identification: This species has black and white lines and an orange submarginal band, gills and rhinophores. It is most similar to *C. africana* which is known from the western Indian Ocean and Red Sea. In that species, the orange band is found right at the edge of the body rather than inside a white band.

Natural History: Found on outer reef walls and slopes, where it feeds on sponges. Mimics *Pseudobiceros* sp. 6 (#344).

Distribution: Australia; New Guinea; Indonesia and Philippines (Madang, Papua New Guinea).

T. M. Gosliner

577. *Chromodoris reticulata* (Quoy & Gaimard, 1832)

Identification: This is one of a complex of species that have a reticulated network of red lines over the surface of the mantle. It lacks marginal spots and has distinct tubercles over the surface of the body.

Natural History: *Chromodoris reticulata* is frequently found under coral rubble at moderately shallow reefs.

Distribution: Tanzania to Vietnam; Tonga; New Guinea; Indonesia; Philippines and Okinawa (Batangas, Luzon Philippines).

T. M. Gosliner
T. M. Gosliner

578. *Chromodoris willani* Rudman, 1982

Identification: This species is similar in appearance to *C. lochi* but is a darker blue, has a larger gill and there are opaque white spots on the gills and rhinophores.

Natural History: Found on exposed reef walls and slopes, where it feeds on sponges.

Distribution: Known from the western Pacific of Vanuatu; Indonesia; Philippines; Guam and Okinawa (Batangas, Luzon, Philippines).

164

579. *Chromodoris* sp.

Identification: This species is part of the *C. bullocki* complex, but is probably a distinct species. It can be recognized by its brilliant purple body color, orange gills and rhinophores and diffuse white marginal band. It may exceed 100 mm in length.

Natural History: This large species is conspicuous when crawling about on living shallow water reefs.

Distribution: Western Australia; Indonesia; Philippines and Okinawa (Batangas, Luzon, Philippines).

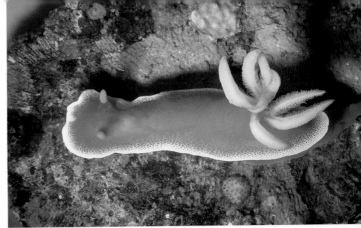

T. M. Gosliner

580. *Glossodoris atromarginata* (Cuvier, 1804)

Identification: The undulating mantle edge, mustard yellow body color and black marginal line on gills and rhinophores distinguish this species, which may reach 100 mm in length.

Natural History: This species is found under coral heads and on sponges inhabiting exposed vertical surfaces on outer reefs.

Distribution: South Africa; Red Sea to the Society Islands (Madang, Papua New Guinea).

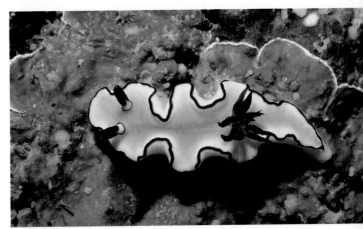

G.C. Williams

581. *Glossodoris cincta* (Bergh, 1889)

Identification: This species can be recognized by its mottled reddish brown and white pigments, and yelllow brown and turquoise marginal bands. Other forms which may be part of a species complex have different colored marginal lines.

Natural History: Found in relatively shallow water on patch reefs, where it may be found in the open or beneath coral rubble.

Distribution: Tanzania; Seychelles to Australia; Fiji; New Guinea; Indonesia; Philippines; Guam and Marshall Islands (Solomon Islands).

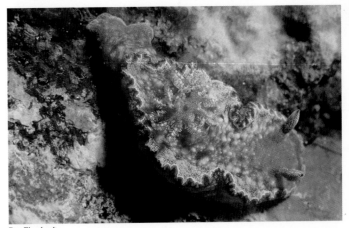

Roy Eisenhardt
Mike Miller

582. *Glossodoris cruentus* Rudman, 1986

Identification: This species has a cream body with blood red spots and broad white and yellow marginal bands. The notum of this species is less convoluted than other members of the genus.

Natural History: Found on reef fronts and drop offs.

Distribution: Australia; New Guinea and Philippines (Davao, Mindanao, Philippines).

T. M. Gosliner

583. *Glossodoris pallida*
(Rüppell & Leuckart, 1828)

Identification: *Glossodoris pallida* is a relatively small member of the genus, up to 30 mm in length. It has a translucent white body with central opaque markings and a yellow marginal band.

Natural History: This species inhabits walls and patch reefs where it is found on the tan sponge shown here. Depth: to 30 m.

Distribution: South Africa; Red Sea to Australia; Fiji; New Guinea; Philippines and Guam (Madang, Papua New Guinea).

T. M. Gosliner

584. *Glossodoris* sp.

Identification: This is one of many undescribed species of *Glossodoris*. It can recognized by its uniformly yellow body color with a broad submarginal color band.

Natural History: The single specimen was collected from a sloping wall on a stretch of rugged limestone coastline.

Distribution: New Guinea (Madang, Papua New Guinea).

585. *Hypselodoris* sp.

Identification: Members of the genus *Hypselodoris* have a high body profile and frequently have longitudinal lines on the dorsal surface. The species depicted here is undescribed and can be recognized by its unique color pattern and large protruding gills.

Natural History: *Hypselodoris* sp. is found on shallow to deep reefs, under coral heads or out in the open. The specimen shown here is next to its brilliantly colored egg mass.

Distribution: New Guinea; Indonesia; Philippines and Okinawa (Batangas, Luzon, Philippines).

Mike Severns
T. M. Gosliner

586. *Miamira sinuata* (Hasselt, 1824)

Identification: *Miamira* is an unusual genus of chromodorids with an irregularly scalloped outline of the body. The more common and widespread of the two species is *M. sinuata*. It has a grayish or greenish body color with blue reticulations.

Natural History: Found on shallow reefs, 5-30 m depth, in association with sponges of the genus *Dysidea*, sometimes together with the opisthobranchs *Sagaminopteron* spp. (#533, 534).

Distribution: Tanzania and Maldives to Hawaiian Islands (Madang, Papua New Guinea).

166

587. *Risbecia tryoni* (Garrett, 1873)

Identification: Species of *Risbecia* can be recognized by a high body profile with a mantle edge that is well extended from the sides of the body. *Risbecia tryoni* has a mottled light and dark tan body with black spots and a blue marginal band. It is similar in color to *Chromodoris leopardus*, but that species has a low rounded body typical of *Chromodoris*.

Natural History: Frequently found crawling in pairs, with one following closely behind the other over shallow reefs. Each animal alternately raises and lowers its head as it crawls.

Distribution: Tanzania; Malaysia; Australia to Okinawa and Society Islands (Madang, Papua New Guinea).

G.C. Williams

588. *Thorunna australis* (Risbec, 1928)

Identification: The genus *Thorunna* contains species with greatly elongated radular teeth with few denticles. Most species are white with colored marginal bands, such as *T. africana*, shown with its commensal copepod (#690). *Thorunna australis* has a rose pink color with white lines and blue and white spots. It is similar in color to some species of *Hypselodoris* and *Noumea*, but lacks contrasting pigment in the central region of the body.

Natural History: This species inhabits shallow patch reefs and is frequently found beneath coral rubble.

Distribution: New Caledonia; New Guinea; Indonesia; Guam; Philippines; Okinawa; Marshall Islands (Madang, Papua New Guinea).

T. M. Gosliner

Porostomata - Dendrodorids and Phyllidiids

This group of dorid nudibranchs lack radular teeth and feed by secreting digestive enzymes into sponges and sucking the partially digested tissue. Dendrodorids have a prominent gill on the back. Phyllidiids have no dorsal gill but a secondary set of gills along the underside of the body margins. The Phyllidiidae are reviewed by Brunckhorst (1993).

589. *Dendrodoris denisoni* (Angas, 1864)

Identification: This species is recognizable by its series of brown and white fleshy tubercles with bright blue spots in depressions between the tubercles.

Natural History: Found on shallow reefs in the open or under rocks, occasionally found in deep water trawls. Depth: 1-120 m.

Distribution: South Africa; Reunion Island to Hawaiian Islands (Dakak, Mindanao, Philippines).

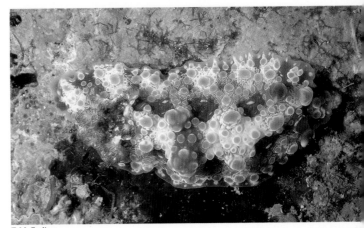
T. M. Gosliner
T. M. Gosliner

590. *Dendrodoris tuberculosa* (Quoy & Gaimard, 1832)

Identification: *Dendrodoris tuberculosa* is an extremely variable species and may reach 200 mm in length. It is almost always brown in color with large, fleshy tubercles.

Natural History: Found under stones during the day and actively crawling over shallow reefs at night.

Distribution: South Africa; Tanzania to Hawai'i (Madang, Papua New Guinea).

167

T. M. Gosliner

591. *Ceratophyllidia africana* Eliot, 1903

Identification: Species of *Ceratophyllidia* can be readily distinguished by the soft, detachable papillae located on the notum. This species has black markings on the papillae. Brunckhorst (1993) considers the western Pacific form, shown here, as a distinct undescribed species.

Natural History: This species is found beneath stones or coral heads on relatively shallow reefs.

Distribution: South Africa; Seychelles; Reunion Island; New Guinea; Marshall Islands (Madang, Papua New Guinea).

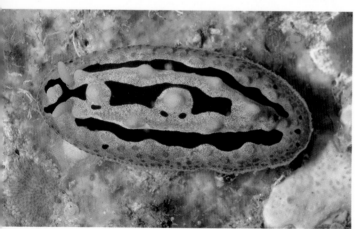

G.C. Williams

592. *Phyllidia babai* Brunckhorst, 1993

Identification: Like other species of *Phyllidia*, this species has a hard body, containing dense calcareous spicules. *Phyllidia babai* is similar to *P. ocellata* and *P. willani*, but has minute, scattered opaque white spots on the notum.

Natural History: Found on shallow patch reefs, usually in the open. Like most species of phyllidiids, this species produces a pungent noxious chemical that is toxic to fish and crustacean predators.

Distribution: Australia; New Guinea and Philippines (Batangas, Luzon, Philippines).

593. *Phyllidia coelestis* Bergh, 1905

Identification: *Phyllidia coelestis* is a common and widespread Indo-Pacific species. It is similar in color to *P. varicosa* but has a Y-shape to the blue ridges and lacks a black line on the sole of the foot. It is usually smaller in size than *P. varicosa* but may reach 60 mm in length.

Natural History: Found on open reef slopes where it is frequently observed on walls in the open. Feeds on sponges to 20 m depth.

Distribution: South Africa; Reunion Island; Aldabra Atoll; Sri Lanka; Malaysia; Thailand; Fiji; Solomon Islands; Australia; New Guinea; Indonesia; Philippines (Madang, Papua New Guinea).

T. M. Gosliner
T. M. Gosliner

594. *Phyllidia elegans* Bergh, 1869

Identification: This species has a pink body with black lines and spots. The tubercles are white or tipped with yellow. The rhinophores are yellow.

Natural History: This species lives in the same habitat as *P. coelestis* and is commonly found on reef faces at moderate depths, 1-40 m.

Distribution: Reunion Island; Red Sea; Australia; Fiji; Solomon Islands; New Guinea; Indonesia; Guam; Philippines; Okinawa; Marshall Islands (Madang, Papua New Guinea).

168

595. *Phyllidia ocellata* Cuvier, 1804

Identification: This species has a yellow to gold body color with black circles or undulating lines often surrounded by white and yellow or white tubercles. It is extremely variable in its color pattern, even from a single locality.

Natural History: This species is found on relatively shallow reefs, almost always conspicuously in the open. Depth: 1-20 m.

Distribution: Tanzania; Reunion Island; Seychelles; Red Sea to Australia; Fiji; Solomon Islands; New Guinea; Indonesia; Philippines; Guam; Okinawa; southern Japan; Marshall Islands (Batangas, Luzon, Philippines).

Mike Miller

596. *Phyllidia varicosa* Lamarck, 1801

Identification: This large species (up to 115 mm in length) is probably the most common and widespread species of *Phyllidia*. It can be recognized by its blue body color, black pigment between the ridges and yellow tubercles and rhinophores. It is also distinguished by a black line on the underside of the foot.

Natural History: Found in the open on patch reefs or reef faces.

Distribution: South Africa; Tanzania; Red Sea; Aldabra Atoll; Madagascar; Reunion Island to Australia; New Guinea; Japan; Hawaiian and Society Islands (Madang, Papua New Guinea).

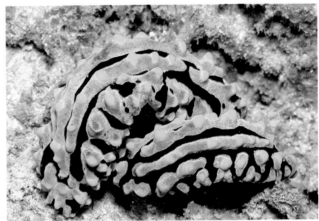
Leslie Newman & Andrew Flowers

597. *Phyllidiella pustulosa* (Cuvier, 1804)

Identification: This species can be recognized by its predominantly black body with pink tubercles arranged in small clusters. Many other species also have pink and black color but differ in the arrangement of lines and tubercles. See Brunckhorst (1993) for detailed differences. *Phyllidiella pustulosa* is by far the most common of these species.

Natural History: Found from shallow water reefs to deep reef slopes. This species, together with *Phyllidia varicosa*, is the most common phyllidiid on Indo-Pacific reefs. Depth: 5-40 m. The flatworm *Pseudoceros imitatus* (#354) mimics this species.

Distribution: Red Sea to Hawaiian Islands (Sapi Island, Sabah, Malaysia).

T. M. Gosliner
Mike Miller

598. *Phyllidiopsis sphingis* Brunckhorst, 1993

Identification: *Phyllidiopsis* differs from other genera in details of the arrangement of its digestive tract. Species of this genus usually have a less rigid body. *Phyllidiopsis sphingis* has blue pigment around the margins of the cream body. There are numerous longitudinal and radiating black lines and spots.

Natural History: Found on moderately steep reef slopes in the open.

Distribution: New Guinea; Guam; Okinawa; Hawaiian Islands (Pupukea, Oahu, Hawai'i).

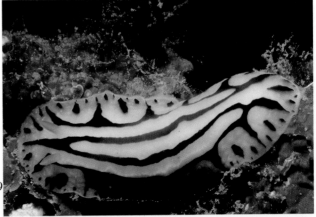

169

Mike Severns

599. *Reticulidia halgerda*
Brunckhorst & Burn, 1990

Identification: The black body, orange ridges and white lines are distinctive. This species may be mistaken for a *Halgerda*, but lacks the circle of gills around the anus.

Natural History: Found on exposed walls in moderately deep water; 14-65 m depths.

Distribution: Australia; Tonga; Fiji; New Guinea; Solomon Islands; Indonesia; Taiwan; Philippines; Okinawa; Marshall Islands (Manado, Sulawesi, Indonesia).

Phanerobranch Dorids

These dorids tend to have a more elongated body than other dorids and the gill can not be retracted into a distinct pocket. Members of this group usually feed upon bryozoans, tunicates or other opisthobranchs.

T. M. Gosliner

600. *Nembrotha kubaryana* Bergh, 1877

Identification: This species is easily recognized by its black body with bright green spots and orange lines.

Natural History: This species feeds upon arborescent bryozoans on the sides of reef walls. Depth: 15-35 m.

Distribution: Australia; New Guinea; Indonesia; Philippines; Belau; Guam; Marshall Islands (Madang, Papua New Guinea).

T. M. Gosliner

Marc Chamberlain

601. *Nembrotha lineolata* Bergh, 1905

Identification: This species is part of a large complex of *Nembrotha* species, many of which are undescribed. It can be distinguished by the cream body with numerous, narrow brown lines and faint red and purple markings on the gills and rhinophores.

Natural History: This species feeds on compound or solitary tunicates and has been observed feeding upon *Clavelina* sp. (#1061), *Rhopalaea* sp. (#1081) and *Oxycorynia fascicularis* (#1068). Depth: 7-20 m.

Distribution: Tanzania to Australia; Fiji; New Guinea; Indonesia; Philippines; Okinawa; Japan (Batangas, Luzon, Philippines).

602. *Nembrotha* sp.

Identification: *Nembrotha* sp. is part of the same species complex as *N. lineolata*. It is white with a dark brown saddle, yellow markings, yellow and purple lines and bright red gills and rhinophores. This species is undescribed.

Natural History: This is one of the most common nudibranchs in the Philippines. It is common on shallow reefs, where it feeds upon tunicates of the genera *Rhopalaea*, *Clavelina* and *Oxycorynia*. Depth: 5-20 m.

Distribution: Philippines and Okinawa (Batangas, Luzon, Philippines).

603. *Roboastra arika* Burn, 1967

Identification: This beautiful species can be recognized by its bright purple color and yellow orange lines surrounded by black. Species of *Roboastra* have radular teeth which differ from those of *Nembrotha* or the other closely related genus *Tambja*.

Natural History: Found feeding on arborescent bryozoans on reef walls. Depth: 25 m.

Distribution: Australia and New Guinea (Madang, Papua New Guinea).

T. M. Gosliner

604. *Okenia* sp. 1

Identification: This species, with its elongate flame red, club-shaped processes, appears more like an aeolid nudibranch than a dorid. However, the circle of gills clearly indicates that it is a dorid. This species is undescribed.

Natural History: Found on the arborescent bryozoan *Tropidozoum cellariiforme* (#903), which inhabits open reef walls, 10-30 m depth.

Distribution: Indonesia; Philippines and Marshall Islands (Bohol, Philippines).

G.C. Williams

605. *Okenia* sp. 2

Identification: This represents another undescribed species of *Okenia* which can be identified by its white body with brown and purple pigment on the back and slender appendages.

Natural History: Found on the underside of the large leafy sponge, *Phyllospongia lamellosa* (#40) and other similar species, where it feeds on crustose bryozoans. Depth: 5-15 m.

Distribution: Only known from Batangas, Luzon, Philippines.

T. M. Gosliner

606. *Gymnodoris ceylonica* (Kelaart, 1858)

Identification: This large (up to 120 mm in length) dorid is distinguished by its white body with large orange spots and lines present on the gills. There are many species of *Gymnodoris*, including numerous undescribed species. Most species are white with yellow or orange markings, and are small, less than 20 mm in length.

Natural History: Species of *Gymnodoris* are voracious predators on other opisthobranchs. *Gymnodoris ceylonica* is commonly found in shallow grass beds and sandy habitats where it commonly feeds upon the sea hare, *Stylocheilus longicauda* (#539). It is active nocturnally. Depth: to 10 m.

Distribution: Aldabra Atoll; Seychelles to Australia; Japan; Guam; Philippines; Marshall Islands (Batangas, Luzon, Philippines).

T. M. Gosliner

171

Mike Miller

607. *Gymnodoris* sp.

Identification: This is another large species (more than 100 mm) of *Gymnodoris* and appears to be undescribed. It can be identified by its orange body with white tubercles and huge gills.

Natural History: Found on shallow reefs, where it is active nocturnally.

Distribution: Philippines (Batangas, Luzon, Philippines).

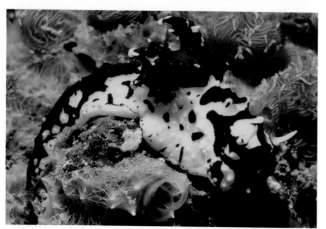

Marc Chamberlain

608. *Notodoris gardineri* Eliot, 1906

Identification: *Notodoris gardineri* is similar to *N. minor* in appearance and size, but has black mottling rather than distinct lines and spots.

Natural History: Also found on the yellow sponge *Leucetta primigenia* (#4) on reef crests and slopes. Depth: 10-20 m.

Distribution: Maldives; Australia; New Guinea; Indonesia and Okinawa (Papua New Guinea).

609. *Notodoris minor* Eliot, 1904

Identification: The bright yellow body of this large (more than 100 mm in length) dorid with black lines and spots makes it instantly recognizable. The body is rigid and hard to the touch.

Natural History: Found feeding on the yellow calcareous sponge, *Leucetta primigenia* (#4), on reef crests and slopes. The egg mass is a bright yellow ribbon. Depth: 10-20 m.

Distribution: Tanzania; Oman to Australia; New Guinea; Indonesia; Philippines and Okinawa (Manado, Sulawesi, Indonesia).

Mike Severns

T. M. Gosliner

610. *Notodoris* sp.

Identification: This species is undescribed and differs from the described species of *Notodoris*, all of which are yellow in color. It may reach 100 mm in length and is green with black markings and large appendages surrounding the gills.

Natural History: Found on the sponge *Leucetta primigenia* (#4) on which it is shown here. It may be found on this sponge together with other species of *Notodoris*. Depth: 10-20 m.

Distribution: New Guinea; Indonesia; Philippines; Belau; Guam and Okinawa (Madang, Papua New Guinea).

Dendronotids

611. *Bornella anguilla* Johnson, 1985

Identification: Most species of *Bornella* are rather drably colored. *Bornella anguilla* has a bright and complex color pattern with orange and black paddle-shaped appendages adjacent to the gills and rhinophores.

Natural History: This species is found on the undersides of rocks and on hydroid colonies. When disturbed it is capable of swimming rapidly by moving its body like an eel. Depth: 5-20 m.

Distribution: South Africa; Australia; New Guinea; Indonesia; Guam; Okinawa; Marshall Islands (Madang, Papua New Guinea).

T. M. Gosliner

612. *Melibe fimbriata* Alder & Hancock, 1864

Identification: This large (more than 200 mm), bizarre looking nudibranch resembles a mass of algae. It has large cerata with a rounded oral hood around its mouth. It is the largest and most widespread Indo-Pacific species of *Melibe*.

Natural History: *Melibe* are specialized predators upon crustaceans, which they trap with their oral hood. Most species are capable of swimming by moving their body from side to side. Many species gain additional nutrition from symbiotic zooxanthellae. Found on shallow sandy bottoms. Depth: 2-15 m.

Distribution: South Africa; Tanzania; into the Mediterranean (where it has migrated through the Suez Canal); Australia; Philippines; Okinawa and Japan (Batangas, Luzon, Philippines).

Mike Miller

613. *Doto ussi* Ortea, 1982

Identification: There are many species of *Doto* known from the Indo-Pacific tropics, most of which are undescribed. This species is cryptic when found on its prey. The cerata are large with numerous low rounded tubercles covering their surface.

Natural History: This species feeds on the stinging hydroid, *Aglaophenia cupressina* (#51). It is also found at the base of the hydroids with its yellow egg masses as well. Depth: 1-15 m.

Distribution: Comoros; Madagascar; New Guinea; Philippines (Batangas, Luzon, Philppines).

T. M. Gosliner

T. M. Gosliner

614. *Marionia distincta* Bergh, 1905

Identification: Many species of *Marionia* are found in the Indo-Pacific tropics, but the systematics are poorly known. This species can be recognized by a series of brown transverse lines and simple unbranched papillae.

Natural History: Species of *Marionia* and *Tritonia* are predators upon soft corals. Many species feed upon xeniids, in particular. This species has not been found in association with any soft coral, but was collected in an area where xeniid soft corals are dominant members of the shallow reef community.

Distribution: Not recorded since its original description from Indonesia; Philippines (Batangas, Luzon, Philippines).

173

615. *Marionia* sp.

Identification: This species has a network of brownish or whitish lines across the surface of the body. The general body color may be yellow or purple.

Natural History: Found under coral heads in shallow water patch reef and rubble communities. This species probably feeds on soft corals.

Distribution: This undescribed species has been found from Madagascar and Philippines (Batangas, Luzon, Philppines).

616. *Tritonia* sp.

Identification: Members of the genus *Tritonia* differ from *Marionia* in lacking chitinous plates in the stomach. This undescribed species is virtually identical to its prey, *Carijoa* sp. 2 (#70).

◀ *Natural History:* The single specimen of this species was found on *Carijoa* sp. 2 in less than one meter of water on a shallow patch reef.

Distribution: East Africa (Msimbati, Tanzania).

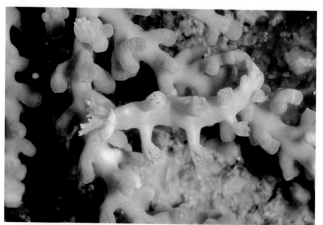

Arminaceans

This is probably an unnatural group of nudibranchs, as members are united by having an oral veil, which has probably originated independently in different Arminaceans. The Arminidae feed on soft corals and penatulaceans, while the group to which *Janolus* belongs, feeds on bryozoans. Several other taxa have been lumped into this group, but it is in need of serious revision.

617. *Dermatobranchus gonatophora* (Hasselt, 1824)

Identification: There are many species of arminids found in the Indo-Pacific tropics. Most are in the genera *Armina* and *Dermatobranchus* and have a series of longitudinal ridges on the body. The body is grey with black and yellow lines and an orange marginal band.

Natural History: Found feeding on the soft coral *Eleutherobia grayi* (#73) in deep sandy habitats. Depth: 36-53 m.

Distribution: Indonesia and Okinawa (Okinawa).

618. *Janolus* sp.

Identification: Species of *Janolus* have bulbous cerata over their body surface and a folded sensory organ between the rhinophores, which resembles a cock's comb. This undescribed species of *Janolus* has an orange body color with opaque white pigment, and black and purple ceratal tips.

Natural History: *Janolus* sp. feeds on an arborescent bryozoan with curved apical branches. It inhabits steep reef walls and pinnacles in areas where strong currents are present. Depth: 20-40 m.

Distribution: Indonesia; Philippines and Okinawa (Batangas, Luzon, Philippines).

174

Aeolidacea - Aeolids

Aeolid nudibranchs lack distinct gills and utilize cerata for respiration and defense. The cerata contain branches of the digestive tract which transport nematocysts acquired from coelenterate prey, to the ceratal tips. The nematocysts are then stored and utilized for the aeolid's defense.

Mike Miller

619. *Flabellina bilas* Gosliner & Willan, 1990

Identification: Indo-Pacific species of *Flabellina* have been reviewed by Gosliner and Willan (1990). *F. bilas* is a recently described species with rhinophores bearing numerous plate-like folds called perfoliations. It is recognized by the medial blue diamond-shaped marking and the blood red rings on the cerata, surrounded by narrower bands of opaque white.

Natural History: Found on hydroids of the genus *Eudendrium*, on steep reef slopes and walls in moderate depths, 20-30 m.

Distribution: Tanzania; New Guinea; Indonesia; Philippines and Marshall Islands (Batangas, Luzon, Philippines).

T. M. Gosliner

620. *Flabellina exoptata*
Gosliner & Willan, 1990

Identification: This species has papillae rather than perfoliations on rhinophores that are orange with yellow spots. Cerata are thick, with a purple subapical ring and a cream apex.

Natural History: Found on relatively shallow reefs and slopes on hydroids of the genus *Eudendrium*. Depth: 1-20 m.

Distribution: Aldabra Atoll to Australia; New Guinea; Indonesia; Philippines and Guam. It may also be known from the Hawaiian Islands, but that record requires confirmation (Madang, Papua New Guinea).

621. *Flabellina macassarana* Bergh, 1905

Identification: This species also has perfoliate rhinophores. It has a pinkish to purplish white body with burnt orange ceratal tips. *Flabellina bicolor* is a similar, but smaller species that has a subapical ceratal band of orange.

Natural History: Found on hydroids of the genus *Eudendrium* on relatively deep reef slopes and walls. Depth: 30 m.

Distribution: Indonesia; Tanzania and Philippines (Batangas, Luzon, Philippines).

Mike Severns
Mike Miller

622. *Flabellina rubrolineata*
(O'Donoghue, 1929)

Identification: Similar to *F. exoptata*, it has papillate rhinophores, and a series of three longitudinal violet lines along the length of the body. Coloration is extremely variable.

Natural History: *Flabellina rubrolineata* also feeds on the hydroid *Eudendrium* sp. (#48) on shallow reefs and walls.

Distribution: Aldabra Atoll; Suez; Australia; Japan; Hawaiian Islands (Batangas, Luzon, Philippines).

175

623. *Cuthona sibogae* (Bergh, 1905)

Identification: *Cuthona sibogae* is one of more than 100 species of this genus inhabiting the Indo-Pacific tropics. Most are undescribed. Most species are small (less than 10 mm in length) and have smooth rhinophores. *Cuthona sibogae* can be recognized by its purple body color with yellow ceratal tips.

Natural History: This species feeds on hydroids of the genus *Sertularella* on relatively shallow reefs, 5-15 m.

Distribution: South Africa to New Guinea; Indonesia and Philippines (Batangas, Luzon, Philippines).

T. M. Gosliner

624. *Phestilla melanobrachia* (Bergh, 1874)

Identification: Species of *Phestilla* are closely related to *Cuthona*, but lack a sac at the tip of the cerata for storing nematocysts. *Phestilla melanobrachia* may be yellow, orange or black, depending upon the color of its prey.

Natural History: This species feeds exclusively upon dendrophyllid corals (#288-291) and is usually found on the underside of detached coral colonies. The white egg masses of the nudibranch are usually deposited directly on the corals.

Distribution: South Africa; Aldabra Atoll; Australia; New Guinea; Japan to Hawaiian Islands (Batangas, Luzon, Philippines).

T. M. Gosliner

625. *Phestilla* sp.

Identification: This undescribed species of *Phestilla* is whitish or brownish in color with a broad body.

Natural History: *Phestilla* sp. feeds exclusively upon corals of the genus *Goniopora* (#235) and is often found under colonies together with its yellowish egg mass. Depth: 1-10 m.

Distribution: Tanzania to Australia; New Guinea; Philippines; Hong Kong; Marshall Islands (Madang, Papua New Guinea).

T. M. Gosliner
Leslie Newman & Andrew Flowers

626. *Caloria indica* (Bergh, 1896)

Identification: This is the most common species of facelinid aeolid in the Indo-Pacific. It can be recognized by the orange head with white and yellow markings and cerata with blue yellow and red markings.

Natural History: Found in relatively shallow reefs where it feeds on hydroids of the genus *Eudendrium* and upon *Pennaria disticha* (#46). Depth: 1-15 m.

Distribution: South Africa and Aldabra Atoll to New Guinea; Australia to Hawaiian Islands (Madang, Papua New Guinea)

176

627. *Phyllodesmium briareus* (Bergh, 1896)

Identification: Species of *Phyllodesmium* are specialized predators upon soft corals. Many species contain symbiotic zooxanthellae for additional nutrition. They also lack sacs for storing nematocysts.

Natural History: This species feeds upon the mat forming soft coral *Pachyclavularia violacea* (#71) in moderately shallow water, 5-20 m.

Distribution: Australia; New Guinea; Indonesia; Malaysia and Philippines (Batangas, Luzon, Philippines).

Mike Miller

628. *Phyllodesmium kabiranum* Baba, 1991

Identification: This beautiful, recently described species is the only *Phyllodesmium* with a reddish body and a white mid-dorsal line.

Natural History: This species is found feeding upon the rust orange soft coral *Heteroxenia* sp. (#131) on large rocky outcrops. Depth: 15-20 m.

Distribution: Okinawa; Malaysia; Indonesia and Philippines (Sipadan, Borneo).

Marc Chamberlain

629. *Phyllodesmium longicirra* (Bergh, 1905)

Identification: This is the largest species of *Phyllodesmium*, which can exceed 120 mm in length. The body is white with golden brown clusters of spots.

Natural History: This species has not been found in association with any prey, but has been seen crawling around in the open on sandy bottoms.

Distribution: Australia and Indonesia (Flores, Indonesia).

Mike Severns
T. M. Gosliner

630. *Phyllodesmium magnum* Rudman, 1991

Identification: This is another large species of *Phyllodesmium*, which may reach 100 mm in length. Its curved, flattened cerata are distinctive.

Natural History: This species is found in relatively shallow sandy habitats where it feeds upon the soft coral, *Sarcophyton* sp. 1 (#81). Depth: 1-10 m.

Distribution: Tanzania; Australia; New Guinea; Indonesia; Philippines; Hong Kong to Guam and Enewetak (Madang, Papua New Guinea).

T. M. Gosliner

Roy Eisenhardt

631. *Pteraeolidia ianthina* (Angas, 1864)

Identification: This dragon-like aeolid is easily recognized by its elongate body shape and perfoliate rhinophores. It is variable in color with different shades of green, blue and brown. It is large and may exceed 100 mm in length.

Natural History: Found commonly in shallow water, often just below the low tide mark on walls. May be found in the open or feeding on hydroids. Adults tend to receive most of their nutrition from symbiotic zooxanthellae. Depth: 1-10 m.

Distribution: Madagascar; Aldabra Atoll; Seychelles to Australia; Japan; Hawaiian Islands (Madang, Papua New Guinea; Solomon Islands).

T. M. Gosliner

632. *Berghia major* (Eliot, 1903)

Identification: Berghia major is a member of the Aeolidiidae, a group of aeolids which feeds primarily upon sea anemones and zoanthids. This species is usually gray or brown with papillate rhinophores. The cerata are curved and flatttened often with blue and yellow markings.

Natural History: Found in shallow water grass beds and algal mats where it feeds upon sea anemones of the genera *Aiptasia* and *Boloceroides* (#177). Depth: 1-5 m.

Distribution: Seychelles and Tanzania; Australia; Japan; Hawaiian Islands and tropical North America (Batangas, Luzon, Philippines).

Leslie Newman & Andrew Flowers

T. M. Gosliner

633. *Cerberilla affinis* Bergh, 1888

Identification: This species can be recognized by its wide foot, black "mask" around the rhinophores and black rings on the cerata.

Natural History: Species of *Cerberilla* are found in sandy habitats where they feed on burrowing coelenterates such as cerianthids. The broad foot is an adaptation for crawling on sandy substrate.

Distribution: Australia; Indonesia; Philippines and Okinawa to Midway Atoll, Hawaiian Islands (Heron Island, Australia).

Scaphopoda - Tusk Shells

Tusk shells are closely related to bivalves but have a single tubular, curved shell. Most species are found burrowing in clean sand and feed on particulate matter in the sediment, using sticky threads that surround the mouth. The tropical species are poorly known.

634. *Fustiaria nipponica* (Yokoyama, 1822)

Identification: This species can be recognized by its distinct curvature, shiny shell with indistinct ribs. The foot is expanded for digging.

Natural History: Found in fine clean sand on slopes in 5-100 m depths.

Distribution: New Guinea; Indonesia; Belau; Philippines; Japan and China (Madang, Papua New Guinea).

178

Bivalvia - Bivalves

The bivalves, which include clams, oysters and mussels, are well known from temperate shores, but not well known by visitors to the coral reefs. The notable exceptions are the giant clams of the genus *Tridacna*. Bivalves are highly modified mollusks. As their name suggests, they have two shell halves; they also lack a radula. The head is reduced and the gills are enlarged and modified for securing prey as well as for respiration. Most bivalves are filter feeders and remove plankton from the water column. Others are deposit feeders and feed upon fine particulate matter situated on sandy or silty bottoms. Some highly specialized bivalves trap small prey by quickly contracting a modified gill. The greatest diversity of bivalves is found burrowed into soft substrate. Often only the siphons are visible at the sand surface. Some species are found in rocky habitats and others may penetrate living coral.

635. *Modiolus philippinarum* (Hanley, 1843)
Philippine Mussel

Identification: The brownish shell with simple, flared aperture distinguishes this species of horse mussel. It is relatively large and may reach 130 mm in length.

Natural History: This species is commonly found living on rocky reefs where it frequently is found protruding from living coral heads.

Distribution: South Africa; Red Sea to Philippines; Marshall Islands (Batangas, Luzon, Philippines).

D.W. Behrens

636. *Pteria crocea* (Lamarck, 1819)
Striped Wing Oyster

Identification: This species of wing oyster can be distinguished by its yellowish color with longitudinal brown or maroon lines. Larger specimens are more brown in color.

Natural History: This winged oyster is commonly found attached to hydroid colonies, in shallow water, especially with *Aglaophenia cupressina* (#51).

Distribution: Tanzania to Philippines (Batangas, Luzon, Philippines).

G.C. Williams
T. M. Gosliner

637. *Pteria penguin* (Röding, 1798)
Penguin Wing Oyster

Identification: This wing oyster can be recognized by its black shell color and large size. The length of the wing may vary considerably. Specimens may reach more than 250 mm in length.

Natural History: This species is commonly attached to antipatharians and may be found from 15 to more than 30 m depth.

Distribution: Indian Ocean; Red Sea to Philippines; Indonesia; New Guinea; Marshall Islands (Batangas, Luzon, Philippines).

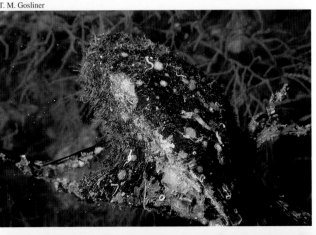

179

G.C. Williams

638. *Pteria tortirostris* (Dunker, 1848)

Identification: *Pteria tortirostris* can be differentiated from other winged oysters by its narrow shell with a much elongated, narrow wing. It is reddish-brown with fine sculpture on the shell.
Natural History: This species has been found attached to gorgonians and the hydroid *Aglaophenia cupressina* (#51).
Distribution: South Africa; Red Sea to Philippines (Batangas, Luzon, Philippines).

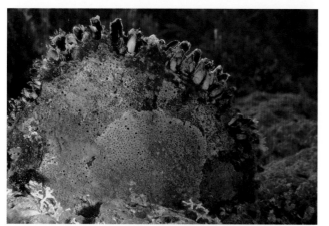

Jack Randall

639. *Pinctada margaritifera* (Linnaeus, 1758)
Common Pearl Oyster

Identification: This species can be distinguished by a series of jagged teeth around the aperture. Younger specimens may also have a series of similar jagged scales on the valves of the shell.
Natural History: Specimens are found attached by byssal threads to coral rubble and rocks in shallow water from intertidal reef flats to depths of about 20 m.
Distribution: South Africa; Red Sea to Tuamotus and Hawaiian Islands (Rangiroa Atoll, Tuamotu Archipelago).

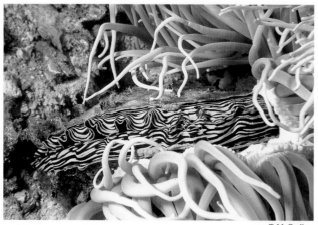

T. M. Gosliner
Laura Losito

640. *Atrina pectinata* (Linnaeus, 1767)
Pen Shell

Identification: The mantle of this pen shell is visible at the opening of the gaping shell valves. The mantle is black with green undulating lines and an orange border. The shell is brown and strongly tapered with scales on the outside of the shell.
Natural History: The animals are largely embedded in coarse sand and rubble with only the top of the shell and exposed mantle visible. They are generally found in shallow water in less than 10 m depth.
Distribution: India to Australia; Indonesia; Philippines and Japan (Puerto Galera, Mindoro, Philippines).

641. *Pedum spondyloideum* (Gmelin, 1791)

Identification: The brightly colored mantle and minute eyes along its margin make this species instantly recognizable. Although this species looks like a mussel, it is a scallop.
Natural History: *Pedum spondyloideum* lives in crevices in colonies of living corals. Only the shell opening and mantle are visible.
Distribution: Madagascar; Red Sea to Solomon Islands; Philippines; Australia; Fiji and Marshall Islands (Solomon Islands).

180

642. *Decatopecten noduliferus* (Sowerby, 1842)

Identification: Like most scallops, this species has a fleshy mantle with numerous tentacles and eyes along the mantle margin.

Natural History: This specimen was trawled in relatively deep water of 30-50 m where it lived on sandy bottoms.

Distribution: South Africa to Hawaiian Islands (Sodwana Bay, Natal, South Africa).

G.C. Williams

643. *Mirapecten rastellum* (Lamarck, 1819)
Scallop

Identification: This species only reaches about 30 mm in shell diameter. It can be identified by the relatively large plates on the shell valves and the smaller sculpture on the shell.

Natural History: This species is found in shallow water of 10-20 m and inhabits the undersides of coral slabs.

Distribution: Red Sea to Australia; New Guinea; Philippines and Marshall Islands (Madang, Papua New Guinea).

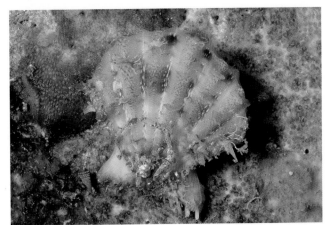
T. M. Gosliner

644. *Spondylus linguaefelis* Sowerby, 1847
Cat's Tongue Thorny Oyster

Identification: This species of thorny oyster can be recognized by a dense covering of long thin spines. The species name means "cat's tongue" and refers to the rough texture of the shell.

Natural History: This species is usually found in relatively deep water below 40 meters, where it is cemented to hard substrate.

Distribution: Hawaiian Islands and Clipperton Island (Maui, Hawaiian Islands).

Mike Severns
Marc Chamberlain

645. *Spondylus squamosus* Schreibers, 1793
Scaly Thorny Oyster

Identification: This species can be recognized by its gray mantle with black lines. The shell has white spines with black markings on their inner side and basally.

Natural History: Found cemented to reef faces and in caves.

Distribution: New Guinea; Indonesia; Philippines and Marshall Islands (Milne Bay, Papua New Guinea).

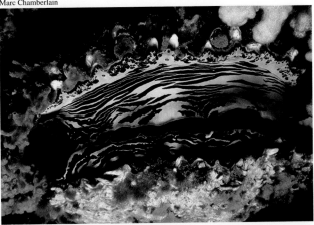

181

646. *Spondylus varians* Sowerby, 1829
Variable Thorny Oyster

Identification: This species can be recognized by its mottled orange mantle and blue eyes. The shell has short white spines covering its surface.

Natural History: This is the most common species of thorny oyster in the western Pacific. It is found cemented to walls and open reef faces in depths ranging from 10-40 m.

Distribution: Australia; New Guinea; Indonesia; Philippines and Marshall Islands (Batangas, Luzon, Philippines).

Lynn Funkhouser

647. *Limaria orientalis* (Adams & Reeve, 1850)
File Shell

Identification: File shells have thin white shells with the red mantle and tentacles protruding. In *Limaria orientalis* the animal is uniformly red. Other species have striped tentacles.

Natural History: File shells are commonly found under rocks in sandy and rubble habitats.

Distribution: New Guinea; Indonesia and Philippines (Manado, Sulawesi, Indonesia).

Mike Severns

648. *Corculum cardissa* (Linnaeus, 1758)
Heart Cockle

Identification: This cockle can be recognized by its flattened shell and heart shape. It has coarser, sharper teeth along the shell margin than closely related species.

Natural History: Found on living coral reefs on the surface of coral rubble to 10 m depth.

Distribution: Malaysia; Solomon Islands; Philippines to southern Japan and Federated States of Micronesia (Sapi Island, Malaysia).

T. M. Gosliner
G.C. Williams

649. *Hyotissa hyotis* (Linnaeus, 1758)
Honeycomb Oyster

Identification: This thickly calcified oyster has jagged triangular teeth along the shell aperture. It differs from *Lopha cristagalli* (#651) in having a more heavily calcified shell which may reach 250 mm in length.

Natural History: This species is attached to rocks or coral on reef faces and walls.

Distribution: South Africa; Red Sea to New Guinea; Indonesia and Philippines (Bohol, Philippines).

650. *Alectryonella plicatula* (Gmelin, 1791)

Identification: This species can be recognized by a series of angular folds that extend from the opening to the base of the shell.

Natural History: *Alectryonella plicatula* is found cemented to the rocky substrate in relatively shallow water.

Distribution: Western Indian Ocean to Philippines (Batangas, Luzon, Philippines)

Nicholas Galluzzi

651. *Lopha cristagalli* (Linnaeus, 1758)
Cock's Comb Oyster

Identification: The cock's comb oyster has a distinctively jagged shell with fine sculpturing. The shell is more lightly calcified than *Hyotissa hyotis* (#649).

Natural History: *Lopha cristagalli* is usually covered with encrusting sponges and is attached to gorgonians or antipatharians.

Distribution: South Africa; Red Sea to Australia; New Guinea; Solomon Islands; Indonesia and Philippines (Madang, Papua New Guinea).

T. M. Gosliner

652. *Amphilepida aurantia* (Deshayes, 1835)

Identification: This species does not even appear to be a bivalve at first glance, since the mantle completely covers the orange shell valves.

Natural History: This species is found under stones and coral rubble.

Distribution: Madagascar to the Red Sea (Tulear, Madagascar).

T. M. Gosliner
T. M. Gosliner

653. *Scintilla* cf. *cuvieri* Deshayes, 1856

Identification: This species has a tuberculate mantle which partially covers the thin, fragile shell. It is small, usually less than 10 mm.

Natural History: This species is common under stones in relatively shallow water.

Distribution: Philippines and Hong Kong (Batangas, Luzon, Philippines).

183

654. *Tridacna crocea* Lamarck, 1819
Crocus Giant Clam

Identification: This species is relatively small reaching a maximum length of about 100 mm. It has very short fluted concentric sculpture on the shell.

Natural History: Tridacna crocea lives in small cracks on shallow reef flats.

Distribution: Malay Peninsula to Australia; New Guinea; Indonesia; Philippines; Okinawa; Fiji and Guam (Madang, Papua New Guinea).

Leslie Newman & Andrew Flowers

655. *Tridacna derasa* (Röding, 1798)

Identification: This is a large species with very low sculpturing on the shell. It may reach 500 mm in length. The base of the shell is near the anterior end rather than at the posterior, as in the remaining species of *Tridacna*.

Natural History: Found in quiet waters, often associated with sandy or silty habitats.

Distribution: Cocos Keeling Atoll to Australia; Fiji; Solomon Islands; New Guinea; Belau; Indonesia and Philippines (Belau).

Dave Zoutendyk

656. *Tridacna gigas* (Linnaeus, 1758)
Giant Giant Clam

Identification: This is the largest species of bivalve in the world, reaching 1.3 m in length! It has a smooth shell with no concentric sculpture.

Natural History: Usually found in isolated offshore reefs. In many places, this species is rare owing to over-harvesting by humans.

Distribution: Malay Peninsula to Marshall Islands and Fiji, but known from fossils as far west as Aldabra Atoll (Manado, Sulawesi, Indonesia).

Terry Schuller
Roy Eisenhardt

657. *Tridacna maxima* (Röding, 1798)
Large Giant Clam

Identification: This species has well developed concentric growth folds and may reach 300 mm in length.

Natural History: Tridacna maxima is common from intertidal reef flats to about 10 m depth. This is the most widespread species of giant clam.

Distribution: South Africa; Red Sea to Marshall and Line Islands, and Pitcairn Island, but absent from the Hawaiian Islands (Heron Island, Great Barrier Reef, Australia).

658. *Tridacna squamosa* Lamarck, 1819
Fluted Giant Clam

Identification: *Tridacna squamosa* can be recognized by the rows of large leaf-like flutes on the surface of the shell. It may reach 400 mm in length.

Natural History: Found on shallow reef flats to depths of about 20 m.

Distribution: South Africa; Red Sea to Samoa; Tonga and Marshall Islands (Solomon Islands).

Dave Zoutendyk

659. *Tridacna tevoroa*
Lucas, Ledua & Braley, 1990
Tevoroa Giant Clam

Identification: This recently described species closely resembles *T. derasa* and *T. gigas* in having relatively low sculpturing on the shell. It differs in having large tubercles on the mantle.

Natural History: Found in clear water on outer slopes of leeward reefs in 20-30 m of water.

Distribution: Fiji and Tonga (Fiji).

Dave Zoutendyk

660. *Hippopus hippopus* (Linnaeus, 1758)
China Clam

Identification: This is a second genus of giant clam. It does not have a large opening at the base of the shell and has brownish markings on the shell surface. It can be recognized by the brownish mantle with fine white lines over the surface. It may reach 400 mm in length.

Natural History: Found on shallow water reef flats and patch reefs.

Distribution: Malay Peninsula to Australia; Okinawa; Fiji; Micronesia and Tonga (Sapwauhfik, Pohnpei).

Dave Zoutendyk
Mike Severns

Cephalopoda - Nautilus, Octopus, Squids and Cuttlefish

661. *Nautilus pompilius* Linnaeus, 1758
Chambered Nautilus

Identification: This species, together with *N. macromphalus* which is restricted to New Caledonia and the Loyalty Islands, represent the only cephalopods with an external shell. Paper nautilus (*Argonauta*) appear to have an external shell. However, paper argonauts are actually octopods and the pearly white "shell" is the egg case of the female.

Natural History: Nautilus pompilius is found from 60-750 m depth. As a consequence of this habitat, it is rarely seen by scuba divers and is frequently trawled or caught in baited traps. The animals migrate to shallower water each evening.

Distribution: Indian Ocean; Andaman Islands; Fiji; Indonesia; Philippines (Manado, Sulawesi, Indonesia).

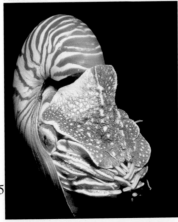

185

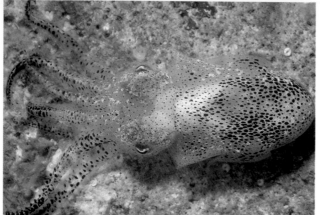

G.C. Williams

662. *Eupyrmna* sp.

Bobtail Squid

Identification: This species can be recognized by its short body with two lobed fins on its tail.

Natural History: This bobtail squid is found in sandy habitats in shallow water (1-10 m depth) where it is active at night.

Distribution: Philippines (Batangas, Luzon, Philippines).

Mike Severns

663. *Sepioteuthis lessoniana* Lesson, 1830

Bigfin Reef Squid

Identification: This squid can be recognized by the fins which extend 90-100% of the total mantle length. Like most squids the color is extremely variable.

Natural History: This species is generally found near the surface or to depths of 100 m. It is frequently seen at the edges of reefs or near anchor lines or underneath boats anchored on reefs.

Distribution: Red Sea; Malay Peninsula; Indonesia; Australia; New Guinea; Philippines and Japan. Also North Island of New Zealand and Hawaiian Islands (Sulawesi, Indonesia).

T. M. Gosliner
Carol Buchanan

664. *Metasepia pfefferi* Hoyle 1885

Flamboyant Cuttlefish

Identification: This cuttlefish is immediately recognizable by its brillant color patterns of deep maroon to black, yellow and red pigment. The colors are more subdued when the animal is not alarmed. This species also has two large dorsal fins and a ventral keel. It reaches a maximum length of about 100 mm.

Natural History: The flamboyant cuttlefish is active during the day and at night on shallow coral reefs. It preys upon small crustaceans, gastropods and fish, which it captures with its two longest tentacles.

Distribution: Australia; Arafura Sea and Philippines (Batangas, Luzon, Philippines).

665. *Sepia apama* Gray, 1849

Australian Giant Cuttlefish

Identification: This is the only large cuttlefish present in most of Australian waters. It overlaps with the next species, *Sepia latimanus*, in the tropical portions of Australia. *Sepia apama* has three large flaps over each eye, while *S. latimanus* has a single eye flap. Individuals may reach 500 mm in length.

Natural History: The Australian giant cuttlefish inhabits rocky and coral reefs and seagrass beds from a few meters to depths of at least 35 m.

Distribution: Endemic to Australia where it is known from all of the temperate waters northward into the tropics. It is apparently absent from the northernmost portions of Australia (Solitary Islands, New South Wales, Australia).

186

666. *Sepia latimanus* Quoy & Gaimard, 1832
Broadclub Cuttlefish

Identification: This is the only large cuttlefish in most of the tropical Indo-Pacific. It may reach 500 mm in length. This species may change its color pattern markedly as illustrated in the three photos.

Natural History: Found on shallow water reefs to depths of at least 30 m, where single individuals or pairs are often observed where it is active during the day.

Distribution: Mozambique; Red Sea to Australia; Solomon Islands; Fiji; New Caledonia; New Guinea; Indonesia; Philippines and Japan (Flores, Indonesia; Solomon Islands; Flores, Indonesia).

Mike Severns

Roy Eisenhardt

Jack Randall

667. *Sepia papuensis* Hoyle, 1885
Papuan Cuttlefish

Identification: This is a small species of cuttlefish (about 60 mm in length) with small papillae situated over the surface of the mantle. It has relatively long tentacles. *Sepia prionota* Voss, 1962 is probably a synonym.

Natural History: This species is common on shallow sandy areas where seagrass and drift algae are abundant. It is active nocturnally.

Distribution: Australia; New Guinea; Indonesia; Philippines (Batangas, Luzon, Philippines).

Marc Chamberlain
Neil Buchanan

668. *Sepia plangon* Gray, 1849

Identification: This small species has a bright reddish color and a relatively smooth mantle.

Natural History: It is found on rocky reefs in shallow water.

Distribution: Australia (Solitary Islands, New South Wales, Australia).

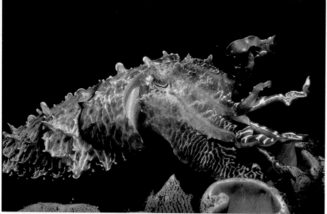
Glenn Barrall

669. *Sepia* sp.
Identification: This small species of cuttlefish is characterized by large branched papillae over its surface. It has relatively short tentacles.
Natural History: This species is found in shallow water sandy habitats at night.
Distribution: Indonesia (Sulawesi, Indonesia).

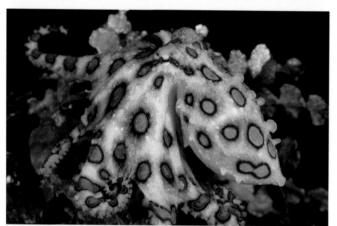
Roy Caldwell

670. *Hapalochlaena lunulata* (Quoy & Gaimard, 1832)
Blue Ringed Octopus
Identification: This species is immediately recognizable by its small size with distinct blue rings that become more intense when the animal is aggravated. This is an extremely toxic species which can inflict a bite fatal to humans. An antivenin must be administered within 20 minutes of being bitten.
Natural History: Found in shallow reefs in the open or underneath rubble. This species is active during the day.
Distribution: Australia; New Guinea; Indonesia and Philippines (Philippines).

Mike Miller
Jack Randall

671. *Octopus luteus* Sasaki, 1929
Identification: This is a fairly large species of *Octopus* that can be recognized by its reddish body color with white spots that can be elevated to form papillae.
Natural History: This species is found on shallow reefs and sandy patches, where it is active at night.
Distribution: Australia; Philippines; Taiwan; Hong Kong (Batangas, Luzon, Philippines).

672. *Octopus cyanea* Gray, 1849
Identification: This is the common large octopus which is active during the day in the Indo-Pacific. This species can be recognized by a black spot surrounded by another thin black ring on the base of the arm web.
Natural History: This species is active during the day and is found on shallow reefs to about 20 m depth.
Distribution: Western Indian Ocean; Red Sea to Hawaiian Islands (Oman).

The following eight taxa are considered as undescribed species:

673. *Octopus* sp. 1

Identification: This is another large "ocellate" octopus with a bright, irridescent blue ring at the base of the arm web. It has been suggested that this octopus may be venomous.
Natural History: Found on shallow coral or coral rubble substrates from 1-54 m depth.
Distribution: New Guinea (Papua New Guinea).

Bud Lee

674. *Octopus* sp. 2

Identification: This species has a reticulated pattern of lines on the mantle and arms
Natural History: Presently nothing is known.
Distribution: Indonesia (Sulawesi, Indonesia).

Mike Severns

675. *Octopus* sp. 3

Identification: Here seen hiding in a bubble shell (*Bulla vernicosa*, #525).
Natural History: This species lives in shallow sandy habitats and is active at night.
Distribution: Philippines (Batangas, Luzon, Philippines).

T. M. Gosliner
Mike Severns

676. *Octopus* sp. 4

Identification: This is another small species of *Octopus* which has a spotted mantle and orange siphon. It may be a member of the *Octopus aeginae* species complex.
Natural History: It is found in coarse sandy habitats where it may partially burrow in cobble.
Distribution: Indonesia (Sulawesi, Indonesia).

189

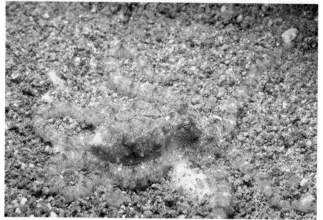

677. *Octopus* sp. 5

Identification: This is another small species of *Octopus*. Until recently it was believed that most small octopuses represented juveniles. More recent study reveals that most of these are mature individuals and many represent undescribed species. It is a member of the *Octopus horridus* species complex.

Natural History: This species is active at night on shallow water sandy habitats. This animal was photographed in less than 1 m depth.

Distribution: Philippines (Batangas, Luzon, Philippines).

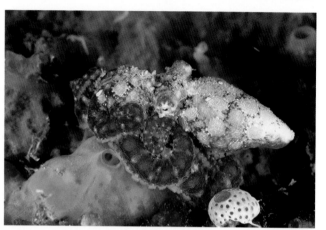

678. *Octopus* sp. 6

Identification: This small species of *Octopus* has a pattern on its mantle that resembles coralline algae or small diatoms that are abundant in most coral reef habitats.

Natural History: This species is found out in the open on living reefs. This may be the only specimen of this species ever recorded.

Distribution: Indonesia (Sulawesi, Indonesia).

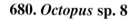

679. *Octopus* sp. 7

Identification: This species can be recognized by its elongate arms with brown transverse bands. It has an elongate papilla above the eye.

Natural History: *Octopus* sp. lives in coarse gravelly sand in shallow water. It may burrow into the sand and can escape quickly by pulling its body below the substrate with one arm which is extended beneath the surface.

Distribution: East Africa to Australia; Indonesia and Philippines (Batangas, Luzon, Philippines).

680. *Octopus* sp. 8

Long Arm Octopus

Identification: The long slender arms of this species are distinctive. It is similar to *Octopus horridus*, but the taxonomy of this species complex needs further study.

Natural History: Found at night in sandy and cobble habitats in a few meters depth.

Distribution: Philippines (Batangas, Luzon, Philippines).

Arthropoda: Sea Spiders and Crustaceans

Pycnogonida - Sea Spiders

Tiny spider-like creatures, they have 4-6 pairs of long segmented legs. The body is divided into a head and segmented trunk. Most species are small, a few millimeters to two centimeters in length. The head bears two pairs of eyes, situated on a rounded knob, and a long cylindrical proboscis. Although the sexes are separate, in several genera the male carries the eggs with special legs, near the head region. Crawling organisms, they are usually found in association with encrusting species. Of the 1200 known species, about 350 are found in the Indo-Pacific (C. Allan Child, pers. comm.). Because this group does not have planktonic larvae, many species are endemic. Refer to Clark (1963) for pycnogonids of Australia.

681. *Anoplodactylus evansi* Clark, 1963

Identification: A dramatically colored species, the base color is bright orange. The leg segments are bright blue, with a thin yellow band at their joints. There are three thin yellow bands on the body. The eyes are easily seen.
Natural History: Like many brightly colored species, its color might be considered aposematic, or a warning color pattern.
Distribution: Tasmania to and subtropical and temperate Australia (South Solitary Island, New South Wales, Australia).

682. *Anoplodactylus* sp.

Identification: The ground color of this species is cream. The head and body bear a deep blue dorsal stripe. The legs have varying degrees of blue on several of the segments.
Natural History: Like other species the eggs are fertilized externally. Males of this genus have cement glands on their femur which are used to cement the eggs together on their special egg-carrying legs called ovigers.
Distribution: Melanesia (Madang, Papua New Guinea).

683. Calipallenid pycnogonid

Identification: The Calipallenidae is a very large family with numerous genera and species living in the Indo-Pacific. This individual is red brown in color, with thick bright yellow bands at the joints in the legs.
Natural History: Both sexes carry eggs on their ovigers.
Distribution: Australia (Pig Island, New South Wales, Australia).

684. *Endeis flaccida* Calman, 1923

Identification: This is a common tropical species of pycnogonid. Thick legged, it has distinctive purple dots between body segments and near the end of each leg segment.
Natural History: Found in association with hydroids.
Distribution: Aden; India; Indonesia; both coasts of Panama; Florida (Batangas, Luzon, Philippines).

Carol Buchanan

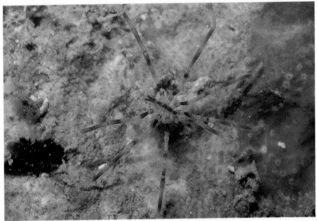

T. M. Gosliner

Carol Buchanan
G.C. Williams

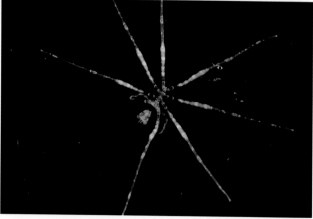

685. *Nymphon* sp. 1

Identification: This is a delicate, very slender-legged species. The body and legs are transparent, with brown and white encrustations. Some legs have chelipeds.

Natural History: The specimen shown here is a female, characteristically carrying eggs on her legs.

Distribution: Australia (South Solitary Island, New South Wales, Australia).

686. *Nymphon* sp. 2

Identification: This very long, smooth-legged pycnogonid is creamy white all over. There are red-brown marks on the legs and at some of the joints in the legs. Legs bear chelipeds.

Natural History: Found on many substrates, including sea fans.

Distribution: Australia (South Solitary Island, New South Wales, Australia).

687. *Nymphopsis* sp.

Identification: Somewhat resembling a decorator crab, this pycnogonid has tall tubercles along its legs. There are at least two species in the Queensland area, making this one difficult to identify precisely. As indicated by the photograph, individuals vary in color from white to cream.

Natural History: The body and legs bear short bristles or cirri which appear to be able to attach detritus, giving the sea spider the decorated appearance.

Distribution: Australia (Byron Bay, New South Wales, Australia).

Ostracoda - Ostracods

Ostracods are very small copepod-like crustaceans. They are characterized by a bivalved carapace which encloses the head, body and limbs. The two long pairs of antennae are used for locomotion. Classification of species is usually based upon features of the carapace valves.

688. Ostracod

Identification: This species' body is enclosed by the typical bivalved shell. It has two long antennae. Compound eyes can be seen on the head portion of the carapace. This particular species of ostracod is easily recognized by the red branching pattern on the thorax.

Natural History: Benthic ostracods such as this one forage on various reef substrata, including sponges and tunicates.

Distribution: Australia (Byron Bay, New South Wales, Australia).

Copepoda - Copepods

These small shrimp-like crustaceans are mostly free-living: planktonic or benthic. Some are parasitic. There are over 300 species in the Indo-Pacific.

689. Parasitic Copepod

Identification: A parasitic species. Head and thorax are fused, abdomen lacks appendages, eyes are fused appearing as one.
Natural History: They attach to the flesh of the host fish, usually on the dorsal surface.
Distribution: Australia (Split Solitary Island, New South Wales, Australia).

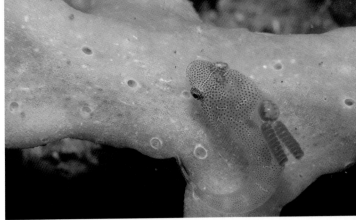

Carol Buchanan

690. Commensal Copepod

Identification: Commensal copepods occur on numerous host species.
Natural History: A free living species, it moves about freely on the mantle of its host. This species is living on a chromodorid nudibranch, *Thorunna africana*.
Distribution: East Africa (Msimbati, Tanzania).

Cirripedia - Barnacles

Although barnacles may be the most obvious and abundant crustacean found on shorelines, because of their calcareous shells they are sometimes mistaken for mollusks. Their larval life history reveals their relationship to other Crustacea. They are distinctive because of their attachment to other objects, and their thick calcareous shells. There are two types: those with shells attached to the substratum (sessile), and those with fleshy stalks (pedunculate).

T. M. Gosliner

691. *Balanus* sp.   Acorn Barnacle

Identification: Shaped like a volcano, the six calcareous plates are tightly fused. The plates are smooth in texture. When submerged, the thoracic feeding appendages or cirri may be observed reaching out of the orifice, in a grasping motion.
Natural History: Primarily an intertidal species, it is found in exposed areas with strong turbulence.
Distribution: Philippines (Batangas, Luzon, Philippines).

692. *Tetraclita formosana* Hiro, 1939
Acorn Barnacle

Identification: Shaped like a volcano, the four calcareous plates are tightly fused, and rugous or ridged in texture. When submerged or subtidal, the feeding appendages or cirri may be observed reaching out of the orifice, in a grasping motion.
Natural History: It is primarily intertidal, restricted to exposed areas with strong turbulence; along continental margins, not found on oceanic islands.
Distribution: Japan; Taiwan; Luzon, Philippines (Batangas, Luzon, Philippines).

G.C. Williams

G.C. Williams

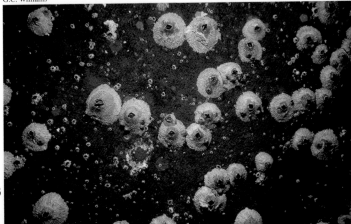

193

693. *Lepas* sp.

Pelagic Goose-neck Barnacle

Identification: The shell of this species is made up of five white plates forming what appears to be a bivalve. The shell, protecting the body and feeding appendages is attached to a fleshy brown stalk.

Natural History: Found attached to floating objects such as buoys, boat hulls and driftwood. Some goose barnacles are capable of limited movement.

Distribution: Melanesia (Solomon Islands).

694. Pyrgomatid Barnacle

Coral Barnacle

Identification: This group of barnacles bores into corals forming a cyst-like structure. The body wall consists of one single fused plate, the typical eight plates being fused into one. Only the thoracic feeding limbs can be seen reaching out of the slit-shaped opening in the surface of the coral, to feed.

Natural History: Forming rounded protuberances in the fire coral *Millepora* (#56), and several species of hard corals.

Distribution: Philippines (Batangas, Luzon, Philippines).

695. Pyrgomatid Barnacle

Coral Barnacle

Identification: Similar to the preceding species, the thoracic limbs are fully extended showing their structural differences.

Natural History: This species (as well as #694) are living on fungiid hard corals.

Distribution: Melanesia (Solomon Islands).

696. *Conopea* sp.

Coral Gall Barnacle

Identification: Coral gall barnacles form globular or deltoid-shaped protuberances on sea fans. Feeding appendages of the barnacle project through a single hole at the apex of the gall-like growth.

Natural History: Commonly associated with ellisellid gorgonians of the genera *Ctenocella* (pictured here, and #155, 156), and the sea whip *Junceella* (#157, 158). The barnacle larval stage settles on the exposed axis of an injured sea fan. During growth, tissues of the coral form around the barnacle.

Distribution: Circumtropical: East Africa to Philippines; Mediterranean; Atlantic and eastern Pacific (Batangas, Luzon, Philippines).

194

Stomatopoda - Mantis Shrimp

Alex Kerstitch

697. *Echinosquilla guerinii* (White , 1861)

Identification: This interesting mantis shrimp gets its generic name from its spiny armored telson. The body color is orange with two white stripes.

Natural History: Found below 15 m in coral rubble and worm tubes. This species almost never leaves its burrow. They are strictly nocturnal and have been observed to feed on small *Odontodactylus* which occur in the same habitat (R. Caldwell, pers. comm.).

Distribution: Seychelles and Mauritius to Indonesia; Fiji; Japan; Hawaiian Island and Marquesas (Hawai'i).

Roy Caldwell

698. *Gonodactylellus affinis* (Man, 1902)

Identification: Maybe the most common relatively deep water gonodactylid. This species is highly polymorphic in color. Often individuals found below 20 m are pink or red, as shown here. See Manning (1994) for taxonomic revision of this genus. Length: to 30 mm.

Natural History: Lives in cavities in coral rubble. Often seen darting over the surface near its cavity. Depth: 5 to 50 m.

Distribution: Kenya; Madagascar; Seychelles; Laccadives to Thailand; Indonesia; Australia; Guam and Society Islands (Queensland, Australia).

699. *Gonodactylus chiragra* (Fabricus, 1781)

Identification: Similar to *Gonodactylus platysoma*, the body is creamy white with green markings. Length: to 100 mm.

Natural History: Found on shallow intertidal reef flats and rocky outcrops, where it is commonly seen foraging at low tide in just a few centimeters of water.

Distribution: South Africa; Red Sea; Maldives; Laccadives; Australia; Guam; Marquesas (Queensland, Australia).

Roy Caldwell
Roy Caldwell

700. *Gonodactylaceus mutatus* (Lancaster, 1903)

Identification: Deep, dark emerald green in color, the legs and antennae fade to red. All specimens have yellow or orange meral spots. See Manning (1994) for taxonomic revision in this genus. Length: to 55 mm.

Natural History: Occurs commonly on reef flats to 5 m deep. Forages at extreme low tides.

Distribution: South Africa; Thailand; Vietnam; Indonesia; Australia; Guam to Hawai'i (Queensland, Australia).

195

Roy Caldwell

701. *Gonodactylus platysoma*
(Wood-Mason, 1895)

Identification: A cream-colored mantis shrimp with an olive green network pattern covering the body. There are two large blue spots on the third and last abdominal segments. The antennae are green with white specks. Length: to 90 mm.

Natural History: Found on shallow intertidal reef flats in cavities in the coral rubble, where it is commonly seen foraging at low tide in just a few centimeters of water.

Distribution: Seychelles to Australia; Guam; Enewetak and the Society Islands (Queensland, Australia).

Roy Caldwell

702. *Gonodactylus smithii* (Pocock, 1893)

Identification: Almost as colorful as *Odontodactylus scyllarus*, this green species bears markings of blue and orange. The antennae are yellow. The species is shown here in full threat posture, displaying raptorial appendages and purple meral spots, enclosed in a white ring.

Natural History: Very common, often seen foraging and mating away from its cavity at low tide (R. Caldwell, pers. comm.). Depth: to 20 m.

Distribution: Red Sea; Zanzibar; Chagos Islands; Maldives; Sri Lanka; Thailand; Vietnam to Australia; Loyalty Islands; Indonesia; Guam; Enewetak; Society Islands (Lizard Island, Australia).

703. *Lysiosquilla* sp. 1

Identification: Like other mantis shrimp, this species has a small carapace, leaving the last four thoracic segments of the body uncovered. This undescribed species is bright orange in color. The eyes are oblong rather than spherical.

Natural History: Depth: to 20 m.

Distribution: Indonesia; Fiji (Sulawesi, Indonesia).

Mike Severns
Marc Chamberlain

704. *Lysiosquilla* sp. 2

Identification: Rarely observed out of its burrow, this undescribed species is extremely cryptic in coloration. It is generally brown in body color, with varying degrees of white encrustations. The eyes, which may range in color from dark green to yellow, are oblong in shape and bear numerous white spots. The claw or second thoracic appendage, is orange in color.

Natural History: Typically a bottom burrower, never observed out of its burrow.

Distribution: Philippines and Indonesia (Indonesia).

196

705. *Lysiosquilla* sp. 3

Identification: The carapace of this undescribed species is brown with some light speckling which is found on the head and antennae. Each thoracic segment has a brown band. The telson is tipped in black. Eyes mottled.

Natural History: Observed out of its burrow crawling on coral rubble bottoms. Depth: 2-50 m.

Distribution: East Africa to Japan (Batangas, Luzon, Philippines).

Mike Miller

706. *Mesacturoides spinosocarinatus* (Fucuda, 1910)

Identification: Almost transparent in color, with a golden hue. The entire body is covered with white patches and brown spots. The telson is densely covered with tall spines. Manning (1994) reassigned this genus to the Family Takuidae. Up to 30 mm in length.

Natural History: Individuals stay near their coral cavities but may be seen darting around close by. Depth: 5-50 m.

Distribution: Thailand; Indonesia; Australia; Guam to Society Islands (Moorea, French Polynesia).

Roy Caldwell

707. *Odontodactylus brevirostris* (Miers, 1884)

Identification: This species is white with orange mottling. To 65 mm in length.

Natural History: A very curious species, often willing to interact with divers. Lives in burrows constructed of small pieces of rubble. Depth: 10-50 m.

Distribution: Seychelles; Sri Lanka; Philippines; Indonesia; New Guinea to Hawai'i (Oahu, Hawai'i).

Roy Caldwell
Mike Severns

708. *Odontodactylus scyallarus* (Linnaeus, 1758)

Identification: Among the most colorful of reef animals. It is easily distinguished from other mantis shrimp by its bright green body color, blue head and red-orange antennae and thoracic limbs. The eyes are spherical in shape and are divided by two parallel lines through their center.

Natural History: Like other mantis shrimp, it is a night feeding carnivore, feeding on other crustaceans, worms, mollusks and fish. A dramatic example of aposematic coloration, this is perhaps one of the most elegantly colored species of crustaceans in the world. Sand and rubble bottoms. Depth: to 70 m.

Distribution: South and East Africa; Madagascar; Maldives to Indonesia; Philippines; New Caledonia; Fiji; Australia; Samoa; Japan and Hawai'i (Indonesia).

197

Mike Miller

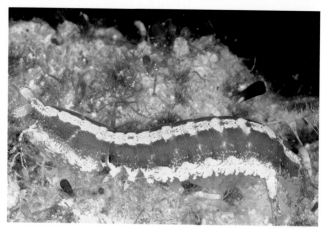

Roy Caldwell

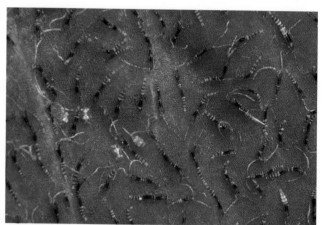

Kathy deWet
Carol Buchanan

709. *Oratosquilla oratoria* (Haan, 1849)

Identification: The body coloration of this species is a translucent grey, with a series of iridescent streaks on the thorax. The telson is distinctive with several of the uropods being blue with black tips. The legs or thoracic appendages are white.

Natural History: A shallow species, it occurs in muddy estuaries, as well as on reefs. Burrows are U-shaped up to a meter long (R. Caldwell, pers. comm.). A nocturnal species.

Distribution: Philippines; China Sea; Taiwan; Japan; Korea; Russia and Hawai'i (Batangas, Luzon, Philippines).

710. *Pseudosquilla ciliata* (Fabricius, 1787)

Identification: Extremely variable in color, taking on the color and pattern of its surroundings. Common colors are green, black, tan, and brown with stripes or mottled color patterns. Length: to 90 mm.

Natural History: Burrows in sand and gravel on coral reefs and sand flats, intertidal to 30 m.

Distribution: Circumtropical except in the eastern Pacific: Red Sea; Thailand; Indonesia; Australia; Guam; French Polynesia; Hawai'i; Bermuda; Caribbean (Queensland, Australia).

Mysidacea - Opossum Shrimp

Small shrimp-like crustaceans, usually found in large numbers in swarms near the bottom. They drift from area to area, and are not commensal on any host species or group of animals.

711. Mysid shrimp

Identification: This species is a typical mysid, having an extremely long slender body and long antennae. The legs are reduced, feather-like and transparent.

Natural History: This is a gregarious group of animals usually occurring in swarms or schools near or on the bottom. They settle on virtually every reef substratum.

Distribution: Indonesia (Indonesia).

Isopoda - Isopods

This group is composed of species with dorso-ventrally flattened bodies, with no carapace. They have seven similar-appearing pairs of legs, providing the name "isopod," meaning similar feet.

712. *Idotea* sp.

Identification: This slender species bears seven pairs of similar legs. It has long, segmented antennae and a simple tapered telson. The body color usually matches the host habitat species of coral or algae.

Natural History: Habitats include various coral and algae.

Distribution: Australia (Marsh Shoal, New South Wales, Australia).

198

713. *Santia* sp.

Identification: This minute species has thick stubby antennae. The uropods are widely separated. The legs hold tightly to the host substratum. The genus is widespread in both tropic and temperate seas. It is probably an undescribed species.

Natural History: Found in dense concentrations on various substrates, predominantly sponges.

Distribution: Indonesia and Philippines (Batangas, Luzon, Philippines).

D.W. Behrens

Bopyrid Isopods - Isopods on Shrimp

714. Bopyrid Parasitic Isopod

Identification: This species is a branchial parasite, living on the gills of shrimp. It can be detected by the "toothache" swelling on the side of the carapace of the host shrimp, in this case, *Periclimenes colemani* (#733).

Natural History: Usually there is only one pair of isopods, but sometimes a host may have a pair on each side. The female is the large, swollen, asymmetrically-segmented animal filling the gill. The male is a tiny, diminutive, symmetrical animal whose sole function in life is to fertilize the many eggs produced by the female.

Distribution: Philippines (Philippines).

Mike Miller

Cymathoid Isopods - Isopods on Fish

715. Cymathoid Isopod
Fish Louse

Identification: This species is typically found in the gill chamber of fish. Resembling a garden variety "pill bug," this species varies in color from cream to grey. Females are larger than males.

Natural History: The specimen shown here has apparently crawled out of the fish's gill and onto the snapping shrimp living in its burrow.

Distribution: Melanesia (Fiji).

Mike Miller

Stephen Smith

Amphipoda - Amphipods and Skeleton Shrimp

716. *Maera mastersii* Haswell, 1879

Identification: This species is typical of most amphipods; the body is laterally compressed and it lacks a carapace. Its appendages are modified for walking. The telson is folded beneath the abdomen.

Natural History: Common, sub-tropical representative of the genus which is found in algal masses at the base of coral colonies.

Distribution: Restricted to subtropical and temperate Australia (South Solitary Island, New South Wales, Australia).

717. Gammarid Amphipod

Identification: A typical example of this group, the body is laterally compressed and it lacks a carapace. Its appendages are quite short. In this particularly distinctive species, each plate of each body segment is outlined with a white line.
Natural History: Occurs in heavily encrusted substrata.
Distribution: Australia (Pig Island, New South Wales, Australia).

Caprellidae - Skeleton Shrimp

718. Unidentified Caprellid

Identification: Amphipods in this group are the "walking sticks" of the ocean. The thoracic limbs have large pincers or claws. The abdomen is reduced and has no appendages in the female, and only one or two in the male. The body is held upright, by the hind legs, similar to that of a praying mantis.
Natural History: This small species is most commonly found on hydroids and gorgonian soft corals.
Distribution: Philippines (Bohol, Philippines).

Decapoda - Shrimp, Prawns, Lobsters and Crabs

Decapods, like other Crustacea have hard, jointed exoskeletons. Their growth involves molting. Decapods fertilize their eggs by copulation, which occurs during molting. The females carry the developing eggs. At some point in the larval period the young hatch and subsequently fend for themselves.

The major differences between shrimp, lobsters and crabs is that the abdomen of shrimps and lobsters is muscular and extends posteriorly, whereas in crabs it is flexed forward under the body. The two major groups of crabs are differentiated by: four conspicuous pairs of walking legs in the Anomura (the fifth pair is hidden) while there are five pairs of legs in the Brachyura.

Penaeidae - Prawns

◀ 719. *Metapenaeopsis* sp. 1

Identification: The rostrum is very short, only slightly passing the eyes. It bears eight small spines on the upper edge and none on the underside. The body is mottled red and white with a slight iridescence. The tail has a blue band and is tipped with yellow.
Natural History: Occurs amongst coral rubble.
Distribution: Philippines (Batangas, Luzon, Philippines).

◀ 720. *Metapenaeopsis* sp. 2

Identification: The rostrum is short, extending just to the eyes. It has six spines, followed by a single head spine. The body is diffusely mottled red and white. The eyes are green.
Natural History: A sandy bottom species, active only at night.
Distribution: Philippines (Batangas, Luzon, Philippines).

721. *Metapenaeopsis* sp. 3

Identification: The translucent white body is covered with irregular white and brown markings. There is a series of white specks down each leg.

Natural History: A nocturnal species found on sandy bottoms.

Distribution: Philippines (Batangas, Luzon, Philippines).

T. M. Gosliner

722. *Heteropenaeus* sp.

Identification: This translucent grey shrimp is covered with black specks which form variably shaped encrustations. The antennae are banded and the legs are colored similarly. The rostrum is shorter than the antennal scale and bears 6 spines on the upper edge.

Natural History: Living in coral rubble, this species is observed most often at night.

Distribution: Philippines (Batangas, Luzon, Philippines).

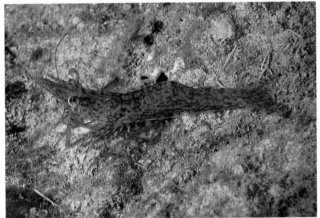

David Reid

723. *Penaeus monodon* Fabricius, 1789

Identification: A large greenish prawn. The rostrum is slightly shorter than the antennal scales. The rostrum bears 7 spines on its upper edge, which are yellow with black tips. Antennal scales have a series of yellow and black marks along their outer edge, giving the appearance of spines along this appendage. The carapace and abdomen have diffuse bands of lighter and darker pigment.

Natural History: This large prawn buries itself in the substratum during daylight.

Distribution: South Africa; Red Sea to Pakistan; Malaysia; Indonesia; Australia; Philippines (Manado, Indonesia).

Mike Severns
Mike Miller

724. *Penaeus latisulcatus* Kishinouye, 1896

Identification: A large gold-colored prawn, with brown rostrum and a large red-brown spot on its side. The swimmerettes and telson are edged with white and blue.

Natural History: Active nocturnally on sand bottoms, where it can be found buried with only its eyes protruding above the substrate.

Distribution: South Africa; Red Sea to Thailand; Philippines; Australia and Japan (Batangas, Luzon, Philippines).

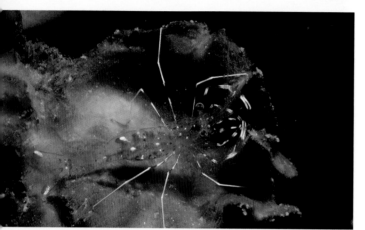

Kathy deWet

Mike Miller

Kathy deWet
D.W. Behrens

Palaemonidae - Palaemonid Shrimps

725. *Leander plumosus* Bruce, 1993
Identification: A slender, distinctively-colored shrimp. Light brown with white and dark brown markings, on the tail, carapace and antennal scales. There are tufts of cirri on the antennal scales and abdomen. The legs and the antennae are transparent. The rostrum is equal in length to the antennal scale. Length: 20 mm.
Natural History: Depth: 15 m.
Distribution: Maldives and Indonesia (Manado, Indonesia).

726. *Urocaridella antonbruunii* (Bruce, 1967)
Identification: Commonly, but incorrectly reported, in the literature as *Leandrites cyrtorhynchus.* A transparent shrimp with a long slender rostrum, which exceeds the length of the antennal scale by about 1/2 its length. The body is covered with white and red spots. The chelipeds and walking legs are red and white banded. Length: 3 cm.
Natural History: Found in holes and caves where it seems to float motionlessly, in the water column.
Distribution: East Africa; Comoro Islands to Australia; Indonesia; Philippines; Belau; New Caledonia and Japan (Batangas, Luzon, Philippines).

Pontoniinae - Commensal Shrimps

More than 240 species of pontoniine shrimp are reported from the waters of the Indo-Pacific. For more information, refer to the more than 100 papers published by Dr. A.J. Bruce on this group of shrimp (see references for a few).

727. *Allopontonia iaini* Bruce, 1972
Identification: At first glance one would believe this shrimp to resemble the galatheid crab *Allogalathea elegans* (#820). However, the rostrum forms a medial ridge, not a flat triangular plate, and the first thoracic legs bear well-developed pincers. Also, the large chelipeds are smooth, not spiny as in the crab.
Natural History: Found on the stinging urchin, *Asthensoma varium* (#997) and on *Salmacis belli* (#1006).
Distribution: Kenya to Australia; Indonesia to the Gulf of California (Lembeh, Indonesia).

728. *Dasycaris zanzibarica* Bruce, 1973
Identification: The transparent body has markings similar in color to that of the sea whip on which it occurs. Its color pattern is unique from the other two shrimp genera living on sea whips. It is distinguished by its large chelipeds and large humps on its thorax and abdomen, and a smooth rostrum. Length: to 15 mm.
Natural History: Lives on the black coral (sea whips) *Cirripathes*, where both male and female are found together. The female is twice as large as the male. Depth: to 41 m.
Distribution: Zanzibar; New Caledonia; Australia; Philippines and Japan (Batangas, Luzon, Philippines).

729. *Hamodactylus* cf. *boschmai*
Holthuis, 1952

Identification: The body is red with white encrustations that form bands across the carapace and abdomen. The telson is yellow. The rostrum bears 2-4 spines. Length: 10 mm.

Natural History: This species is found on a red gorgonian coral, which it matches in color. Found to a depth of 30 m.

Distribution: Kenya; Zanzibar; Madagascar; Indonesia; Solomon Islands; Australia; New Caledonia (Solomon Islands).

Carol Buchanan

730. *Periclimenes amboinensis* (Man, 1888)

Identification: Variably colored to match its host crinoid. The body coloration is banded similar to the banding of the crinoid. Colors include red or brown with white bands and red with yellow and orange banding. Eye stalks thick and white.

Natural History: Found in association with feather stars, crinoids. Shown here on *Oxycomanthus bennetti* (#916).

Distribution: Indonesia; New Caledonia; Australia; Solomon and Marshall Islands (Marshall Islands).

Charlie Arneson

731. *Periclimenes brevicarpalis*
(Schenkel, 1902)

Identification: Distinguished by large white spots, the spot on the head giving the appearance of a wart or hunchback. The tail has five large black spots with orange centers. Length: 4 cm.

Natural History: Found in association with sea anemones, especially *Cryptodendrum adhaesivum* (#189).

Distribution: East Africa; Mozambique; Red Sea to Vietnam; New Guinea; Indonesia; Belau; Australia; Line Islands; Singapore and South China Sea; Marshall Islands (Indonesia).

Mike Severns

732. *Periclimenes* cf. *ceratophthalmus*
Borradaile, 1915

Identification: This species has a long slightly curved rostrum, with a series of small teeth along the upper edge. Color varies widely depending on the host species. Its base coloration is made up of white granulations. Usually has several thin stripes the length of the body. The antennae are very short.

Natural History: Associated with crinoids in the genera *Himerometra*, *Dichrometra* and *Lamprometra* (#913, 914). Depth: to 43 m.

Distribution: Kenya; Zanzibar; Seychelles and Maldive Islands; Indonesia; Belau; Solomon Islands; Federated States of Micronesia; Australia (South Solitary Island, New South Wales, Australia).

Carol Buchanan

203

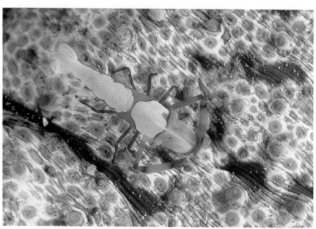

Mike Severns

733. *Periclimenes colemani* Bruce, 1975

Identification: The white to yellowish colored body is covered with large purple spots. The legs and chelipeds are banded. When found in pairs, the female is usually larger, up to 2 cm in length.

Natural History: Found exclusively on the venomous sea urchin, *Asthenosoma varium* (#997). The shrimp occupies a patch which it has cleared of tube feet and spines. They are able to move amongst the poisonous spines and pedicellaria without being harmed.

Distribution: Australia; Indonesia; Philippines (Indonesia).

734. *Periclimenes holthuisi* Bruce, 1969

Identification: Transparent with deep blue-purple spots which contain varying degrees of white at the center. Eyes bright red. The legs are transparent. The chelipeds are banded blue-purple. Some specimens have red streaks on the abdominal hump.

Natural History: Found on a number of hosts, including bubble and mushroom coral, sea anemones and the upside down jellyfish *Cassiopea.*

Distribution: East Africa; Red Sea; Sri Lanka to Vietnam; Maldives; New Caledonia; Indonesia; New Guinea; Australia; Philippines; Solomon Islands; Marshall Islands; Hong Kong; South China Sea and Japan (Solomon Islands).

Carol Buchanan

735. *Periclimenes imperator* Bruce, 1967

Identification: The head has a flattened, broad duck-bill appearance, created by the very wide lamina of the antennal scale. The tiny rostral spines (30-36) give a smooth appearance. Color varies widely depending on the host. Most have a broad creamy white dorsal stripe extending from rostrum to tail. The chelipeds are tipped in purple. The ornamental coloration on the body, whether dispersed evenly over the body as on specimens living on *Hexabranchus,* or enclosed with an outline as in those living on cucumber hosts, is always made up of tiny white specks.

Natural History: This species is found on several hosts, including the nudibranch *Hexabranchus* (#566), where it is found near the gill, feeding on fecal pellets; the sea cucumbers, *Stichopus, Bohadschia* and *Synapta* and the sea star, *Gomophia* (#952).

Distribution: South and East Africa; Red Sea; Mozambique; Comoros and Seychelles; Indonesia; Australia; Japan; Hawai'i; (Lembeh Strait, Indonesia).

Mike Severns
Roy Eisenhardt

736. *Periclimenes inornatus* Kemp, 1922

Identification: Body transparent with a slight tinge of copper orange. The antennal scale is very small. The rostrum bears 7-8 spines on the dorsal surface, and 0-2 below.

Natural History: Found on giant anemones, *Stichodactyla* or *Heteractis.* Depth: to 20 m.

Distribution: Kenya; Zanzibar; Seychelles; Comoros; Maldive and Andaman Islands; Ryukyu Islands; Indonesia; Australia; Fiji; Solomon Islands; South China Sea, Federated States of Micronesia (Solomon Islands).

737. *Periclimenes kororensis* **Bruce, 1977**

Identification: Transparent to orangish body with a conspicuous white spiny head and eyes. The chelipeds are very long and transparent. Length: to 4 cm.

Natural History: Lives on mushroom coral and various sea anemones, where its abdomen or tail is hidden.

Distribution: Belau; Philippines; Australia; Marshall Islands (Batangas, Luzon, Philippines).

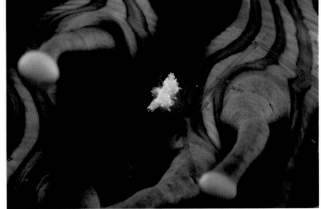
Mike Miller

738. *Periclimenes magnificus* **Bruce, 1979**

Identification: Body transparent. The carapace and the first four segments of the abdomen have a band of white specks outlined in red. The tail and hump in the abdomen are white. The chelipeds and antennae are white, the legs red. The antennal scale is broad and white. The rostrum bears 7-8 spines on the dorsal surface, 1-2 below.

Natural History: Found on the scleractinian coral, *Catalaphyllia* (#281) and the anemone, *Dofleinia armata*. Depth: 3-29 m.

Distribution: Australia; Indonesia; Philippines; South China Sea (Batangas, Luzon, Philippines).

Marc Chamberlain

739. *Periclimenes ornatus* **Bruce, 1969**

Identification: The transparent body bears a pattern of fine red reticulations. The legs, chelipeds and telson are specked with deep purple and white.

Natural History: Lives on the anemones, *Gyrostoma* and *Parasicyonis*.

Distribution: Kenya to Australia; Hong Kong; Japan; Marshall Islands (South Solitary Island, New South Wales, Australia).

Carol Buchanan
Robert Herrick

740. *Periclimenes psamathe* **(Man, 1902)**

Identification: A transparent species, it bears a red spot on the hump in the abdomen. There are six spines on the dorsal surface of the rostrum.

Natural History: Often found with black coral, *Antipathes*, melithaeid gorgonians, and the sea anemones, *Heteractis* and *Actinodiscus*.

Distribution: East Africa; Maldives; Chagos; New Guinea; Indonesia; Australia; Belau; New Caledonia; South China Sea (Belau).

David Reid   Laura Losito

741. *Periclimenes soror* Nobili 1904

Identification: The body is thinner, but the coloration similar to specimens of *P. imperator,* bearing numerous white specks over its body. The rostrum is large, with 10-11 teeth along the upper edge. The body color varies with that of the host, tan to deep purple. As in the photograph, on *Linckia laevigata* (#954), the shrimp is blue.

Natural History: This species is found on several species of sea stars including *Acanthaster, Culcita, Choriaster* and *Linckia.*

Distribution: East Africa; Rea Sea; Seychelles; Vietnam; Malaysia; Indonesia; Australia; Philippines; Fiji to Society and Tuamotu Islands; Hawai'i and Japan to Gulf of California (Batangas, Luzon, Philippines).

Terry Schuller

742. *Periclimenes tenuipes* Borradaile, 1898

Identification: A transparent species with a very long rostrum and long chelipeds. The rostrum bears 8-10 spines on its upper edge, and 6-9 spines below. There is a white line connecting the eyes. The chelipeds are tipped in orange.

Natural History: Generally a free living species, living in holes and crevices. It is sometimes found on sea anemones.
Depth: to 25 m.

Distribution: East Africa; Red Sea; Sri Lanka; Andaman Islands; Indonesia; Philippines; Australia; Belau and Marshall Islands (Batangas, Luzon, Philippines).

743. *Periclimenes venustus* Bruce, 1990

Identification: Similar to *P. holthuisi* in color, but with a large white spot on the abdominal hump, the center of which is pinkish. The rostrum has 5-7 spines on the dorsal surface and 0-2 ventrally.

Natural History: Found on anemones to a depth of 5 m.

Distribution: Ryukyu Islands; Philippines; Australia (Batangas, Luzon, Philippines).

Mike Miller
Mike Miller

744. *Periclimenes* sp. 1

Identification: Apparently an undescribed species, this transparent *Periclimenes* species has a large white spot on the abdominal hump, red marks and specks on the body and banded chelipeds. The tail is quite colorful, the telson being white with a red tip, the inner uropods being white to lime green, while the outer uropods are white with blue tips.

Natural History: Little is known about this free-living species.

Distribution: Philippines (Batangas, Luzon, Philippines).

745. *Periclimenes* sp. 2

Identification: Another undescribed species it is transparent with extremely long chelipeds. The color of the chelipeds is cream yellow with large brown leopard spots. There is an bridge of color between the eyes, composed of a deep red-brown path with a line of gold specks. Closely related to *Periclimenes tenuipes*.
Natural History: This is a poorly known species.
Distribution: Philippines (Philippines).

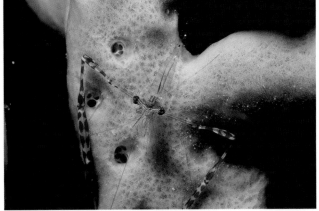
Bruce Watkins

746. *Pliopontonia furtiva* Bruce, 1973

Identification: This is the sole pontoniine shrimp commensally associated with corallimorph anthozoans. Its body is transparent with white bands and white and yellow spots. The legs and pincers are also banded.
Natural History: Occurs on corallimorpharians such as *Discosoma*.
Distribution: East Africa; Indonesia; Australia; Philippines; Solomon Islands; Japan (Solomon Islands).

Roy Eisenhardt

747. *Pontonides unciger* Calman, 1939

Identification: This shrimp is easily identified by its unique color pattern, which mimics the polyps of the black coral on which it lives. Similar in appearance to *Dasycaris zanzibarica* (#728), this species has a well-developed rostrum bearing 8-10 spines, and lacks the large humps on the thorax and abdomen. The body is usually ochre, with lighter patterns. Females are reported to be lighter in color. Length: 15 mm.
Natural History: Usually on the black coral *Cirripathes* (#293-295).
Distribution: Red Sea to Borneo; Indonesia; Philippines; Australia; Japan to Hawai'i (Hawai'i).

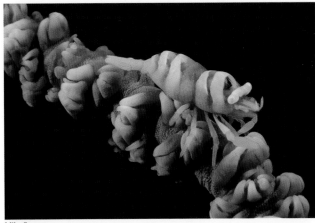
Mike Severns
D.W. Behrens

748. cf. *Pontonides* sp.

Identification: Found on black corals with the other *Pontonides*, this species is transparent, its body covered only with luminescent orange specks. Known only from this and two other photographs, one taken in the Philippines, the other in Hawai'i. Length: to 10 mm.
Natural History: Found on spiral black coral. Active at night.
Distribution: Philippines and Hawai'i (Batangas, Luzon, Philippines).

Mike Severns

749. *Stegopontonia commensalis* Nobili, 1906

Identification: This species has a long flattened body, adapted to living between sea urchin spines. The color is dark purple to black, with three thin white stripes the length of the body.

Natural History: Found most commonly among the spines of the sea urchins *Diadema, Echinothrix diadema* (#1003) and *Astropyga radiata* (#999). Unlike *P. colemani*, this species lives solely on the spines, never walking on urchin's test.

Distribution: Kenya; Mauritius; Seychelles; New Caledonia; Australia to Tuamotu Islands and Hawai'i (Hawai'i).

Mike Severns

750. *Vir philippinensis* Bruce & Svoboda, 1984

Identification: Reported incorrectly in several publications as a *Periclimenes*, this transparent species is easily identified by its purple antennae, and the purple line down each leg and cheliped and down the center of the body, as well as its exclusive occurrence on bubble coral.

Natural History: Commonly occurring on bubble coral, *Plerogyra sinuosa* (#282).

Distribution: Ryukyu Islands; Philippines; Australia (Philippines).

Alpheidae - Snapping Shrimp

Alpheids have received much attention due to their commensal relationships, particularly with various species of gobies. These associations are believed to be mutually beneficial. The common name, snapping shrimp, comes from the shrimp's ability to produce a snapping or cracking sound with its pincers.

751. *Alpheopsis yaldwyni*
Banner & Banner, 1973

Identification: A white snapping shrimp with 7-9 broad red stripes on the body running from the antennal peduncle to the telson.

Natural History: Occurs in coral rubble.

Distribution: Australia to Japan (Heron Island, Great Barrier Reef, Australia).

Leslie Newman & Andrew Flowers
Glenn Barrall

752. *Alpheus bellulus* Miya & Miyake, 1969

Identification: Body white with a red-brown reticulated pattern. The chelipeds and legs are banded white and brown.

Natural History: Lives in association with the goby species, *Amblyeleotris steinitzi.*

Distribution: East Africa; Thailand; Indonesia; Solomon Islands; Japan (Indonesia).

753. *Alpheus djeddensis* Coutiere, 1897

Identification: Similar to *A. djiboutensis*, however the greenish brown mottling is lighter and more diffuse. Antennae are yellow.

Natural History: Found on coral rubble bottoms, in association with the goby *Amblyeleotris*.

Distribution: Red Sea to Solomon Islands (Solomon Islands).

Roy Eisenhardt

754. *Alpheus djiboutensis* Man, 1909

Identification: One of the snapping shrimps that lives in association with gobies, it bears the typical enlarged chelipeds which are used to generate the snapping sound divers sometimes hear underwater. This species has olive green mottling on its white body.

Natural History: Lives in association with the goby species *Amblyeleotris steinitzi*.

Distribution: Djibouti; Thailand; Indonesia (Manado, Indonesia).

Terry Schuller

755. *Alpheus ochrostriatus*
Karplus, Szlep & Tsurnamal, 1981

Identification: Another goby associate, this species has thin longitudinal yellow stripes on carapace and abdomen.

Natural History: Lives in association with the goby species, *Amblyeleotris steinitzi*.

Distribution: Red Sea; Solomon Islands; Indonesia; Fiji; Japan (Indonesia).

Glenn Barrall
Alex Kerstitch

756. *Alpheus paracrinitus* Miers, 1881

Identification: This species is easily identified by its orange and white body banding.

Natural History: Commensal associations unknown.

Distribution: Circumtropical: Red Sea; Coral Sea; Japan and Baja California; Caribbean (Coral Sea).

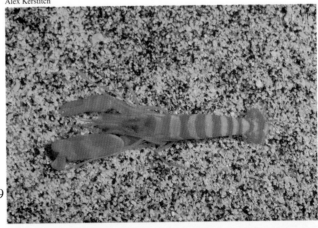

757. *Alpheus randalli* Banner, 1981

Identification: The body is banded red and white on the dorsal surface, with yellow ventrally. The chelipeds are white with two thick red bands.

Natural History: Like several other *Alpheus* species, it lives in association with the goby species, *Amblyeleotris steinitzi.*

Distribution: Indonesia; New Guinea; Marquesas Islands (Indonesia).

758. *Alpheus rubromaculatus*
Karplus, Szlep & Tsurnamal, 1981

Identification: Whitish with irregularly shaped brown marks distributed over the entire body.

Natural History: Found on sandy bottoms in association (shown here) with the white-capped goby, *Lotilia graciliosa.*

Distribution: Endemic to Red Sea (Red Sea).

759. *Alpheus* sp. 1

Identification: This species is undescribed, and closely related to *A. diadema* (Y. Miya, pers. comm.). A unique and easily identifiable species, it is white to salmon pink in color with a large, dark red bullseye on the side of the second abdominal segment. Chelipeds are purple in some specimens.

Natural History: Not yet observed with gobies.

Distribution: Sri Lanka and Philippines (Philippines).

760. *Synalpheus carinatus* (Man, 1888)

Identification: One of the smaller species of snapping shrimp, its body is translucent, colored only by a thin white line down the rostrum and the center of the carapace. In some specimens brown coloration may occur around eyes and on the antennae.

Natural History: Occurs on crinoids, such as *Comanthina schlegeli* (#909).

Distribution: Malaysia; Indonesia; Australia; Federated States of Micronesia; Marshall and Gilbert Islands (Marshall Islands).

761. *Synalpheus stimpsonii* (Man, 1888)

Identification: Another crinoid associate, this species has a white body which bears varying degrees of dark, black reticulations and markings. Unlike other species which closely resemble the color of their host species, this snapping shrimp does not, often contrasting in color with the red and yellow crinoid individuals on which it is found.

Natural History: Lives on crinoids *Comanthina* and *Oxycomanthus.*

Distribution: Singapore; Thailand; Australia; Indonesia; Philippines; Marshall and Gilbert Islands and Japan (South Solitary Island, New South Wales, Australia).

Carol Buchanan

Stenopodidae - Cleaner Shrimp

762. *Microprosthema validum* Stimpson, 1860
Robust Shrimp

Identification: Almost lobster-like in appearance, this shrimp has very large flattened chelipeds. The white body is encrusted with varying amounts of golden-brown pigmentation. The carapace has numerous spines, as do the edges of the chelipeds and abdomen. Length: 25 mm.

Natural History: Lives beneath rocks and coral rubble to 20 m.

Distribution: Arabian Sea; Australia (South Solitary Island, New South Wales, Australia).

Carol Buchanan

763. *Stenopus hispidus* (Oliver, 1811)
Banded Coral Shrimp

Identification: Goy and Randall (1986:99) provide a key to the six described Indo-West Pacific species of *Stenopus.* The third pair of thoracic legs form very large chelipeds, and the body is covered with spines. The red and white bands on the body and chelipeds are distinctive. Reaches 5 cm in length.

Natural History: This common cleaner shrimp waits in crevices or out in the open at its cleaning station, waving its long white antennae to attract fish. As fish approach to be cleaned, the shrimp touches them with the antennae until the fish becomes quiet, allowing the shrimp to move about its body in search of parasites.

Distribution: South Africa; Red Sea; Malaysia; Vanuatu; Philippines to Hawai'i and Caribbean (Sipadan Island, Borneo).

Marc Chamberlain
Dan Gotshall

764. *Stenopus pyrosonotus* Goy & Devaney, 1980
Fountain Shrimp

Identification: Similar in shape and spination to *S. hispidus,* this species has white legs, antennae, chelipeds and telson. The abdomen is transparent with a red stripe down the back.

Natural History: This species is found to depths of 15 m in crevices. Like *S. hispidus,* they are thought to be a cleaner shrimp. When fecund, the females carry light blue eggs.

Distribution: Mauritius; New Guinea; New Caledonia; Hawai'i (Maui, Hawai'i).

211

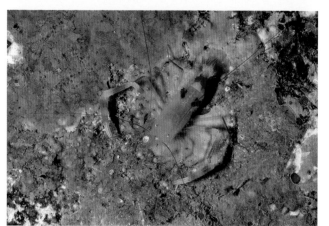

Mike Severns

765. *Stenopus tenuirostris* Man, 1888
Identification: Probably the most colorful *Stenopus*, the body and legs are purple with blue encrustations. The abdomen and chelipeds are banded in a characteristic repetitious white-red-yellow-red-white pattern. Length: 2 cm.
Natural History: This shrimp can be found picking parasites off various species of reef fish at specific cleaning stations on reef faces.
Distribution: South Africa to Indonesia (Indonesia).

766. *Stenopus zanzibaricus* Bruce, 1976
Identification: Similar to *Stenopus tenuirostris,* but body and first leg segments are yellow. The abdomen and chelipeds have the same pattern as all other *Stenopus*. Antennae are red. Length: 3 cm.
Natural History: Usually found in pairs.
Distribution: Tanzania to Indonesia (Indonesia).

Leslie Newman & Andrew Flowers

Hippolytidae - Hump-backed Shrimp

767. *Hippolyte commensalis* Kemp, 1925
Identification: Typical in shape with a broad antennal scale. The body is white and entirely covered with a reticulating pattern of olive green or brown.
Natural History: This species derives protection from predators by its remarkably close resemblance to xeniid soft corals.
Distribution: Andaman and Nicobar Islands; Australia (New South Wales, Australia).

Carol Buchanan
Junji Okuno

768. *Koror misticius* Clark, 1989
Identification: Similar in color pattern to *Parhippolyte uveae* (#772), below, to which it is closely related. This species has orange stripes on a translucent body. Other differences require microscopic examination.
Natural History: Found primarily in submarine caves, in rubble.
Distribution: Seychelles; Belau; Japan; Ryukyu Islands (Ryukyu Islands).

212

769. *Lysmata amboinensis* (Man, 1888)
Hump-back Cleaner Shrimp
Identification: This hump-backed shrimp has a long straight rostrum bearing 5-6 evenly spaced spines on the upper edge. The body is orange with a broad red stripe and a thin inner white stripe down the back. The telson is white and the uropods bear two white spots. The antennae are white with a red stripe. There is some dispute whether this species is synonymous with *L. grabhami* which resides in Atlantic waters.
Natural History: This shrimp has been observed to be a cleaner species on reef fishes.
Distribution: Red Sea to Australia; Indonesia; Hawai'i; Society Islands (Bali, Indonesia).

Jack Randall

770. *Lysmata debelius* Bruce, 1983
Scarlet Cleaner Shrimp
Identification: Deep red in color, this species has white legs and antennae. There are several series of distinctive white spots on each side of the carapace.
Natural History: Usually in pairs, they aggressively defend their territory. Depth: below 20 m.
Distribution: Maldives; Indonesia; Philippines; Society Islands; Japan (Coral Sea).

Alex Kerstitch

771. *Lysmatella prima* Borradaille, 1915
Identification: A yellow colored shrimp with a red line running the length of the body. The legs and antennae are yellow.
Natural History: Lives on sandy bottoms. Depth: 10-80 m.
Distribution: Maldive and Andaman Islands; Malaysia; Japan (Izu Pennisula, Japan).

Junji Okuno
Alex Kerstitch

772. *Parhippolyte uveae* (Borradaille, 1899)
Identification: The rostrum of this species is very short only slightly passing the eyes. It has very long legs and antennae. The long slender white body has 5-6 red and white bands on the abdomen. The antennae, legs, and abdominal appendages are red with white encrustations. Closely related to *Koror misticius* (#768). Length: 3 cm.
Natural History: Free living, found primarily in anchialine pools.
Distribution: Aldabra Atoll; Philippines; Loyalty Islands; Fiji; Hawai'i (Philippines).

213

Mike Severns

773. *Saron marmoratus* (Oliver, 1811)

Identification: Also very cryptic in coloration, the body is brown with numerous green spots over the body. These spots contain varying numbers and intensities of white specks. Beginning at the base of the large spiny rostrum there is a series of tufts of long cirri down the back. The antennae, legs and chelipeds are banded with two-tone brown and white speckled bands. Length: 4 cm.

Natural History: Nocturnal, usually found in protected lagoons of fringe reefs.

Distribution: Mozambique; Kenya; Red Sea to Australia; Philippines; Coral Sea; Society Islands; Marquesa and Hawaiian Islands and Japan (Hawai'i).

Junji Okuno

774. *Saron neglectus* Man, 1902

Identification: Similar in shape to *S. marmoratus*, it has long chelipeds. This species is greenish brown with aggregations of bluish spots encircled with white rings. The chelipeds and legs are banded with white and blue lines.

Natural History: Lives in coral rubble and within the branches of branching corals.

Distribution: Red Sea; Madagascar; Seychelles to New Caledonia; Japan and Ryukyu Islands (Ryukyu Islands).

Alex Kerstitch
Glenn Barrall

775. *Saron rectirostris* Hayashi, 1984

Identification: A very cryptic species, the body resembles a piece of dead coral, being cream colored with brown pore-like spots. It has a huge rostrum with spines on both the upper and lower edge. The legs and tail are blue. Length: 3-4 cm.

Natural History: A nocturnal species.

Distribution: Indonesia; Vanuatu (Vanuatu).

776. *Saron* sp. 1

Identification: Indescribably beautiful, the rostrum of this species is long and upwardly curved. This species has a very characteristic series of long cirri along the rostrum and down the top of the body, as well as along the lower margin of the abdominal plates and carapace. The body is white with uniquely shaped orange marks, with dark orange center regions. The legs and antennae are banded. Length: to 120 mm.

Natural History: Observed living on typical reef face habitats to 33 m deep.

Distribution: Flores and Ambon, Indonesia; Philippines (Biak, Indonesia) .

214

777. *Saron* sp. 2

Identification: This undescribed species has small bundles of sparse cirri on the back, relative to other species. It also has bundles of cirri on the front chelae. The general body color is greenish gray mottled with paler blotches and numerous small white speckles. There are less than ten large "bull's eyes" on each side of the carapace, with black, yellow, orange and white concentric rings. The bases of some walking legs and pleopods are purple, edged in red. The remaining walking legs have green and white alternating stripes. The chelae have purple and red stripes, as well.

Natural History: Found in coral rubble on disturbed reef.

Distribution: Known only from Indonesia (Sumatra, Indonesia).

Jim Black

778. *Thor amboinensis* (Man, 1888)
Ambonian Shrimp

Identification: A small brown shrimp, it has a saddle of large opalescent spots on its thorax and abdomen. These spots are encircled by a thin purple outline. Similarly colored spots occur at the base of and on the tail, on the abdominal plates, and on the eye stalks. Individuals hold their tail almost vertically.

Natural History: Commensal on corals and a number of anemone species. Usually observed in pairs, but you may find as many as six or eight individuals on a single host. Female nearly twice the size of the male.

Distribution: Kenya; Madagascar; Andaman and Nicobar Islands; Indonesia; Australia; Belau; Philippines; Marshall and Ryukyu Islands; Easter Island; Caribbean (Indonesia).

Mike Severns

779. *Tozeuma armatum* Paulson, 1875

Identification: An extremely elongated species, the rostrum is nearly 1/3 its total length. The body is transparent with irregular bands, which provides camouflage on its host.

Natural History: Occurs on black coral where its shape and coloration made it almost impossible to discern.

Distribution: Red Sea; Malaysia; Australia; Indonesia; Japan (Indonesia).

Mike Severns

Gnathophyllidae - Bumblebee and Harlequin Shrimp

780. *Gnathophylloides mineri* Schmitt, 1933

Identification: This tiny species is cigar shaped, similar to the sea urchin spines on which it occurs. The shrimp is white all over, except for a dark purple stripe up each side of the body. Some specimens have a lighter stripe on the back, made up of a series of fine lines. The chelipeds are held linearly to complete the minicry of the urchin's spine.

Natural History: Occurs on sea urchins, on individual spines.

Distribution: Circumtropical: Zanzibar; Seychelles; Australia; Tonga; Hawai'i and western Atlantic from Florida to Yucatan and Grenadines (Hawai'i).

Mike Severns

215

T. M. Gosliner

781. *Gnathophyllum americanum* Guerin-Meneville, 1855

Bumblebee Shrimp

Identification: A blunt headed shrimp with short rostrum. The largest cheliped is nearly as long as the body. The body is white, banded with a series of black-tan-black bands. The tail and chelipeds have orange markings.

Natural History: Associated with echinoderms; observed on asteroids, sea cucumbers and urchins.

Distribution: Circumtropical: South Africa; Red Sea to Tuamotu Archipelago; Hawai'i; Japan; western Atlantic; Gulf of Mexico; Caribbean Sea; Canary Islands (Aldabra Atoll).

Mike Severns

782. *Hymenocera picta* Dana, 1852

Harlequin Shrimp

Identification: This species has well-developed mouth parts and chelipeds for preying on starfish. Easily identified by its large broad chelipeds and its bold colorful mottling. Occurring on a white background, the contrasting color markings vary from burgundy to lavender and grey.

Natural History: This most colorful species preys on several species of sea stars, including *Fromia, Nardoa, Linckia* and *Acanthaster*. The species has been observed to eat the sea star from tip of the leg to the central disc, thus keeping the star alive for as long as possible. The large chelipeds are used for display only. Often found in pairs.

Distribution: South and East Africa; Red Sea to Australia; Philippines; Indonesia; Tuamotus; Hawai'i and Panama (Hawai'i).

Rhynchocinetidae - Hinge-beak Prawns

This group of shrimps has been greatly confused and mis-identified in the current literature. In general, hinge-beak prawns can be distinguished from other shrimp and prawns by the large eyes and the extreme upward angle of the rostrum. Males have very large pincers. Classification of the group is reviewed by Okuno & Takeda (1992) and Okuno (1994a & b).

Mike Miller

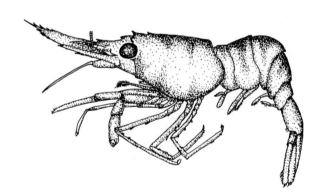

783. *Rhynchocinetes brucei* Okuno, 1994

Identification: The rostrum is hinged, rather than rigid to the carapace, in this group of shrimps. In this species the rostrum is roughly equal in length to the antennal scale. The body is transparent with a reticulated pattern of lines which are made up of two lines, which are either red or dark grey in color, with a thin broken white line between. The remaining spaces are filled with white spots. Its most prominent feature is the dark medial spot on the third abdominal somite.

Natural History: A rocky reef inhabitant, found in rubble, caves and crevices. Depth: to 30 m.

Distribution: Australia; Philippines; Hong Kong (Philippines).

216

784. *Rhynchocinetes concolor* Okuno, 1994

Identification: The rostrum of this red bodied hinged-beak prawn exceeds the length of the antennal scale. It has a characteristic white band on the second abdominal segment followed by a white "U"-shaped mark extending from the second segment down the sides of the 3rd and 4th abdominal segments.

Natural History: This species feeds at night, and hides by day.

Distribution: Zanzibar; New Guinea; Australia; Ryukyu Islands; Japan (Papua New Guinea).

Marc Chamberlain

785. *Rhynchocinetes conspiciocellus* Okuno & Takeda, 1992

Identification: Similar to *R. brucei* (#783) in color with the spot on the third abdominal segment, but occurs only in Japan. Other differences are more difficult to discern without close microscopic examination of the arthrobranchs, or small gills on the legs.

Natural History: Found in coral and rock rubble.

Distribution: Japan; Ryukyu Islands (Ryukyu Islands).

Junji Okuno

786. *Rhynchocinetes durbanensis* Gordon, 1936

Identification: Similar in coloration to and commonly reported incorrectly as *R. uritai* (#791), this species differs by its very long rostrum The rostrum bears spines along the entire upper edge, culminating with 9-10 tightly spaced spines at its tip and 16-18 spines along the underside. The species also differs by the bold red and white patterns on the body. The white inner line of the red stripe is not broken and very bold, as wide as the red lines. The angle of the main stripes on the abdomen also differs from *R. uritai*. Length: 9 mm. See Okuno and Takeda (1992) for details.

Natural History: Lives deep in crevices and holes. Usually occuring in large numbers together.

Distribution: South Africa to New Guinea; Indonesia; Philippines; Belau; Ryukyu Islands (Belau).

Jack Randall
Junji Okuno

787. *Rhynchocinetes hendersoni* Kemp, 1925

Identification: This species does not have the typical red color of this group of shrimp, but is brown mottled. The rostrum and legs are banded with white.

Natural History: Hides in coral rubble by day and feeds actively at night.

Distribution: India; Indonesia; Fiji; Japan (Izu Peninsula, Japan).

217

788. *Rhynchocinetes hiatti*
Holthuis & Hayashi, 1967

Identification: The abdomen and posterior portion of the carapace bear bold red-white-red vertical stripes. Between these stripes it is yellow orange.

Natural History: Occurs in coral rubble.

Distribution: Australia; Taiwan; Guam; Marshall Islands (Guam).

Junji Okuno

789. *Rhynchocinetes rugulosus* Stimpson, 1860

Identification: Similar to *R. brucei* (#783) in color, but lacks the bold spot on the third abdominal segment. This region is ornamented instead, only with red and white markings. Length: 60 mm.

Natural History: A bold and almost inquisitive species, it is also capable of quick retreats from predators.

Distribution: Australia; Indonesia; Japan; Loyalty Islands to leeward Hawai'i (Australia).

Dave Tarrant

790. *Rhynchocinetes striatus*
Nomura & Hayashi, 1992

Identification: This hinged-beak prawn has a very long upturned rostrum. It has 5 spines on the head and only 2 on the upper edge of the tip of the rostrum. There are about 12 spines on the underside of the rostrum. It has a transparent body with red and white zebra stripes forming bands across the carapace and abdomen. The legs are red with a white stripe along the upper edge.

Natural History: Occurs in coral rubble.

Distribution: Philippines; New Caledonia; Taiwan; Ryukyu Islands; Japan; Enewetak (Fiji).

Mike Miller
Junji Okuno

791. *Rhynchocinetes uritai* Kubo, 1942

Identification: Body translucent with irregular lines over the entire body. These lines vary from red to black. There are white spots, with no broken white lines, within the interspaces of the bolder lines. There are no white oblique bars on the abdomen.

Natural History: Occurs in coral rubble.

Distribution: Ryukyu Islands; Japan; Korea (Hachijo-jima, Japan).

218

792. *Rhynchocinetes* sp.

Identification: A transparent hinged-beak, the body is covered with red blotches. The legs have irregular red and white speckled bands. The rostrum has about 4 head spines, three at the tip and about eight spines on the under side. Its description by J. Okuno is in press.

Natural History: Found on coral flats and reef faces.

Distribution: Zanzibar to Malaysia; Indonesia and the Marquesas (Sipadan, Borneo).

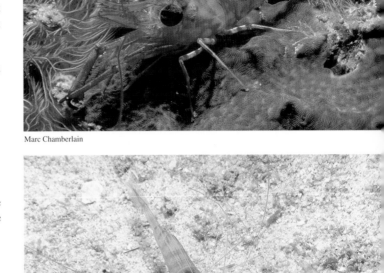

Marc Chamberlain

Pandalid Shrimps and Others

793. Pandalid sp.

Identification: The body is transparent with an orange tinge created by numerous specks. Clearings in this pigmentation give the appearance of stripes down the abdomen.

Natural History: Found in the open on sand bottoms.

Distribution: Malaysia (Sipadan Island, Borneo).

Marc Chamberlain

794. Processid sp.

Identification: A clear bodied, penaeid-like shrimp, it is covered with opalescent pink and blue specks. The rostrum is about 1/3 the length of the antennal scale. This species could be either a *Processa* or a *Nikoides*.

Natural History: A sandy bottom species.

Distribution: Philippines (Batangas, Luzon, Philippines).

Marc Chamberlain
Alex Kerstitch

Palinura - Lobsters
Nephropidae - Soft Lobsters

795. *Enoplometopus debelius* Holthuis, 1982

Identification: This white-bodied, soft lobster is covered with red spots. There are two prominent spiny ridges on the carapace. The legs and antennae are yellow orange. The chelipeds are pink, tipped in yellow-orange. Length: 10 cm.

Natural History: Occurs in caves and crevices of reef faces.

Distribution: Indonesia to Hawai'i (Indonesia).

219

796. *Enoplometopus occidentalis* (Randall, 1840)

Hairy Lobster

Identification: A red-bodied soft lobster, its body is covered with white spots, each enclosed in a red circle. It has long white tipped cirri along the edges of the chelipeds, the medial ridge of the carapace, and uniformly dispersed on the abdomen. It also has two prominent spiny ridges on the carapace. The legs and antennae match the body color. Length: 12 cm.

Natural History: Occurs in caves and crevices of reef faces.

Distribution: East Africa to Indonesia; Philippines and Hawai'i (Hawai'i).

Andy Sallmon

797. *Hoplometopus holthuisi* (Gordon, 1968)

Identification: This red-orange soft lobster is characterized by a series of white spots over its body and a large white "O" on each side of its carapace. Length: 12 cm.

Natural History: Occurs in caves and crevices of reef faces.

Distribution: East Africa to Hawai'i (Kona, Hawai'i).

Marc Chamberlain

Palinuridae - Spiny Lobsters

798. *Justitia longimanus* H. Milne Edwards, 1837

Long-Handed Lobster

Identification: One of the most colorful spiny lobsters, this species can be distinguished by its unique red pattern on the yellow background. The antennae, legs and chelipeds are red, orange and white banded.

Natural History: Occurs in caves at the outer edge of the reef to a depth of 35 m. Feeds at night.

Distribution: Circumtropical: Mauritius to Hawai'i; Bermuda; Caribbean (Hawai'i).

Mike Severns
Mike Severns

799. *Palinurellus wieneckii* (Man, 1881)

Mole Lobster

Identification: This lobster, while not possessing the typical large spines on the body of the spiny lobster, is covered with tiny tightly set spines over the carapace and abdomen. Its color is uniformly orange. Length: 15 cm.

Natural History: One of the smaller species, it has no commercial value, except in the aquarium trade.

Distribution: Thailand; New Guinea; Solomon Islands; Okinawa; Marshall Islands; Hawai'i (Hawai'i).

800. *Panulirus marginatus*
(Quoy & Gaimard, 1825)

Identification: A typical spiny lobster in appearance, it is distinguished by a white line on each abdominal segment, white spots on the carapace and dark legs and antennae.
Length: to 41 cm.
Natural History: Spiny lobsters are scavengers, feeding on a variety of organic material.
Distribution: Endemic to the Hawaiian Islands (Hawai'i).

Mike Severns

801. *Panulirus penicillatus* (Oliver, 1791)

Identification: This typical spiny lobster has a series of distinct white spots, encircled with black, down its spotted abdomen. Although some variation in color exists, all specimens have a number of white lines down each leg. The antennae are dark in color.
Natural History: The most common spiny lobster in the Indo-Pacific. Of commercial importance the species migrates into shallow water seasonally, during the summer months.
Distribution: South and East Africa; Red Sea; Maldives; Australia; Marshall Islands; Japan; Hawai'i; Society Islands; Clipperton Island; Galápagos; Mexico (Raine Island, Great Barrier Reef, Australia).

802. *Panulirus versicolor* (Latreille, 1804)
Painted Crayfish

Identification: This spiny lobster is boldy patterned. The abdominal bands are wider that those of *P. marginatus* (#800). The white and tan pattern on the carapace is distinctive. The antennae are white, and conspicuous, and the legs have several bold white or blue stripes.
Natural History: Nocturnal, hiding in crevices during the daylight.
Distribution: South Africa; Red Sea; India; Lakshadweep; Australia; Federated States of Micronesia; Belau (Coral Sea).

Jack Randall

Scyllaridae - Slipper Lobsters

Slipper lobsters differ from spiny lobsters by having dorso-ventrally flattened bodies. The antennae are flattened into a shield-shaped structure, and are not long and whip-like as in spiny lobsters.

Alex Kerstitch
Mike Severns

803. *Arctides regalis* Holthuis, 1963
Regal Slipper Lobster

Identification: One of the most brightly colored of the slipper lobsters, this species is most easily identified by the black central spots on the frontal antennal plates. The body is generally red in color with characteristic black markings and encrustations. There are small spines on the margin of the carapace, and strong spines on the abdomen. Length: 15 cm.
Natural History: Found in caves in groups of up to ten animals.
Distribution: Endemic to Hawaiian Islands (Hawai'i).

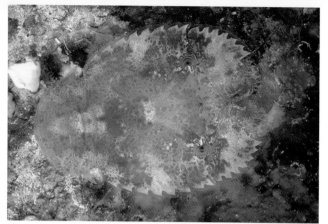
T. M. Gosliner

804. *Ibacus sp.*

Identification: This is a small, extremely flattened species of slipper lobster. It is delicate in appearance bearing a series of very large spines around the margin of the entire body. The spines are white, the remainder of the body is pigmented with brown and green specks and encrustations.

Natural History: Found under a coral head at 10 m depth.

Distribution: New Guinea (Madang, Papua New Guinea).

Terry Schuller

805. *Parribacus antarcticus* (Lund, 1793)

Identification: Like other slipper lobsters, the second antennae form flat shield-shaped structures, giving the head a flat paddle shape. The spines at the margin of the body are covered with numerous cirri. The body is covered with wart-like bumps. It is cream in color with irregular brown encrustations. There are numerous encircled spots along the margin of the carapace and head structures. Length: 20 cm.

Natural History: Depth: to 15 m.

Distribution: Circumtropical: East Africa to Australia; Japan; Hawai'i; Caribbean (Manado, Indonesia).

Mike Severns
Mike Severns

806. *Scyllarides haanii* (Haan, 1841)

Identification: A rugid, leathery slipper lobster. It has no spines along the margin of the carapace. There are several strong spines on the abdomen. The body is covered with small warts. Its color is cream and brown.

Natural History: Very docile, easy to handle.

Distribution: Indonesia; Taiwan; Hawai'i; Japan (Hawai'i).

807. *Scyllarus sp.*

Identification: One of about 40 species of *Scyllarus*, it is a very cryptic species, mottled tan and brown. This species lacks the characteristic spines or warts on the carapace, seen on so many slipper lobsters. The legs are banded and there are tufts of cirri on the carapace and at the sutures in the abdomen. The dark spot on the first abdominal segment is found in several species.

Natural History: This species hides in coral rubble.

Distribution: Indonesia (Sulawesi, Indonesia).

Anomura - Hermit Crabs, Squat Lobsters and Porcelain Crabs

Hermit Crabs

Hermit crabs are walking decapods that have chelipeds or claws on the first pair of walking legs, and have the fourth or fifth pair of walking legs reduced in size. They have large soft abdomens requiring protection from predators, hence the need to seek refuge in abandoned snail shells and worm tubes. Tudge (1995) reviews the tropical Australian species. McLaughlin and Lemaitre (1993) review the genus *Paguritta*.

Mike Severns

808. *Aniculus maximus* Edmondson, 1952

Identification: The exposed appendages of this species are covered with bristles. They are yellow and uniquely striped with white and thin dark bands.
Natural History: Outer edges of reefs to over 35 m.
Distribution: Seychelles; Ryukyu Islands; Hawai'i; Marquesas Islands (Hawai'i).

809. *Calcinus minutus* Buitendijk, 1937

Identification: This small white species' body is covered with small orange pits. The last segment of the walking legs is orange tipped with black. Eyes are black, set on white to pale orange stalks.
Natural History: Found in branching corals, such as *Acropora* and *Pocillopora*. Occupies a variety of shells.
Distribution: Red Sea; Australia; Indonesia; Taiwan; Vietnam; Malaysia; New Guinea; Philippines; Soloman Islands; Ryukyu Islands; Japan (Solomons).

Roy Eisenhardt

810. *Coenobita perlatus*
H. Milne Edwards, 1837
Land Hermit Crab

Identification: This large hermit crab is orange in color, and is covered with blunt white spines. Its black eyes are set on thick tapering stalks.
Natural History: Living high on the shoreline, often hiding in burrows, this species is capable of staying out of water for long periods of time. Spawns underwater.
Distribution: Australia; Philippines; Belau; Marshall Islands; Society Islands; Tuamotus (Coral Sea, Australia).

Mike Miller

Leslie Newman & Andrew Flowers

811. *Dardanus gemmatus*
(H. Milne Edwards, 1848)

Identification: A red and white mottled species. The chelipeds are smoother than most other species, and bear a few cirri along the inner margin.
Natural History: Lives at the outer edge of reefs to depths of 15 m.
Distribution: Australia; New Guinea; Taiwan to Hawai'i and Marquesas Islands (Papua New Guinea).

223

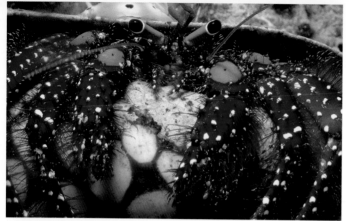

Ben Tetzner

812. *Dardanus guttatus* (Oliver, 1812)

Identification: This species is a deep maroon in body color, covered with white spots. It is easily identified by the large sky blue spots on the first segment of the legs and chelipeds. The eye stalks do not taper and are olive green in color. The eyes are usually black.

Natural History: Found in shells with narrow apertures, such as cone snails.

Distribution: South Africa to Australia; Indonesia; Samoa; South China Sea (Flores, Indonesia).

813. *Dardanus lagopodes* (Forskål, 1775)

Identification: The body of this small species is mottled tan and maroon. It is uniformly covered with dark bristles which are tipped in white. The eye stalks and antennae are yellow.

Natural History: Commonly lives in shells with wide apertures.

Distribution: East Africa; Red Sea to Australia; Solomon Islands; Society and Tuamotu Islands; Marshall Islands and Japan (Solomon Islands).

Roy Eisenhardt

814. *Dardanus megistos* (Herbst, 1804)

Identification: This large hermit crab is easily identified by its red-orange color with white spots and spines outlined with black. This dark outlining is the origin of numerous long thick brown bristles.

Natural History: Has been observed to feed on mollusks by breaking open the shells.

Distribution: South Africa; Madagascar to Australia; Indonesia; China; Japan; Hawai'i; Tuamotu and Society Islands (Sulawesi, Indonesia).

Jack Randall
Laura Losito

815. *Dardanus pedunculatus* (Herbst, 1804)

Identification: Probably the most common species of hermit crab, it is identified by its spiny chelipeds, light colored bristles, red and white striped eye stalks and green eyes. The color often varies in intensity, some specimens being lighter or darker, leading some to believe that there may be more than one species.

Natural History: The shells used by this species are nearly always covered with small sea anemones. This relationship provides camouflage and protection for the crab, giving the anemone the opportunity to feed in many differing locations as the crab moves. When disturbed the anemone expels long stinging threads called acontia. Depth: to 40 m.

Distribution: South Africa; Madagascar; Vanuatu; Korea to Society Islands (Vanuatu).

816. *Dardanus* sp.

Identification: An extremely hairy species. The hairs are red, tipped in white. The eyes are small and spherical with long thin eye stalks, which bear longitudinal stripes.
Natural History: Shell typically covered with anemones.
Distribution: Melanesia (Fiji).

Bruce Watkins

817. *Paguritta gracilipes* Melin, 1939
Coral Hermit Crab

Identification: This species of coral-inhabiting hermit crab is characterized by a larger right cheliped, and feather-like antennae. There are longitudinal stripes on the chelipeds and a few large curved spines on the dorsal edge and a series of numerous smaller spines on the ventral edge (McLaughlin and Lemaitre, 1993).
Natural History: In polychaete tubes associated with corals, and in holes in hydrocorals and hexacorals.
Distribution: Australia; Philippines; Vietnam; Solomon Islands; Japan (Solomon Islands).

Roy Eisenhardt

818. *Paguritta harmsi* (Gordon, 1935)
Coral Hermit Crab

Identification: A small coral-inhabiting species, the right cheliped is always larger than the left. The antennae are very long and feather-like with evenly set lateral branches. The chelipeds have a characterisitc brown reticulated pattern and reddish brown tips. It is similar to *P. scottae*, whose reticulating pattern is blue.
Natural History: This hermit crab prefers sessile habitats, choosing the vacant holes of tube worms on the surface of corals, as its enclosure. It uses its long feather-like antennae to strain plankton out of the water to feed on.
Distribution: According to McLaughlin and Lemaitre (1993) known only from Christmas Island, now also recorded from the Solomon Islands (Solomon Islands).

Marc Chamberlain
Marc Chamberlain

819. *Trizopagurus strigatus* (Herbst, 1804)

Identification: The body of this hermit crab is flat in appearance. It is easily identified by its bright yellow and red banding. Its pincers are tipped in black.
Natural History: Found in cone shells.
Distribution: East Africa; Red Sea to Australia; New Guinea; Hawai'i and Japan (Papua New Guinea).

Carol Buchanan

Debbie Zmarzly

Galatheidae - Squat Lobsters

This group of anomurans has oval bodies with a sharp rostrum. They get their lobster name because of the long chelipeds. Some species are commensals on corals and crinoids.

820. *Allogalathea elegans* (White, 1847)

Identification: The color varies depending on the crinoid host, but the pattern remains similar; in most cases, longitudinal stripes down the carapace and chelipeds. The exception is red individuals living on red hosts. The rostrum is long and tapering with a series of tiny spines along each edge. Length: 2 cm.

Natural History: Found in association with feather stars. The color of the individual mimics the host. The female of the species is larger. A shrimp species, *Allopontonia iaini* (#727), closely resembles this crab.

Distribution: Australia; Philippines; Indonesia; Enewetak (South Solitary Island, New South Wales, Australia; Marshall Islands).

Mike Severns

821. *Galathea* sp. 1

Identification: This bright red species has very long, thin chelipeds. The body is covered with bristles. The chelipeds have a series of sharply pointed spines, as does the head region. There is a single distinctive white spot on each leg and cheliped.

Natural History: Found in crevices.

Distribution: Indonesia (Sulawesi, Indonesia).

822. *Galathea* sp. 2

Identification: This species has strong sharp spines and bristles on the surface of the chelipeds. The cheliped and claw are shorter than the previous species, and the red body color is mottled with white markings.

Natural History: Lives in small holes on coral substrata. Several individuals are usually found together on a single outcrop.

Distribution: Philippines (Batangas, Luzon, Philippines).

Terry Schuller
Kathy deWet

823. *Lauriea siagiani* Baba, 1994

Identification: This elegantly colored squat lobster, is brilliant pink in color with purple stripes on the carapace and legs. The chelipeds have numerous sharp spines and purple spots. The entire animal is densely covered with white and pink bristles.

Natural History: Found inside the central cavity of *Xestospongia* (#35, 36).

Distribution: Indonesia; Philippines (Manado, Indonesia).

Porcellanidae - Porcelain Crabs

Superficially similar to crabs, they are in fact more closely related to squat lobsters. The presence of long antennae, the small last pair of walking legs, and the presence of a tail, although tucked under the carapace, distinguish them from true crabs.

824. *Neopetrolisthes maculata* (H. Milne Edwards, 1837)
Spotted Porcelain Crab ▶

Identification: Has typical large flat chelipeds. The body is white with reddish brown spots and specks. This coloration varies widely from location to location. The large white mouth parts extend far in front of the head.

Natural History: Lives in the shelter of sea anemones. The crabs keep from being captured and eaten by the anemone by moving around among the tentacles. Individuals can enter and leave the mouth of the anemone freely.

Distribution: Australia; Indonesia; Philippines; China (Ai Island, Banda, Indonesia).

825. *Petrolisthes lamarckii* (Leach, 1820)

Identification: Ground color is greyish tan with brown mottling.

Natural History: Intertidal to shallow subtidal under rubble.

Distribution: South Africa; Somalia to India; Australia; Guam; Tuamotu, Line and Marshall Islands (Heron Island, Great Barrier Reef, Australia).

826. *Porcellanella picta* Stimpson, 1858

Identification: A small commensal porcelain crab, the body is translucent white, sometimes having white markings and encrustations.

Natural History: Found living on the sea pens *Pteroeides,* (#172) and *Veretillum* (#161). Both sexes are usually present, the female being larger.

Distribution: Indonesia; Philippines; Japan (Manado, Indonesia; Batangas, Luzon, Philippines).

827. Unidentified Porcelain Crab

Identification: This tiny species, with typical large, flattened chelipeds, and long antennae, is easily identified by its blue and yellow-orange markings.

Natural History: Found on a pinkish-orange sponge, with a brittle star of the same coloration.

Distribution: Indonesia (Manado, Indonesia).

Jack Randall

Leslie Newman & Andrew Flowers

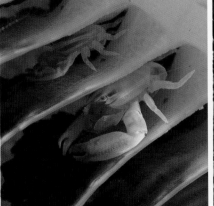 

Mike Severns
Mike Severns

D.W. Behrens

227

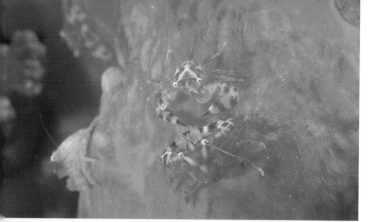

828. Unidentified Porcelain Crab

Identification: A tiny symbiotic species, its color makes it hard to differentiate from its host. The long antennae and large flat chelipeds indicate it is a porcelain crab. The white body has varying patterns of red and violet. The chelipeds are tipped in purple.

Natural History: Numerous individuals of this species may be found on a single soft coral colony of *Dendronephthya*.

Distribution: Indonesia (Sulawesi, Indonesia).

Hippidae - Mole Crabs

829. *Emerita pacifica* (Dana, 1852)

Mole Crab

Identification: This sand dwelling crab differs from the spanner crab (#834) in that the pincer is not at a right angle to the claw. The eyes are set in sockets in the carapace and it has a pair of long antennae which are feather-like.

Natural History: Typical sand crab, rarely seen, usually buried in clean, coarse sand.

Distribution: Western India Ocean to Indonesia; Samoa; Hawai'i (Sulawesi, Indonesia).

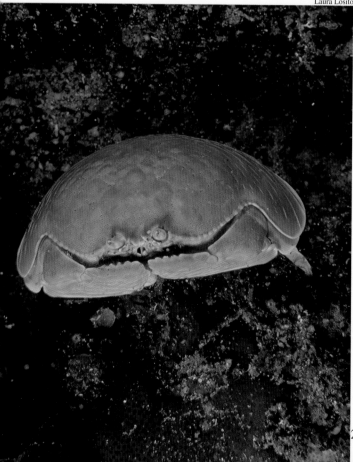

Brachyura - True Crabs

True crabs have dorso-ventrally flattened bodies, and the carapace is fused with the ventral plate. The abdomen is tucked tightly underneath the carapace. The antennae are small. The first pair of thoracic legs are developed into strong pincers, or chelipeds.

Calappidae - Box Crabs

830. *Calappa calappa* Linnaeus, 1758

Identification: Named box crabs because they can tuck their legs into the body giving the appearance of a neat box. It has smooth lateral extensions of the carapace which cover the legs. The domed carapace varies in color from creamy white, with uniformly spaced brown spots, to a uniform tan.

Natural History: This species feeds by crushing mollusks with its powerful chelae.

Distribution: Zanzibar; Mauritius; New Guinea; Indonesia; Philippines; Japan; Hawai'i (Philippines).

831. *Calappa hepatica* (Linnaeus, 1758)

Identification: The carapace has a warty texture, with six postero-lateral spines. The color is white with gold and brown markings. The lateral extensions of the carapace in this species bear several large spines.

Natural History: Like *Calappa calappa* above, this species is a mollusk eater.

Distribution: South Africa; Red Sea; Philippines; Japan; Wake Islands; Hawai'i and Society Islands (Batangas, Luzon, Philippines).

G.C. Williams

832. *Calappa lophos* (Herbst, 1782)

Identification: The dome shaped carapace is smooth in texture. The carapace is golden with white lines, while the chelipeds are white with irregular brown spots and markings.

Length: to 40 mm.

Natural History: Occurs on reef flats. The hooked claw, used to break open mollusks, is easily seen in this photo.

Distribution: South Africa; Tanzania; Sri Lanka; Thailand; Australia; Philippines; China; Japan to Panama (Philippines).

Lynn Funkhouser

833. *Calappa* sp.

Identification: This is a rugose, wart-covered species. The color is reddish brown. The chelipeds bear a series of spines along the anterior edge of the claw, and a couple of large spines on the posterior edge.

Natural History: Like other box crabs, the right cheliped has a toothed claw which is used to break open gastropod and hermit crab shells.

Distribution: Hawaiian Islands (Hawai'i).

Raninidae - Spanner Crabs

Mike Severns
Mike Severns

834. *Ranina ranina* (Linnaeus, 1758)
Spanner Crab

Identification: This species has an elongated carapace, which does not cover the abdomen. The pincer is where this crab gets its name, being "spanner-shaped" at a right angle to the flattened end of the moveable joint. Unlike sand crabs, it has a short discrete antennae.

Natural History: Usually observed partially buried in the sand.

Distribution: South Africa; Reunion Island; Mauritius to Indonesia; South China Sea; Japan and Hawai'i (Hawai'i).

Mike Severns

Dorippidae - Dorippid Crabs

835. *Dorippe frascone* (Herbst, 1785)
Identification: Small spherical crab with long white eye stalks and slender sharp chelipeds. The body is often encrusted with debris, similar to a decorator crab.
Natural History: This group of crabs uses its third and fourth legs to hold debris such as anemones, on its back.
Distribution: Mozambique; East Africa to Australia; Indonesia; Malaysia; Japan (Indonesia).

836. *Ethusa* sp.
Identification: A light brown, flat bodied decorator. The eyes are tear-drop shaped. The hair-like structures are easily seen on the carapace and legs.
Natural History: This is a typical example of this species carrying a jellyfish on its back with the third and fourth legs.
Distribution: Indonesia (Flores, Indonesia).

Jack Randall

Dromiidae - Sponge Crabs

McLay (1993) provides the most recent review of the species currently assigned to this family.

837. *Cryptodromia* cf. *coronata* Stimpson, 1858
Identification: Matching its sponge cover (*Desmacella* sp. #18) almost identically, this red dromiid crab bears white eyes that mimic the white polyps of *Nausithoe* (#306) boring in the sponge, which it carries.
Natural History: Similarly, in this species, the sponge worn by the crab does not die after being torn from the parent colony, but continues to live and grow, covering the entire carapace of the crab. Depth: to 47 m.
Distribution: New Caledonia; New Guinea; Indonesia; Samoa; Japan; Society Islands (Indonesia).

Mike Severns
Bruce Watkins

838. *Cryptodromia* cf. *fukuii* (Sakai, 1936)
Identification: This tiny dromiid is concealed within the sponge it carries. The body is semi-transparent.
Natural History: Much like a hermit crab, this species lives in the cavity it has created in the sponge and carries it with its hind legs. A shallow water species.
Distribution: New Caledonia; Philippines; Japan (Philippines).

839. *Dromia dormia* (Linnaeus, 1763)

Identification: Similar to *Lauridromia dehaani*, the species has a furry domed carapace. The lateral spines on the carapace are larger and sharply pointed.

Natural History: One of the larger sponge crabs. Like *Lauridromia dehaani*, it is often found carrying a sponge with its rear legs.

Distribution: South and East Africa; Red Sea; Seychelles; Mauritius; Indonesia; New Caledonia; Philippines; China; Japan; Hawai'i (Hawai'i).

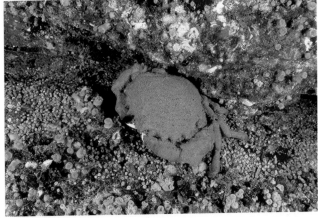
Mike Severns

840. *Lauridromia dehaani* (Rathburn, 1923)

Identification: The high dome shaped carapace with a furry appearance, and pinkish-white tipped chelipeds distinguish this species.

Natural History: Usually carrying a piece of sponge with its rear thoracic legs. Depth: to 80 m.

Distribution: South Africa; Aden; India; Indonesia; Japan; Hong Kong to Sal y Gomez Islands in the eastern Pacific (Batangas, Luzon, Philippines).

Hymenosomatidae - Spider Crabs

841. *Trigonoplax* sp.

Identification: The carapace is triangular, and the legs and chelipeds long and slender; minute size.

Natural History: Occurs in coral rubble.

Distribution: This species is known only from Australia (Australia).

G.C. Williams

Majidae - Spider or Decorator Crabs

These true crabs have a triangular-shaped carapace and long slender legs giving them the name "spider." Many of the species camouflage themselves by attaching organisms and other materials to their carapace and legs. Referred to as decorator crabs when they do this, the materials are held in place by hair-like projections on the surface of the body.

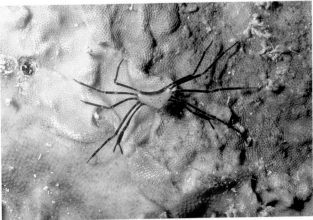
Leslie Newman & Andrew Flowers
Terry Schuller

842. *Achaeus japonicus* Haan, 1839

Identification: The long hairs on the body of this species enable it to attach small particles for camouflage. When fully covered only the eyes may protrude from the covering.

Natural History: This species may take on very different appearances depending on the quantity and type of material used to cover itself. It is most often found in association with coelenterates, particularly bubble coral (#282), *Dendronephthya* and *Parazoanthus*. Length: to 80 m.

Distribution: Fiji; Solomon Islands; Indonesia; Belau; Japan (Manado, Indonesia; Russell Group, Solomon Islands).

Carol Buchanan

231

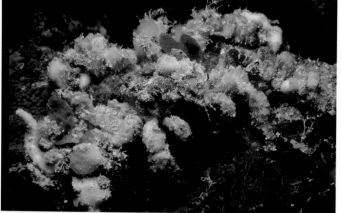

© Franklin J. Viola

843. *Camposcia retusa* Latreille, 1829

Identification: A master of camouflage, this species has been observed with many organisms, algae and other ocean-born material attached to it. It is often identified only by its black tear-drop shaped eyes, protruding from its decoration.

Natural History: This species may be one of the most artistic decorators in the Indo-Pacific, usually choosing several species and colors of material for decoration. Depth: 10-15 m.

Distribution: South Africa to Indonesia; Malaysia; Philippines; Belau; Japan (Sipadan, Borneo).

Bruce Watkins

844. *Cyclocoeloma tuberculata* Miers, 1880

Identification: As indicated by its name, the surface of the carapace is covered with densely packed tubercles. These tubercles aid in attaching anemones, tunicates and other materials to the body. The chelipeds and legs are covered with brown spots, and there are two black bands on each leg segment.

Natural History: A striking decorator crab, it is usually covered with anemones.

Distribution: Western Pacific: Malaysia; Australia; Philippines; Indonesia; Ryukyu Islands (Manado, Indonesia).

T. M. Gosliner

845. *Hoplophrys oatesii* Henderson, 1893

Identification: The body appears as a mass of large spines. The legs and chelipeds are also covered with large sharp spines. Color is translucent with red and white lines and patterns, mimicking the color of the *Dendronephthya* soft coral it lives on.

Natural History: Found in association with *Dendronephthya* soft corals.

Distribution: Australia; Philippines; Indonesia; New Guinea; New Caledonia; Fiji; Japan (Madang, Papua New Guinea).

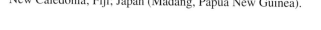

G.C. Williams

846. *Huenia heraldica* Haan, 1837
Arrowhead Crab

Identification: This species has a long triangular carapace. The eyes appear as small lumps on the edges of the long tapering rostrum. The chelipeds are long and the same color as the body and first segment of the walking legs. There may be a white stripe along the edge of the carapace. Length: 2 cm.

Natural History: Collected in 30 m of water associated with coral rubble and gorgonians.

Distribution: South Africa to India; Australia; Indonesia; Malaysia; Korea; Ryukyu Islands; Japan and Hawai'i (Natal, South Africa).

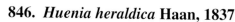

232

847. *Huenia* sp.

Identification: Green in color, its body is shaped similarily to the segments of the alga *Halimeda*. The carapace has a suture-like line connecting the eye sockets. Length: 3 cm.
Natural History: Found in association with the green algae, *Halimeda*.
Distribution: Australia (Heron Island, Great Barrier Reef, Australia).

Leslie Newman & Andrew Flowers

848. *Hyastenus bispinosus* Buitendijk, 1939

Identification: A small oval species, it is light in color, and often sparsely decorated with hydroids and marine detritus.
Natural History: A common decorator crab.
Distribution: Indonesia and Melanesia (Fiji).

Ben Tetzner

849. *Micippa* sp.

Identification: A small decorator crab, its eyes are spherical, and carapace knobby. The body color is red.
Natural History: Another common decorator species that utilizes anemones for cover.
Distribution: Melanesia (Fiji).

Mike Miller

Carol Buchanan

850. *Naxioides taurus* (Pocock, 1980)

Identification: The rostrum of this spider crab is bifurcated into two long slightly curved horns. The body is oval with a small spine at the anterior end of the carapace. The body is tan colored and the legs and chelipeds lightly mottled.
Natural History: This spider crab often decorates itself with a sparse cover of hydroids and other miscellaneous small items.
Distribution: Andaman Islands; Indonesia; Philippines; Fiji; Solomon Islands (Solomon Islands).

233

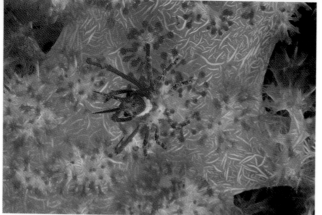

Marc Chamberlain

851. *Naxioides* sp.
Identification: The rostrum of this species is also bifurcated into two long horns. The coloration is striking, being primarily red. Some specimens have a bright yellow sickle-shaped mark on the carapace.
Natural History: Occurs on a variety of reef surfaces.
Distribution: Malaysia (Sipadan, Borneo).

Nicholas Galluzzi

852. *Schizophrys aspera*
(H. Milne Edwards, 1834)
Identification: This spider crab is easily identified by the series of large spines along the edge of the carapace and on the first two segments of the chelipeds. The carapace is covered with low tubercles. The color is reddish brown all over.
Natural History: Found to depths of 15 m.
Distribution: South and East Africa to Australia; Fiji; Malaysia; Japan; Hong Kong and Hawai'i (Sipadan, Borneo).

Carol Buchanan
Mike Severns

853. *Xenocarcinus conicus*
(H. Milne Edwards, 1934)
Identification: Another long triangular-bodied spider crab. The color varies greatly from silver-grey to bright red. There is usually a broad white stripe down the center of the carapace. There may be varying degrees of banding and mottling on the legs. The eyes are set in shallow notches at the edge of the carapace.
Natural History: Lives on black corals, hydroids and gorgonians.
Distribution: Australia; New Guinea; Indonesia; Japan (Byron Bay, New South Wales, Australia).

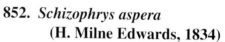

854. *Xenocarcinus tuberculatus* White, 1847
Identification: A distinctive spider crab, its smooth shiny carapace has several large broadly tapering spines. The rostrum is long and tapering. The color is brown, with light tips on the spines. The legs and chelipeds are lightly banded.
Natural History: Lives on black coral.
Distribution: South Africa to Australia; Indonesia; Hong Kong; Taiwan; China; Japan (Indonesia).

234

Latreillidae - Arrow Crabs

855. *Latreilla valida* Haan, 1839
Arrow Crab

Identification: An extemely long, legged arrow crab. The body is oval, with several large spines. Characteristically the legs are banded red and white, and the fifth legs are held erect over the body in the posture shown here.
Natural History: Depth: to 85 m.
Distribution: South Africa; Seychelles; India; Indonesia; Philippines (Sulawesi, Indonesia).

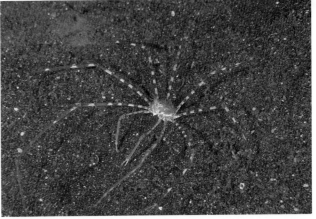
Mike Severns

856. *Eplumra phalangium* (Haan, 1839)
Arrow Crab

Identification: The carapace is triangular with a long tapering rostrum. A typical arrow crab the legs are about four times the length of the body. The color pattern on the carapace is a series of converging red and yellow stripes. The fifth legs end in hooked dactyls. Previously, *Latreilla phalangium* in older literature.
Natural History: Depth: to 85 m.
Distribution: Australia; Philippines; Japan (Philippines).

Portunidae - Swimming Crabs

Swimming crabs have broad carapaces which are often armed with long, sharp lateral spines. The fifth, or last pair of legs, are flattened to form a paddle for swimming and burrowing in the sand.

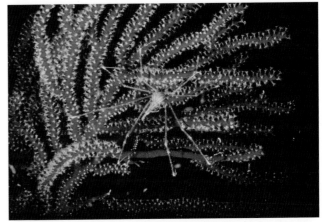
James Hargrove

857. *Charybdis erythrodactyla* (Lamarck, 1818)
Red Legged Swimming Crab

Identification: A colorful swimming crab, the carapace bears five large lateral spines, the anterior-most two separated by smaller spines, and six inter-orbital spines. The chelipeds bear large spines, similar to *C. hawaiiensis*. It has characteristic blue blotches on the carapace.
Natural History: Feeds at night. Depth: 5-10 m.
Distribution: Red Sea to Australia; Japan; Marshall, Tuamotu, Society, Hawaiian and Marquesas Islands (Hawai'i).

Mike Severns
Jack Randall

858. *Charybdis hawaiiensis* Edmondson, 1954
Identification: A tannish to red swimming crab with black tipped chelipeds and spines over the body. Specifically, there are five lateral spines on the carapace, eight inter-orbital spines and large sharp spines on the surface of the chelipeds. The tips of the pincers have a ridge of small spines down the posterior edge. The eyes are dark with green specks.
Natural History: On coral reef faces. Depth: 10-20 m.
Distribution: Hawaiian, Tuamotu and Society Islands (Oahu, Hawai'i).

235

Mike Severns

859. *Charybdis* sp.

Identification: The lateral spines are short, but sharp. The four antero-lateral spines are separated by a smaller spine. There are six inter-orbital spines, the medial two being longest. The carapace has several ridges, highlighted in white. All spines are tipped in white.

Natural History: Observed scurrying on reef flats and faces.

Distribution: Hawaiian Islands (Hawai'i).

Carol Buchanan

860. *Lissocarcinus laevis* Miers, 1886

Identification: Brightly patterned, white and brownish red. The legs are white. It has four inter-orbital spines. To 3 cm.

Natural History: Found living in association with sea anemones and soft corals.

Distribution: South Africa; Madagascar; Sri Lanka; Andaman Islands; Indonesia; Australia; Japan; Hawai'i (South Solitary Island, New South Wales, Australia).

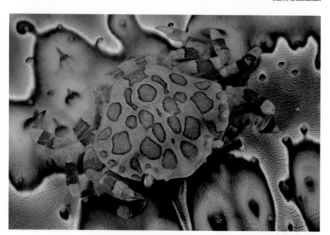

Mike Severns
Jack Randall

861. *Lissocarcinus orbicularis* Dana, 1852
Harlequin Crab

Identification: The color of this species is quite variable. The white and brown patterns may be reversed. The inter-orbital space is smooth, without spines. The lateral spines are weak. Length: 3-4 cm.

Natural History: A free swimming species, it is commensal on cucumbers and sea anemones.

Distribution: South Africa; Tanzania; Red Sea; Seychelles; Sri Lanka; Australia; Indonesia; New Guinea; Fiji; Japan; Hawai'i (Indonesia).

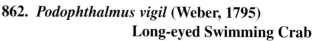

862. *Podophthalmus vigil* (Weber, 1795)
Long-eyed Swimming Crab

Identification: This swimming crab is easily identified by its long eyestalks. The carapace is angular, forming a single lateral spine. There is a series of small spines on the first segment of the chelipeds, with several larger spines at the joint. The color is tan with brown patterns and spotting.

Natural History: A mud bottom dweller, often attracted to a diver's light. Depth: to 70 m.

Distribution: South Africa; Red Sea to Australia; Indonesia; Samoa; Japan; Society Islands (Maumere, Flores, Indonesia).

863. *Portunus pelagicus* (Linnaeus, 1766)
Blue Swimming Crab

Identification: A typical swimming crab with its large sharp lateral spine, this species has eight smaller antero-lateral spines. There is a series of small, sharp spines on the chelipeds. It is olive green with lighter lines and mottling. The legs and tips of chelipeds are blue.

Natural History: Intertidal to 65 m, in mud and soft bottoms.

Distribution: South Africa; Tanzania to Australia; Philippines; northern New Zealand; Japan and Society Islands (Philippines).

Mike Miller

864. *Thalamita picta* Stimpson, 1858

Identification: This small swimming crab has a long slender lateral spine, eight antero-lateral spines and three inter-orbital spines. The chelipeds bear several sharp spines. The color is white, green and brown mottling with some red highlights.

Natural History: Occurs in coral rubble.

Distribution: South Africa; Tanzania; Seychelles; Madagascar; Red Sea to Australia; Philippines; Japan; Tuamotu and Hawaiian Islands (Batangas, Luzon, Philippines).

Mike Miller

865. *Thalamita* sp.

Identification: The lateral spines of this species are very long, slender, and slightly curved. There are seven small irregular antero-lateral spines. The body is tan with white and brown encrustations.

Natural History: A sand bottom species, it is most active at night.

Distribution: Philippines (Batangas, Luzon, Philippines).

Marc Chamberlain
D.W. Behrens

866. Portunid sp.

Identification: Its long legs and chelipeds give it the appearance of a spider crab. The hind legs are characteristically flattened into a paddle. The chelipeds are strongly spined, and have long slender pincers. The color varies from orange with white specks to brown, and has lighter bands on the legs.

Natural History: Very active at night, scurrying on sandy bottoms.

Distribution: Philippines (Batangas, Luzon, Philippines).

237

Xanthidae - Dark-Finger Coral Crabs

This group is referred to as coral crabs because of their association with reef corals. All members have black tipped claws, despite the variation in body forms. The species are variably colored. In general they have broad fan-shaped carapaces, and well-separated eyes. Several species live commensally on corals.

Mike Miller

▶ **867.** *Atergatis subdentatus* **Haan, 1835**
Identification: The broad domed carapace has an orange-peel texture. The entire body is orange in color with a dark orange spot in the center of the carapace.
Natural History: Found on reef flats and faces living in coral crevices.
Distribution: Philippines and Japan (Philippines).

868. *Carpilius maculatus* **(Linnaeus, 1758)**
Identification: The carapace is relatively smooth. There are four blunt inter-orbital spines. The brownish grey body has a series of large brown spots.
Natural History: Slow and clumsy, it inhabits rocky areas. Depth: 3-35 m.
Distribution: South Africa; Red Sea to Australia; Philippines; Malaysia; Taiwan; Japan; Society Islands; Hawai'i (Sipadan, Borneo).

Marc Chamberlain

869. *Etisus dentatus* **Herbst, 1785**
Identification: This large red crab has large hooked chelipeds. There is a series of 10-12 irregularly sized and spaced lateral spines. The legs are covered with hair. Length: to 18 cm.
Natural History: The large chelipeds are used to break open mollusk shells. Depth: to 15 m.
Distribution: Red Sea to Indonesia; Japan; Marshall Islands; Society, Tuamotu and Hawaiian Islands (Bali, Indonesia).

Jack Randall
Jack Randall

870. *Etisus sp.*
Identification: A large brown species. The carapace bears seven strong lateral spines. The black-tipped chelipeds are also heavily spined, as are the walking legs, which bear rows of slender sharp spines.
Natural History: Found in coral rubble.
Distribution: Marshall Islands (Enewetak, Marshall Islands).

871. *Lybia edmondsoni*
Takeda & Miyake, 1970
Pompom Crab

Identification: A delicate crab, it gets its common name from the anemones it carries in its claws, resembling two pompoms. There are two tufts of cirri on the head and carapace. The body is salmon with reticulating white lines. The legs are tan with thin maroon bands interspersed with white specks. The antennae are maroon and the eyes are red and white striped. Length: 1-2 cm.

Natural History: Carries anemones *Bunodeopsis* sp. (#178) in its claws, which it waves at predators in defense.

Distribution: Endemic to the Hawaiian Islands (Hawai'i).

Mike Severns

872. *Lybia tessellata* (Latreille, 1812)
Boxer Crab

Identification: It has a distinctive brown and cream checkered pattern on its carapace. The pattern is outlined in dark brown. This species also has small tufts of cirri on the carapace. The legs are colored similarly to *Lybia edmondsoni*. Length: 1-2 cm.

Natural History: Like *Lybia edmondsoni*, it defends itself by carrying small anemones *Bunodeopsis* sp. (#178), in its claws.

Distribution: Mozambique; Seychelles; New Guinea; Indonesia; Philippines; Taiwan; Marshall Islands (Batangas, Luzon, Philippines).

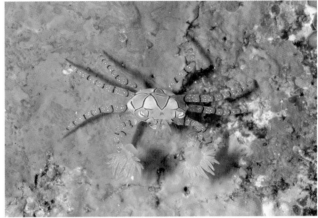
Leslie Newman & Andrew Flowers

873. *Lophozozymus incisus*
(H. Milne Edwards, 1834)

Identification: The carapace and chelipeds of this species are lumpy in appearance. There are dense cirri along the sutures of the body and claws. The carapace has a series of low lateral spines. The inter-orbital space is free of spines. The body is creamy orange with brown and pink patches. The eye is almost human-like.

Natural History: Occurs in crevices on reef slopes.

Distribution: Australia; Indonesia; Philippines; Japan; Hawai'i (Batangas, Luzon, Philippines).

Mike Miller
Leslie Newman & Andrew Flowers

874. *Neoliomera insularis*
(Adams & White, 1849)

Identification: Bright red xanthid with smooth carapace. The lateral spines are low, giving the carapace an angular appearance.

Natural History: Rubble areas to 10 m depth.

Distribution: Australia; New Caledonia; New Guinea; Philippines; Japan (Madang, Papua New Guinea).

Mike Severns

875. *Polydectus cupulifer* (Latreille, 1825)

Identification: Although a xanthid, not a majid or dromiid, this species utilizes other animal material to camouflage its body. The crab appears fuzzy, adorning itself with living yellow anemones.

Natural History: Found under rubble to depths of 15 m.

Distribution: Madagascar; Mauritius; Red Sea; Singapore; Japan; Hawai'i; Tuamotu Islands (Philippines).

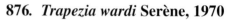

Mike Severns

876. *Trapezia wardi* Serène, 1970

Identification: An unmistakable coral crab, white with red spots.

Natural History: Lives symbiotically with *Pocillopora* corals.

Distribution: Red Sea; Maldives; Sri Lanka; Vietnam; Japan; Samoa; Hawai'i (Maui, Hawaiian Islands).

G.C. Williams
Mike Miller

877. *Xanthias maculatus* Sakai, 1961

Identification: This species is readily distinguished by the bright purple to white spots on it entire body. The spots are outlined with a reddish-brown line. The dome shaped carapace is free of spines.

Natural History: Depth: to 30 m.

Distribution: Previously known only from Japan; recorded here from South Africa (Natal, South Africa).

878. *Zosymus aeneus* (Linnaeus, 1758)

Identification: A white xanthid with dark brown irregularly shaped markings over its body. The few lateral spines are reduced to smooth bumps.

Natural History: This species is poisonous if eaten. Found in shallow water.

Distribution: South Africa; Red Sea; Andaman Islands; New Caledonia; Samoa; Japan; Hawai'i; Tuamotu and Marquesas Islands (Japan).

879. Unidentified Xanthid sp.

Identification: This white and brown mottled xanthid bears 4-5 moderate lateral spines. The carapace is very wide. The legs are banded white and brown.
Natural History: Found under coral rubble.
Distribution: Australia (Heron Island, Australia).

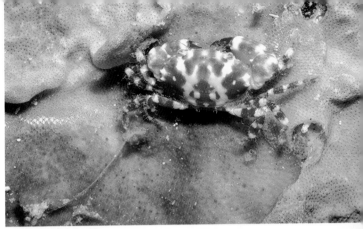
Leslie Newman & Andrew Flowers

Eumedonidae - Urchin Crabs

880. *Zebrida adamsii* White, 1847
Urchin Crab

Identification: White and dark brown striped and banded, the carapace forms four large, forward-facing tapered plates. The chelipeds also, have several angular plates or spines. Each leg bears several spines. The last segment of the leg forms a hook to hold onto sea urchin spines. To 25 mm.
Natural History: Found in association with several species of sea urchins. Depth: 5-10 m.
Distribution: Sri Lanka; India; Thailand; Australia; Indonesia (Indonesia).

Mike Severns

Parthenopidae - Elbow Crabs

881. *Harrovia elegans* Man, 1887

Identification: Having an angular carapace, the rostrum is bifurcated, forming two horns. The body is black with yellow stripes and outline.
Natural History: Found on the crinoids *Comaster* and *Comanthus*. Depth: to 10 m.
Distribution: Somalia; Pakistan; Singapore; Indonesia; Philippines; Taiwan; Korea; Japan; Marshall Islands (Batangas, Luzon, Philippines).

Mike Miller
Lynn Funkhouser

882. *Lambrus* sp.

Identification: The carapace is triangular and the chelipeds are long, broad, flattened and bear several rows of large spines. This crab is distinctive because of the large tubercles covering the body and the transparent hairs along the body margins.
Natural History: A sandy substrate inhabitant.
Distribution: Philippines (Philippines).

241

Mike Severns

Nicholas Galluzzi

Grapsidae - Shore Crabs

883. *Plagusia depressa* (Lamarck, 1818)
Identification: A typical grapsid shore crab, it has a flattened body and long flattened legs. The eyes are large, notched into the front of the carapace. The legs have a row of spines along the front edge. Length: 4 cm.
Natural History: Occurs intertidally to subtidally. Fast moving, it scurries around rapidly from rock to rock.
Distribution: South Africa; Red Sea; Australia; Japan; China; Hawai'i (Hawai'i).

884. *Percnon planissimum* (Herbst, 1804)
Flat Rock Crab
Identification: With its highly flattened body and distinctive coloration, this crab is easily identified. In males the chelipeds are red, the legs have a yellow line, the eyes and antennae are yellow, and the frontal region bears a horizontal white line. Color in females may be a less dramatic mottled or dark olive green. Always has a line down the center of the carapace and spines on the front edge of the legs. Length: to 8 cm.
Natural History: Extremely fast, its flattened body makes it perfectly adapted for scurrying in rock crevices and under sea urchins.
Distribution: South Africa; Red Sea; Sri Lanka; Andaman Islands to Australia; New Zealand; Japan; Hawai'i; Society Islands (Kona, Hawai'i).

Lophophorates:
Phoronida - Phoronids
Brachiopoda - Lamp Shells
Bryozoa - Moss Animals

These three groups of animals are placed together since they have a ring of unbranched tentacles surrounding the mouth. The tentacles are used for suspension feeding and removal of plankton

David Reid

from the water column. Generally, the body is divided into two or three distict regions.

Phoronida - Phoronid Worms

Phoronids are a small group of lophophorates with 17 species in two genera known worldwide. All species have an elongate worm-like body devoid of segments and a U-shaped digestive tract. Most species live in tubes and many reproduce asexually as well as sexually.

885. *Phoronis australis* Haswell, 1883
Identification: This is one of the larger species of phoronids, with a lophophore consisting of two distinct spirals. The lophophore is black or white in color.
Natural History: This species is found in association with several species of cerianthid tube anemones. The body of the phoronid penetrates the tissue of the cerianthid, but is not parasitic as no nutrients are obtained from the anemone directly. These phoronids are hermaphroditic and are found from the intertidal zone to 36 m depth.
Distribution: Circumtropical and warm temperate: Madagascar to India; western Australia and Queensland; Philippines; China and Japan; Canary Islands and the Mediterranean (Batangas, Luzon, Philippines).

Brachiopoda - Lamp Shells

Brachiopods were once thought to be mollusks, owing to the fact that they have a bivalved shell. Unlike molluscan bivalves, the shells of brachiopods are dorsal and ventral rather than lateral to the body. Inside the brachiopod valves is a large lophophore for filter feeding. Brachiopods were much more diverse prior to the Cretaceous Period. The greatest diversity of living species is found in cold temperate and polar waters. Few species are known to inhabit the Indo-Pacific tropics.

886. *Frenulina sanguinolenta* (Gmelin, 1817) ▶

Identification: This species can be recognized by its alternating red and yellowish zigzag markings and fine punctate sculpture. *Frenulina cruenta* Cooper, 1973, from the western Indian Ocean is likely a synonym.
Natural History: This species is found attached by its peduncle to the undersides of dead coral heads and rubble in areas of strong current, such as passes in barrier reefs. Depth: 3-40 m.
Distribution: Mozambique; Madagascar; Tanzania to Australia; New Caledonia; New Guinea; Indonesia to the Hawaiian and Society Islands (Madang, Papua New Guinea).

T. M. Gosliner

Bryozoa - Moss Animals

Bryozoans are colonial animals composed of numerous small individuals called zooids. Marine bryozoans are divided into three major groups, cyclostomes, ctenostomes and cheilostomes. In many ctenostomes the zooids are born, on chitinous tubes. This group is well represented in temperate environments while tropical members are less conspicuous. Cyclostomes have thickly calcified skeletons and relatively small colonies. The openings for the zooids lack an operculum. The majority of species are cheilostomes, which are also calcareous, but have an operculum closing the aperture of the zooids.

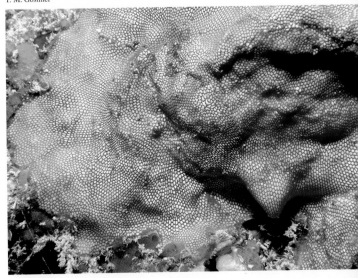
Leslie Newman & Andrew Flowers
T. M. Gosliner

887. *Alcyonidium* sp. ↗

Identification: This species has widely separated zooids with minute openings. The white pigment is present on only some of the zooids and is solid rather than spotted.
Natural History: Found on rock and reef surfaces in the open, in shallow water.
Distribution: Australia (Great Barrier Reef, Australia)

888. *Bantariella verticillata* (Heller, 1867) ▶

Identification: These ctenostomatous bryozoans form small or massive colonies composed of chitinous tubes bearing zooids. This species forms stellate aggregations of zooids that are separated from each other by a thin tube which contains a strand of tissue. Seen here with the nudibranch *Thorunna daniellae*.
Natural History: This species is found under small bits of dead coral on shallow patch reefs.
Distribution: Tropical and warm temperate: New Guinea; Bonin Islands and Indonesia; temperate western and eastern Atlantic (Madang, Papua New Guinea).

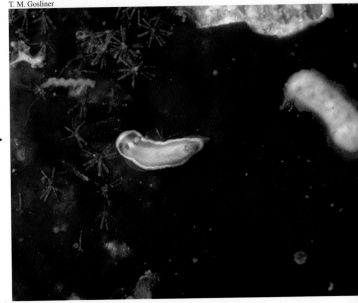

243

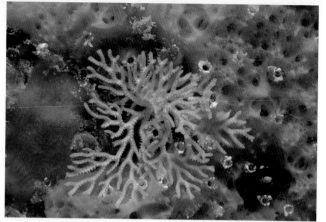

Marc Chamberlain

889. cf. *Idmidronea* sp.

Identification: This is a fairly large, highly-branched cyclostome with whitish colonies.

Natural History: Found on the surface of living reef among sponges and corals.

Distribution: Hawaiian Islands (Kona, Hawai'i, Hawaiian Islands).

T. M. Gosliner

890. *Lichenopora* sp.

Identification: *Lichenopora* species are small cyclostomes with round whitish or bluish colonies with tubular zooids. A colony can be seen on either side of the *Elysia* sp., a sacoglossan opisthobranch.

Natural History: Colonies of *Lichenopora* can be found on the undersides of coral rubble on shallow patch reefs.

Distribution: New Guinea (Madang, Papua New Guinea).

Dave Zoutendyk
Leslie Newman & Andrew Flowers

891. *Iodictyum axillare* (Ortman, 1889)

Identification: Species of *Iodictyum* can be generally distinguished by the reddish purple pigment of the colony. They are largely restricted to the western Pacific. The generic name comes from their resemblence to the color of iodine. They are members of a group of bryozoans, the Phidoloporidae, which have the colonies organized in a reticulated network. *Iodictyum axillare* can be distinguished by its relatively open colonies which are pale red or purple in color.

Natural History: This species is found associated with living corals on outer barrier reefs and walls where there is strong current. Commensal zancleid hydroids are visible as white dots scattered around the colony.

Distribution: Pohnpei; Indonesia; Japan (Sapwauhfik, Pohnpei).

892. *Iodictyum buchneri* Harmer, 1934

Identification: This species is characterized by its mottled coloration and open, cup-shaped colony growth.

Natural History: Found on living reefs in relatively shallow water.

Distribution: Australia; Philippines and southern China (Lizard Island, Great Barrier Reef, Australia).

893. *Iodictyum sanguineum* (Ortman, 1889)

Identification: This species is deep red on the top side and white underneath. It appears to be more thickly calcified than other species.

Natural History: Found on the surface of living reefs and walls where there is strong current.

Distribution: Australia; Indonesia and Japan (Heron Island, Great Barrier Reef, Australia).

Leslie Newman & Andrew Flowers

894. *Reteporella graeffei* (Kirchenpauer, 1869)

Identification: This species is thickly calcified, but delicate in construction. Unlike species of *Iodictyum*, it is white in color and has few, if any, interconnections between branches.

Natural History: Found on outer sides of barrier reefs in areas of relatively strong current.

Distribution: Sri Lanka to Australia; Indonesia; Philippines; Japan; Gilbert Islands and Tonga (Lizard Island, Great Barrier Reef, Australia).

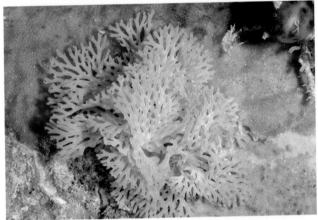
Leslie Newman & Andrew Flowers

895. *Reteporellina denticulata* (Busk, 1884)

Identification: This species is similar to the preceding one but is coarser, with shorter, thicker branches.

Natural History: Found on the surface of living reefs among sponges and corals.

Distribution: Tanzania and Seychelles; Australia; Tonga; Japan and Hawaiian Islands (Heron Island, Great Barrier Reef, Australia).

Leslie Newman & Andrew Flowers
Jack Randall

896. *Reteporellina / Reteporella* sp.

Identification: This species forms large yellowish colonies with a few perpendicular connections between branches.

Natural History: It is found in areas with reasonably strong current from 10-20 m depth.

Distribution: Australia and Hawaiian Islands (Maui, Hawaiian Islands).

897. *Rhynchozoon larreyi* (Audouin, 1826)

Identification: This species forms a large crustose colony which may exceed 10 cm in diameter.

Natural History: Found in the open on rock walls in 10-20 m of depth. Commensal zancleid hydroids are visible as white polyps emerging from the bryozoan colony.

Distribution: Red Sea; Sri Lanki; Australia; Indonesia; Mediterranean (Batangas, Luzon, Philippines).

898. *Triphyllozoon inornatum* Harmer, 1934

Identification: There are about 15 species of *Triphyllozoon* described from the Indo-Pacific tropics. Most are poorly differentiated and difficult to determine from living animals.

Natural History: Species of *Triphyllozoon* are found on walls in areas which are exposed to strong currents, from the intertidal zone to more than 40 m depth. This specimen bears hydroid polyps with capitate tentacles. Zancleid hydroids frequently grow in a commensal relationship with tropical phidoloporid and celleporariid bryozoans.

Distribution: Western Pacific: New Guinea; Indonesia and Philippines (Batangas, Luzon, Philippines).

899. *Bugula dentata* (Lamoroux, 1816)

Identification: This is another delicately branching species with minute spines along the colony.

Natural History: This species is found on walls on outer barrier reefs in 15-30 m depth. It is preyed upon by several polyceratid nudibranchs, including *Roboastra rubropapillosa,* shown here.

Distribution: South Africa to Australia; New Guinea; Indonesia; Singapore; Japan; Atlantic: Cape Verde Islands; Madeira (Madang, Papua New Guinea).

900. *Canda* sp.

Identification: This is a delicately branched species with many small cross-connections between adjacent branches.

Natural History: It is shown here on an outer reef, growing with an encrusting bryozoan and a didemnid tunicate.

Distribution: Malaysia (Sipadan Island, Borneo).

901. *Scrupocellaria* sp.

Identification: This brown species is highly branched with flattened blades and is only lightly calcified.

Natural History: Found on walls in 20-30 m of water in areas of moderate current. This specimen has small epiphytic spirorbid polychaetes growing on the colony.

Distribution: Philippines (Batangas, Luzon, Philippines).

T. M. Gosliner

902. *Tricellaria / Scrupocellaria* sp.

Identification: This is a large bushy species that characteristically has curved ends to the branches.

Natural History: Found on rocky walls from 10-30 m depth in areas of strong current.

Distribution: Philippines (Batangas, Luzon, Philippines).

T. M. Gosliner

903. *Tropidozoum cellariiforme* Harmer, 1957

Identification: Colonies of this species are bright red in color and are evenly dichotomously branched. The zooid blades are flattened and zooids emerge from the edges of each side of the branch.

Natural History: This species is found on walls of reef margins and drop offs. The bryozoan is preyed upon by the undescribed nudibranch *Okenia* sp. (#604).

Distribution: This species has been found from Indonesia; Philippines and Marshall Islands (Balacasag Reef, Bohol, Philippines).

T. M. Gosliner

T. M. Gosliner

904. *Celleporaria sibogae* Winston & Heimberg, 1986

Identification: This recently described bryozoan can be easily recognized in the living state. The colonies have brown, crustose zooids.

Natural History: Found forming a crustose colony on flat rocky surfaces in areas with strong current in 10-30 m depth. As in species of *Iodictyum* (#891), *Rhynchozoon* (#897) and *Triphyllozoon* (#898), this species has elongate polyps of zancleid hydroids growing commensally. The bryozoan colony actually encloses the hydroid with calcification.

Distribution: Northern Australia; New Guinea; Indonesia; Malaysia; Philippines; Belau; Chuuk (Batangas, Luzon, Philippines).

247

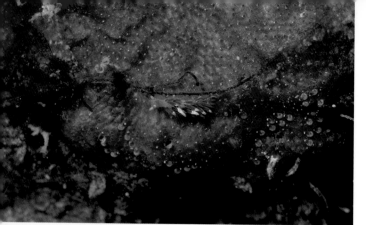

Leslie Newman & Andrew Flowers
T. M. Gosliner

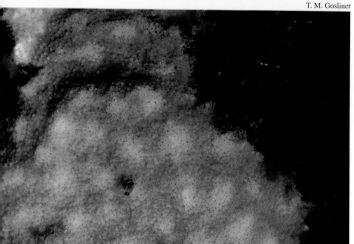

T. M. Gosliner

905. cf. *Calyptotheca* sp.

Identification: This species forms a thin calcareous crust. The zooids are ovoid with minute papillae on their surface.

Natural History: This species is found on the under surface of coral rubble and dead coral heads. The swollen white brood chambers of the ovicells are visible on individuals at the lower end of the colony. The undescribed nudibranch, *Cuthona* sp. is crawling on the colony.

Distribution: New Guinea (Madang, Papua New Guinea)

906. Cheilostome sp. 1

Identification: This species has orange zooecia with scattered opaque white pigment on the outer zooecial surface.

Natural History: Found growing on the upper and lower surfaces of rocks and coral on shallow reefs.

Distribution: Northern Australia and New Guinea (Lizard Island, Great Barrier Reef, Australia).

907. Cheilostome sp. 2

Identification: This species is yellow with regular patches devoid of pigment.

Natural History: Found on the underside of rocks and dead coral heads on shallow patch reefs.

Distribution: Philippines (Batangas, Luzon, Philippines).

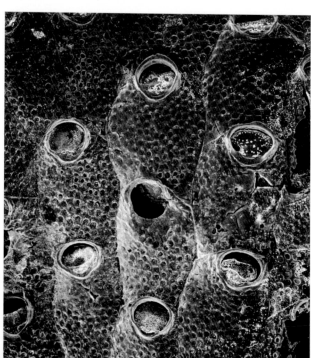

Echinodermata - Feather Stars, Sea Stars, Brittle Stars, Sea Urchins, and Sea Cucumbers

Crinoidea - Crinoids or Feather Stars

A considerable diversity exists among crinoids; several species often occurring together. Best observed at night, they are plankton feeders. The group on the whole is difficult to identify without close inspections of both the pinnules and cirri. See Zmarzly (1984, 1985) concerning the ecology, distribution and zoogeography of the central Pacific members of this group.

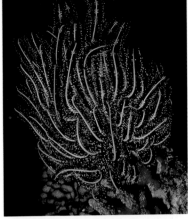

Charlie Arneson

Juvenile

908. *Comantheria briareus* (Bell, 1882)

Identification: This is a large crinoid, which appears as a dense mass of arms and lacks cirri. The color of this species is consistent. Black pinnules with white tips, alternate with white pinnules. Yellow is also present on the pinnules.
Natural History: Found on living and dead coral projections during both day and night.
Distribution: Australia; Indonesia; Philippines; Japan; Belau and Marshall Islands (Kwajalein, Marshall Islands).

G.C. Williams

909. *Comanthina schlegeli* (Carpenter, 1881)

Identification: This species is variable in its color, mostly ranging from yellow to orange and brown to black. The average number of arms is 130.
Natural History: Commonly found on live and dead coral, often under or on the sides of ledges. It is active during the day and night.
Distribution: Maldives; western Australia; Solomon Islands; New Guinea; Indonesia; Philippines; Marshall Islands (Madang, Papua New Guinea).

910. *Comaster gracilis* (Harlaub, 1890)

Identification: This and the following species lack cirri. Individuals are solid reddish to dark brown. *Comaster multifidus* has arms that are three times the length of those of *C. gracilis*.
Natural History: This species is found under rubble during the day and emerges onto low relief in areas with strong current or is found in more exposed areas in lagoons where current is weak.
Distribution: Maldives; Bay of Bengal; Indonesia; Japan; Gilbert and Marshall Islands (Kwajalein, Marshall Islands).

Charlie Arneson
Charlie Arneson

911. *Comaster multifidus* (Müller, 1841)

Identification: Comaster multifidus is orange to orange-brown with more elongate arms than the preceding species.
Natural History: Partially emergent from crevices during both day and night.
Distribution: Australia; Indonesia; Philippines and Marshall Islands (Kwajalein, Marshall Islands).

249

Charlie Arneson

912. *Comissa* cf. *pectinifera* (Clark, 1911)

Identification: Specimens from the Marshall Islands have more elongate arms than specimens from other localities. On this basis, their systematic status remains uncertain. Orange-brown or dark brown with light green.

Natural History: Nocturnal on living reefs, where it occupies living coral heads such as *Lobophyton* and only partially extends its arms.

Distribution: Maldives; Christmas Island; Indonesia; and possibly the Marshall Islands (Kwajalein, Marshall Islands).

Debbie Zmarzly

913. *Lamprometra palmata* (Müller, 1841)

Identification: In this and related species in the Mariametridae, the mouth is located near the center rather than peripherally. Species in this family are virtually imposible to distinguish without examining specimens.

Natural History: A nocturnal species which emerges from coral rubble to feed at night. Members of the Mariametridae are largely nocturnal.

Distribution: India; Maldives; Marshall and Hawaiian Islands (Kwajalein, Marshall Islands).

914. *Lamprometra* sp.

Identification: This species is largely reddish with a bit of white and yellow. It appears to have thicker cirri than the preceding species.

Natural History: Found on living coral heads at night.

Distribution: New Guinea (Madang, Papua New Guinea).

Leslie Newman & Andrew Flowers

Charlie Arneson

915. *Liparometra regalis* (Carpenter, 1888)

Identification: This species has about 40 arms, which are reddish with white bands or white with reddish bands arranged in an inverted "V."

Natural History: Strictly nocturnal, where it occupies exposed coral perches, but is completely hidden during the day.

Distribution: Only Tonga and Marshall Islands (Kwajalein, Marshall Islands).

916. *Oxycomanthus bennetti* (Müller, 1841)

Identification: This extremely common species has an average of 100 arms and is yellow, orange-brown, brown or black. It has longer (2-3x) basal attachment cirri than the preceding species.

Natural History: *Oxycomanthus bennetti* inhabits live and dead coral faces, well elevated above the reef surface, in relatively shallow water, where strong currents prevail. It is active during the day and at night.

Distribution: Bay of Bengal to Philippines and Marshall Islands (Batangas, Luzon, Philippines).

Mike Miller

917. cf. *Oxymetra* sp.

Identification: This species of Mariametridae cannot be positively identified to genus or species. It is strikingly colored.

Natural History: Found on living coral heads in shallow water at night.

Distribution: Philippines (Batangas, Luzon, Philippines).

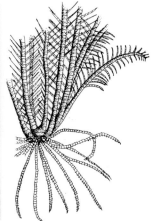
Mike Miller

918. *Stephanometra* cf. *indica* (Smith, 1876)

Identification: This species has alternating maroon and white markings. It is possibly *Stephanometra indica*, but its identity cannot be certain in the absence of a specimen.

Natural History: Found on living coral reefs where it is emergent at night.

Distribution: East Africa; Red Sea to Fiji and Marshall Islands (Kwajalein, Marshall Islands).

Charlie Arneson
Leslie Newman & Andrew Flowers

919. *Petasometra clarae* (Hartlaub, 1890)

Identification: This species has 10-31 arms with elongate, widely spaced pinnules. It is brownish with darker pigment.

Natural History: Found partially exposed on shallow, living reefs during the day.

Distribution: Australia; New Guinea and Indonesia (Madang, Papua New Guinea).

251

Charlie Arneson

920. Crinoid sp.

Identification: A beautiful species with orange pinnules and purple ribs and cirri. Neither the genus nor species of this individual is known.

Natural History: Nocturnal where it sits on living coral heads.

Distribution: Marshall Islands (Kwajalein, Marshall Islands).

Asteroidea - Sea Stars or Starfish

The sea stars are a large and diverse group of echinoderms. Most are predatory on a wide variety of other organisms. Many are specialists on particular groups and others are opportunistic scavengers and grazers. The most notorious asteroid is the crown-of-thorns, a voracious predator of hard corals.

T. M. Gosliner

921. *Luidia maculata* Müller & Troschel, 1842

Identification: This species is a member of a group that can be distinguished by having 7-9 arms which are elongate and slender. It has a reddish brown mottling on the top surface.

Natural History: *Luidia maculata* is found in relatively shallow water to 90 m depth. It inhabits sandy substrates and is active nocturnally.

Distribution: Mozambique; East Africa; Red Sea to Australia; Philippines; southern Japan (Batangas, Luzon, Philippines).

922. *Luidia magnifica* Fisher, 1906

Identification: This species is similar to the preceding one but has 9-10 rays. It also differs in details of the spination of the body. This species may be synonymous with *L. mauritiensis* from Mauritius.

Natural History: This sea star is nocturnal and inhabits coarse sandy slopes and flats in 18-133 m depth.

Distribution: Philippines and Hawaiian Islands (Maui, Hawaiian Islands).

Mike Severns
Marc Chamberlain

923. *Astropecten polyacanthus* Müller & Troshel, 1842

Identification: This brown sea star, with prominent marginal plates, can be recognized by elongate, white marginal spines along the lateral edges of each ray. Members of this family have pointed tube feet without a terminal sucker, an adaptation for living in soft substrate.

Natural History: This species inhabits sandy bottoms from 1-40 m depth. It feeds primarily upon bivalves.

Distribution: South Africa; Red Sea to New Caledona; Indonesia; Phlippines to Hawaiian Islands (Batangas, Luzon, Philippines).

924. *Astropecten* sp. 1

Identification: This species could not be positively identified. It has long sharp spines and appears to be similar to *Astropecten andersoni.*

Natural History: This species inhabits sandy areas in relatively shallow water and is active nocturnally.

Distribution: Indonesia (Sulawesi, Indonesia).

Mike Severns

925. *Astropecten* sp. 2

Identification: This sea star has relatively thin spines with a pair of strong spines at the tip of each ray.

Natural History: This species was dredged off the coast of South Africa in 52 m depth.

Distribution: South Africa (Sodwana Bay, Natal, South Africa).

G.C. Williams

926. *Iconaster longimanus* (Möbius, 1859)

Identification: *Iconaster longimanus* has large marginal plates along the rays with smaller plates in the region of the disk. It is one of the more brightly colored species of the family Goniasteridae.

Natural History: This species is found on deep reefs and slopes in association with coral rubble at 30-85 m depth.

Distribution: Saudi Arabia to Australia; Indonesia; and Philippines (Manado, Sulawesi, Indonesia).

Mike Severns
G.C. Williams

927. *Stellaster childreni* (Gray, 1840)

Identification: *Stellaster childreni* has long slender arms and a margin with large, prominent ossicles.

Natural History: It has been trawled in moderately shallow habitats on the outer edges of coral reefs, in relatively soft substrate.

Distribution: South Africa; Mozambique; Red Sea to Australia; Indonesia; Philippines; southern Japan (Sodwana Bay, Natal, South Africa).

253

928. *Pentagonaster duebeni* Gray, 1847

Identification: This species can be immediately recognized by its white surface with large red marginal and aboral plates.

Natural History: This species inhabits shallow temperate and subtropical rocky reefs and appears to feed upon sponges and bryozoans.

Distribution: Western Australia: New South Wales and southern Queensland (Coral Sea, Queensland, Australia).

Alex Kerstitch

929. *Tosia queenslandensis* Livingstone, 1932

Identification: The bright red-orange body color, small size (30 mm diameter) and large marginal plates clearly distinguish this species.

Natural History: This sea star is found on shallow water living reefs. Depth: 5-20 m.

Distribution: Australia and New Caledonia (Lizard Island, Great Barrier Reef, Australia).

Leslie Newman & Andrew Flowers

930. *Bothriaster primigenius* Döderlein, 1916

Identification: This genus and species were recently considered to represent unidentifiable juveniles. The present specimen closely resembles the type material described by Döderlein. The fine pentagonal outline on the disk, with fine radiating lines, appears distinctive.

Natural History: This species has been found in shallow water on patch reefs. Depth: 2-45 m.

Distribution: Described from Indonesia and now found in the Philippines (Batangas, Luzon, Philippines).

G.C. Williams
Roy Eisenhardt

931. *Choriaster granulatus* Lütken, 1869

Identification: This is a large species of sea star with broad rays, usually pinkish gray with orange or rust mottling in the central region.

Natural History: *Choriaster granulatus* is a common resident of shallow, living reefs. *Choriaster* appears to be a scavenger on dead animal material. Depths: 5-40 m.

Distribution: East Africa; Red Sea to Australia; Fiji; Solomon Islands; New Caledonia; New Guinea; Philippines and Guam (Solomon Islands).

932. *Culcita coriacea* Müller & Troschel, 1842

Identification: This is one of three species of cushion stars, all of which are found in the Indo-Pacific tropics. It is characterized by small pores over most of the aboral and marginal areas with a few small spines.

Natural History: Like other species of *Culcita*, *C. coriacea* is found on living coral reefs and gains at least part of its nutrition from coral polyps, as well as algae.

Distribution: Known only from the Red Sea (Oman).

Jonathan Mee

933. *Culcita novaeguineae*
Müller & Troschel, 1842

Identification: This is the most widespread and common species of cushion star. It is extremely variable in its coloration and ranges from red to green and brown.

Natural History: Like the preceding species, *C. novaeguineae* is a common inhabitant of shallow patch, barrier and fringing reefs. The commensal shrimp, *Periclimenes soror* (#741) is commonly found on the oral or aboral surface of this species.

Distribution: Andaman Islands; western Pacific to Hawaiian and Society Islands (Hawai'i).

Andy Sallmon

934. *Culcita schmideliana* (Retzius, 1805)

Identification: The specimen of this cushion star is a juvenile. Juveniles are characterized by having a more stellate shape and larger marginal plates. They have often been described as distinct species in different genera, owing to their difference from the adults.

Natural History: Juveniles and adults of this species are commonly found on shallow water reefs and reef flats. It is a predator on scleractinean corals, octocorals, sponges and ascidians.

Distribution: Madagascar; Reunion Island; Mauritius and Seychelles (Nosy Bé, Madagascar).

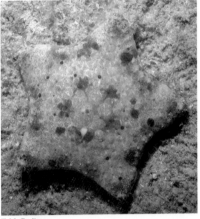
T. M. Gosliner
G.C. Williams

935. *Halityle regularis* Fisher, 1913

Identification: The red body color with triangular plates covering the aboral surface of the animal are distinctive.

Natural History: Little is known about the natural history of this species. The specimen shown here has a pair of parasitic eulimid snails partially embedded in its aboral surface. Depth: 3-90 m.

Distribution: South Africa; Madagascar to western Australia; Philippines and New Caledonia (Sodwana Bay, Natal, South Africa).

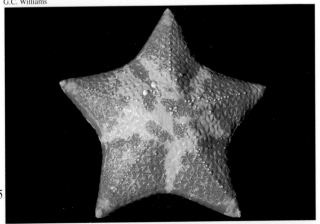

255

T. M. Gosliner

936. *Monachaster sanderi* (Meissner, 1892)

Identification: This species can be recognized by its small body size and gray body color with large marginal plates.

Natural History: This sea star is found on shallow patch reefs and coral rubble habitats. Depth: to 68 m.

Distribution: Mozambique; Madagascar; Tanzania to the Gulf of Suez (Nosy Bé, Madagascar).

Mike Severns

937. *Pentaceraster cumingi* (Gray, 1840)

Identification: Pentaceraster cumingi has a broad body with thick, heavy rays. The body color is reddish orange with scattered, conical spines on the aboral surface.

Natural History: This species is found from shallow to moderately deep water, usually on rocky or coral reefs. It presumably feeds upon sponges. Depth: 5-183 m.

Distribution: Hawaiian Islands to Gulf of California; Peru and Galapágos Islands (Maui, Hawaiian Islands).

T. M. Gosliner
T. M. Gosliner

938. *Pentaceraster alveolatus* (Perrier, 1875)

Identification: This species can be identified by the elongate marginal spines which are longer and more well developed at the apices of the arms. Large, isolated conical spines are present on the aboral surface.

Natural History: Pentaceraster alveolatus is found on shallow grass flats on reef platforms. Depth: 1-25 m.

Distribution: Western Australia; New Caledonia; Indonesia; Melanesia and Philippines (Bohol, Philippines).

939. *Pentaster obtusatus*
(Bory de St. Vincent, 1827)

Identification: The dense rounded tubercles of this species distinguish it from other members of this family. The body is white with numerous red tubercles.

Natural History: Pentaster obtusatus is found on grass flats in shallow water, where it presumably feeds on sponges.

Distribution: Western Pacific: Indonesia; New Britain; Philippines and Okinawa (Bohol, Philippines).

940. *Pentaster tyloderma* Fisher, 1913

Identification: This species has been recently considered to be a synonym of the preceding species. However, the smaller body size, different color and larger, sparser tubercles clearly distinguish this species.

Natural History: Pentaster tyloderma is found in shallow, silty habitats, often with seagrasses or algae.

Distribution: Known only from Philippines (Batangas, Luzon, Philippines).

T. M. Gosliner

941. *Protoreaster lincki* (Blainville, 1830)

Identification: The gray body color with brilliant, sharp red spines is distinctive for this species.

Natural History: This species is an inhabitant of shallow sandy habitats and grass beds. It probably feeds on sponges.

Distribution: Mozambique; East Africa; Madagascar to western Australia (Mombasa, Kenya).

Jack Randall

942. *Protoreaster nodosus* (Linnaeus, 1758)

Identification: This species is similar to the preceding one, but has fewer, thicker spines that are black rather than red.

Natural History: Protoreaster nodosus can be a common inhabitant of shallow grass beds and other sandy habitats. It probably feeds on sponges. Depth: 1-30 m.

Distribution: Seychelles to Australia; New Guinea; New Caledonia; Philippines; Belau and Japan (Bohol, Philippines).

T. M. Gosliner
G.C. Williams

943. *Ferdina sadhensis* Marsh & Campbell, 1991

Identification: This species resembles a species of *Fromia*, but has more robust and shorter rays. This species was described from five specimens from Oman. This represents the first record of this species since its original identification.

Natural History: This species was trawled from relatively shallow water on living reefs. Depth: to 18 m.

Distribution: Oman in the Red Sea; Southeastern Africa (Sodwana Bay, Natal, South Africa).

944. *Fromia elegans* Clark, 1921

Identification: About a dozen species of *Fromia* have been recognized from the Indo-Pacific tropics. This species can be identified by its aboral plates which vary in size. This specimen is abnormal with four rather than five rays.

Natural History: This sea star is found on shallow reefs in the open. Some species of *Fromia* have been observed to feed upon sponges, though more information about the generality of this feeding behavior is needed.

Distribution: Australia; Solomon Islands; Indonesia and Philippines (Solomon Islands).

Laura Losito

945. *Fromia ghardaqana* Mortensen, 1938

Identification: The yellow and red beaded aboral surface is distinctive.

Natural History: This species in found in the open on living reefs.

Distribution: Red Sea and Mauritius (Mauritius).

Jack Randall

946. *Fromia indica* (Perrier, 1869)

Identification: This species is similar in appearance to *F. elegans*, but has denser aboral tubercles over the surface. It is usually red with black lines between the tubercles. The specimen shown here is unusual in having seven rays, rather than the typical five.

Natural History: *Fromia indica* is found on shallow living reefs and in coral rubble.

Distribution: South Africa; India; Indonesia; western Australia; Fiji; New Caledonia to southern Japan (Sodwana Bay, Natal, South Africa).

T. M. Gosliner

Leslie Newman & Andrew Flowers

Glenn Barrall

947. *Fromia milleporella* (Lamarck, 1816)

Identification: This species usually has a uniformly red coloration with prominent pores visible over the aboral surface. Specimens from the Red Sea may have additional blue or white spots. It lacks large tubercles and has a more flattened appearance than other members of the genus.

Natural History: Like other members of the genus, this species inhabits shallow living reefs.

Distribution: South Africa; Red Sea to Marshall and Society Islands (Madang, Papua New Guinea; Red Sea).

258

948. *Fromia monilis* (Perrier, 1869)

Identification: This is one of the most striking species of *Fromia* with cream to yellow pigment and a bright red disk area. It has low tubercles except around the margins and distinct pores which are readily visible in the disk area.

Natural History: This is one of the most common species of *Fromia* in the western Pacific, where it is found on shallow reefs in the open. Depth: 1-35 m.

Distribution: Andaman Islands to Australia; New Guinea; Indonesia; Philippines and Japan (Batangas, Luzon, Philippines).

Marc Chamberlain

949. *Fromia nodosa* Clark, 1967

Identification: This species is similar in appearance to *F. monilis*, but has larger, prominent tubercles along the midline of each ray.

Natural History: *Fromia nodosa* is found in the same general habitat as other members of the genus.

Distribution: Seychelles; Maldives; New Guinea and Marshall Islands (Madang, Papua New Guinea).

G.C. Williams

950. *Fromia* sp. 1

Identification: This beautiful specimen could not be identified positively with any described species. It has prominent tubercles on the margins and few, low tubercles over the remainder of the aboral surface.

Natural History: This species is found in relatively shallow water on living reefs.

Distribution: Solomon Islands (Solomon Islands).

Roy Eisenhardt
G.C. Williams

951. *Fromia* sp. 2

Identification: This species is similar to the preceding one, but has more numerous tubercles on the aboral surface.

Natural History: Found in shallow water on living reefs.

Distribution: New Guinea (Madang, Papua New Guinea).

259

Marc Chamberlain

952. *Gomophia egyptiaca* Gray, 1840

Identification: The isolated purple tubercles surrounded by a white ring clearly distinguish this species. Each tubercle is tipped with a short spinose papilla.

Natural History: Known from shallow water living reefs. It has been observed feeding on colonial ascidians and sponges. Depth: 15-50 m.

Distribution: East Africa; Red Sea to Australia; New Caledonia; Guam; Belau and Society Islands (Belau).

Nicholas Galluzzi

953. *Leiaster speciosus* Martens, 1866

Identification: This species can be recognized by its red color and evenly spaced rows of low, regular tubercles.

Natural History: Known from shallow water living reefs. Depth: 10-30 m.

Distribution: Australia; New Caledonia; Malaysia; Indonesia; Philippines and Okinawa (Sipadan, Borneo).

G.C. Williams

Alex Kerstitch
G.C. Williams

954. *Linckia laevigata* (Linnaeus, 1758)

Identification: This species is variable in color, ranging from bright blue to green, pink or yellow.

Natural History: This is one of the most common and obvious shallow water species of sea stars in most of the Indo-Pacific. It appears to be a scavenger and also to feed upon algae and microbes. It may have specimens of *Periclimenes soror* (#741) on its oral surface. The commensal shrimp are generally the same color as the sea star. The parasitic snail, *Thyca crystallina* (#489) may be found on the oral side of the animal partially imbedded in the ambulacral grooves. Depth: to 60 m.

Distribution: South Africa; Tanzania (but absent from the Red Sea) to Philippines; Melanesia; Marshall and Society Islands (Batangas, Luzon, Philippines; Vanuatu).

955. *Linckia multifora* (Lamarck, 1816)

Identification: This species is smaller than the preceding one and has mottled red, blue and yellow colors.

Natural History: Like the preceding species, *L. multifora* is a common inhabitant of shallow water reefs. It is has been suggested that this species may be a suspension feeder and that it may also feed upon microscopic algae and microbes. The parasitic snail, *Thyca crystallina* (#489) is also found on this sea star. Depth: to 40 m.

Distribution: South and East Africa; the Red Sea to Hawaiian and Society Islands (Madang, Papua New Guinea).

956. *Nardoa frianti* Koehler, 1910

Identification: Species of *Nardoa* can be recognized by the presence of tubercular warts over the aboral surface. *Nardoa frianti* has prominent tubercles along the margins of the rays as well as on the aboral surface.

Natural History: This species is found on shallow water living reefs. Depth: 3-45 m.

Distribution: Andaman Islands to Philippines; Federated States of Micronesia and New Caledonia (Batangas, Luzon, Philippines).

Jerry Allen

957. *Nardoa novaecaledoniae* (Perrier, 1875)

Identification: This species can be recognized by red or yellow color with numerous low tubercles of variable size.

Natural History: Found in shallow water on living reefs. Depth: 1-5 m.

Distribution: New Caledonia; Australia; New Guinea and Philippines (Madang, Papua New Guinea).

G.C. Williams

958. *Nardoa rosea* Clark, 1921

Identification: This species has prominent rounded tubercles on the nonmarginal portions of the rays and disk. It has both white and orange tubercles.

Natural History: Like other members of this genus, *N. rosea* inhabits shallow reefs.

Distribution: Known only from Queensland, Australia (Lizard Island, Great Barrier Reef, Australia)

Leslie Newman & Andrew Flowers
T. M. Gosliner

959. *Neoferdina cumingi* (Gray, 1840)

Identification: Species of *Neoferdina* have large marginal plates with bright purple or red spots. Usually some of the other plates on the aboral surface have similar circles of bright pigment. *Neoferdina cumingi* has a single longitudinal row of purple plates in the center of each ray.

Natural History: Little is known about the natural history of *Neoferdina* except that species are found in the open on shallow, living reefs. Depth: 5-30 m.

Distribution: Christmas Island to Australia; New Caledonia; New Guinea; Philippines to Society Islands (Madang, Papua New Guinea).

261

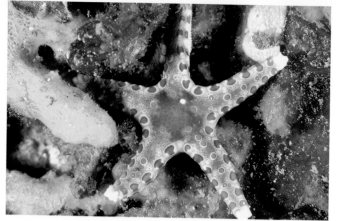

Mike Severns

960. *Neoferdina glyptodisca* (Fisher, 1913)

Identification: This species has extremely large marginal plates with a few scattered red plates on each ray. This sea star had been known only from the holotype specimen. The specimen photographed here closely approximates the holotype and comes from the type locality (Sulawesi).

Natural History: Virtually nothing is known about the natural history of this species. Depth: 44 m.

Distribution: Known only from Sulawesi, Indonesia (Lembeh Strait, Sulawesi, Indonesia).

Carol Buchanan

961. *Neoferdina insolita* Livingstone, 1936

Identification: This species was originally described as lacking large, colorful plates on the non-marginal portions of the rays. Jangoux (1973) doubted the validity of this species. However, the animal photographed here closely resembles the type specimen depicted by Livingstone and is tentatively identified as *N. insolita*.

Natural History: Nothing is known about the natural history of this species.

Distribution: New Guinea to Australia (Solitary Islands, New South Wales, Austrlaia).

962. *Ophidiaster confertus* Clark, 1916

Identification: This species can be recognized by its uniformly orange color with distinct rows of quadrangular plates.

Natural History: Found on shallow water rocky reefs.

Distribution: Queensland and New South Wales; Norfolk and Lord Howe Islands, Australia (Solitary Islands, New South Wales, Australia).

Stephen Smith
Leslie Newman & Andrew Flowers

963. *Asterina sarasini* (Loriol, 1897)

Identification: The extremely granular surface and short arms distinguish this species of shallow water sea star. Other members of this family may have shorter or longer rays. The species related to *A. sarasini* are poorly understood and may represent a single species or a species complex.

Natural History: *Asterina sarasini* is a common inhabitant of shallow water reefs.

Distribution: Members of this species or species complex are known from Sri Lanka; Andaman Islands; Hong Kong; Australia and New Guinea (Madang, Papua New Guinea).

964. *Acanthaster planci* (Linnaeus, 1758)
Crown-of-Thorns

Identification: This sea star, with 10-20 rays and elongate spines, is immediately distinctive. The spines can inflict a painful wound to divers when the skin is punctured by the crown-of-thorns.

Natural History: The crown-of-thorns is known to be a voracious predator upon scleractinean corals. In some areas of the south and western Pacific, population explosions of this species have resulted in the local decimation of living reefs. Depth: to 18 m.

Distribution: Western India across the entire Indo-Pacific to the Pacific coast of Mexico and the Galápagos Islands (Bohol, Philippines).

G.C. Williams

965. *Mithrodia bradleyi* Verrill, 1870

Identification: This species can be recognized by its spiny arms with a reticulated network of interconnecting plates. The taxonomy of this genus remains confused. Two or three species are known. *Mithrodia clavigera* is known from the western Indian Ocean to the western Pacific. The short spined species, *M. fisheri*, is restricted to the Hawaiian Islands. *Mithrodia bradleyi* is known from the eastern Pacific and has been considered as a synonym of *M. clavigera* by recent workers. The animal photographed here from Hawai'i does not resemble specimens of *M. fisheri* but is virtually identical to specimens of *M. bradleyi* from the eastern Pacific.

Natural History: Virtually nothing is known about the natural history of this species.

Distribution: Gulf of California; Colombia and Galápagos Islands, also from Hawaiian Islands (Maui, Hawaiian Islands).

Mike Severns

966. *Thromidia catalai* Pope & Rowe, 1977

Identification: This species has a fine network of interconnecting plates. The aboral surface is white with burnt orange tips. In specimens from Hawai'i, the orange pigment may cover most of the rays; only the disk region is white.

Natural History: This species is found on moderately shallow reefs where it is observed in the open. Depth: 10-45 m.

Distribution: New Caledonia; New Guinea; Indonesia; Philippines; Okinawa to Hawaiian Islands (Sulu Sea).

Fred McConnaughey
Nicholas Galluzzi

967. *Echinaster callosus* Marenzeller, 1895

Identification: This species can be recognized by its rows of rounded tubercles along the rays. Near the tips the tubercles are united to form an elevated band spanning each ray.

Natural History: Found in shallow water on living reefs.

Distribution: East Africa; Red Sea to New Caledonia; New Guinea; Indoneisa; Philippines and southern Japan (Batangas, Luzon, Philippines).

263

Mike Severns

968. *Echinaster luzonicus* (Gray, 1840)

Identification: *Echinaster luzonicus* is extremely variable in its color, ranging from yellow, brown to red. It may be mottled or uniformly colored.

Natural History: This is a common species in shallow water communities. It may reproduce asexually by fission. The commensal platyctene, *Coeloplana astericola* (#308) may be found on the aboral surface of this species.

Distribution: Western Pacific: New Caledonia; Australia; New Guinea; Indonesia; Philippines; southern Japan and Society Islands (Flores, Indonesia).

Ophiuroidea - Basket Stars, Brittle Stars & Serpent Stars

969. *Astroboa nuda* (Lyman, 1874)
Basket Star

Identification: The highly branched rays are divided into multiple bifurcations. The animal is uniformly whitish in color with no prominent tubercles along the margin of the rays.

Natural History: *Astroboa nuda* is commonly found at night with its rays fully extended into areas of current. It removes plankton from the water column by filter feeding. During the day it is found underneath large coral heads and in crevices.

Distribution: East Africa; Madagascar to Australia; New Guinea; Indonesia and Philippines (Batangas, Luzon, Philippines).

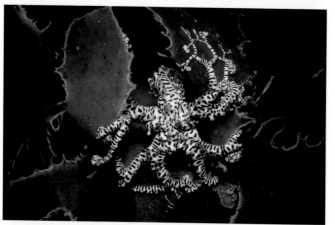

Mike Miller

970. *Conocladus australis* (Verrill, 1876)
Basket Star

Identification: The black and white coloring and less extensive branching of the rays distinguish this species from other basket stars.

Natural History: A shallow water filter feeder.

Distribution: Temperate and subtropical portions of eastern Australia to western Australia and Tasmania (Nelson Bay, New South Wales, Asutralia)

Neil Buchanan
Laura Losito

971. Basket Star

Identification: This species can be distinguished by the large paired tubercles along the margins of the rays and the reddish body color.

Natural History: Little is known about the natural history of this species, except that it is a filter feeder.

Distribution: Melanesia (Vanuatu).

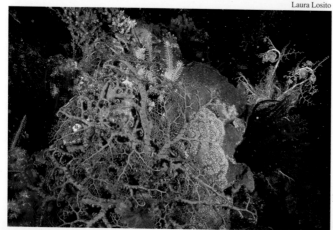

972. *Ophiomaza cacaotica* Lyman, 1871

Identification: The distinct yellow and black pigment and relatively smooth skin distinguish this species. The spines of this specimen are shorter than in most specimens.

Natural History: *Ophiomaza cacaotica* is a commensal on crinoids. The black and yellow color makes it extremely cryptic on its host. It is found in shallow living reefs.

Distribution: Western Indian Ocean to Australia; New Guinea; and southern Japan (Madang, Papua New Guinea).

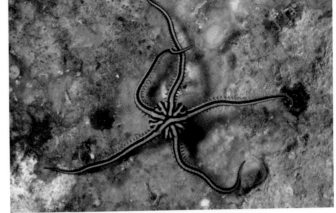

T. M. Gosliner

973. *Ophiomyxa australis* Lütken, 1869

Identification: This species can be readily distinguished by an apparently smooth disk and lack of dorsal plates on the arms. Actually, the plates are covered by a soft layer of tissue, making it appear that spines are absent.

Natural History: *Ophiomyxa australis* has been collected on shallow and deep reefs, where it is found actively crawling in the open. Depth: 2-50 m.

Distribution: South Africa; Madagascar to Marshall and Gilbert Islands (Sodwana Bay, Natal, South Africa).

G.C. Williams

974. *Ophiopteron* cf. *elegans* Ludwig, 1888

Identification: This uniformly tan species has brilliant blue markings on the disk, which continue on the arms as longitudinal lines. It is distinguished in having basal arm spines with webbing between them.

Natural History: Found under coral rubble on shallow reefs.

Distribution: Maldives to Australia; Indonesia; Philippines and Japan (Batangas, Luzon, Philippines).

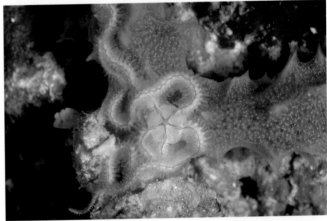

T. M. Gosliner
Marc Chamberlain

975. *Ophiothrix foveolata* Marktanner & Turneretscher, 1887

Identification: This species can be distinguished by its whitish disk and rays with prominent red on the disk, spines and rays. There is a longitudinal red line along the middle of each ray.

Natural History: Like other members of this genus, *Ophiothrix foveolata* is found in the open on living shallow water reefs.

Distribution: Mozambique; Solomon Islands and northern Australia (Solomon Islands).

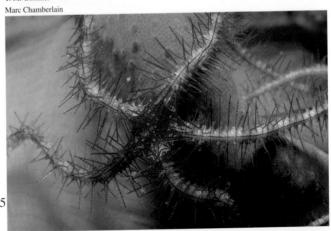

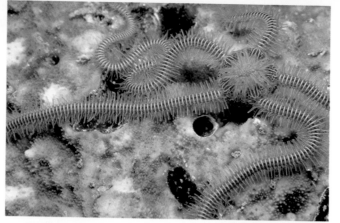

Leslie Newman & Andrew Flowers

976. *Ophiothrix nereidina* (Lamarck, 1816)

Identification: The distinctive green disk and blue transverse markings on the rays distinguish this species of long-spined brittle star. There are about fifty species of this diverse genus represented in the Indo-Pacific tropics.

Natural History: Found in the open on shallow water reefs.

Distribution: Madagascar; East Africa to Australia; New Guinea; Philippines and southern Japan and Marshall Islands (Madang, Papua New Guinea).

Kathy deWet

977. *Ophiothrix purpurea* Martens, 1867

Identification: This species has a reddish color on the disk and ray spines. It also has a distinct black longitudinal line down the center of each ray surrounded by a yellow line on either side.

Natural History: Found in the open on shallow water reefs.

Distribution: Madagascar; Red Sea to Australia; New Caledonia; Indonesia and Philippines (Bunaken, Indonesia).

Ben Tetzner
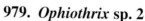
Greg Ochocki

978. *Ophiothrix* sp. 1

Identification: This species resembles *O. purpurea*, but has transparent spines with black rather than red stripes. It also has two additional longitudinal lines on each arm.

Natural History: Found in the open on shallow water reefs.

Distribution: Melanesia (Fiji)

979. *Ophiothrix* sp. 2

Identification: This species is also similar to *O. purpurea*, but has a single maroon spot in the center of each arm segment.

Natural History: Found in the open on shallow water reefs.

Distribution: Indonesia (Indonesia)

266

980. *Ophiothrix* sp. 3

Identification: The presence of maroon spots on the disk and a pair of maroon spots on each arm segment distinguish this species of *Ophiothrix*.

Natural History: Found in the open on shallow water reefs.

Distribution: Philippines (Batangas, Luzon, Philippines).

D.W. Behrens

981. Ophiotrichid Brittlestar

Identification: This strikingly colored brittle star has yellow, blue and black bands on the disk and has extremely elongate arms.

Natural History: Found under dead coral heads on shallow reef tops. Depth: 6 m.

Distribution: New Guinea (Madang, Papua New Guinea).

T. M. Gosliner

982. *Ophiomastix variabilis* Koehler, 1905

Identification: This species with, elongated arms, has a black body, and a series of yellow markings on the disk and arms. Members of this genus have some arm spines modified into paddle-like structures.

Natural History: Found on shallow and deep reefs on the surface of living corals.

Distribution: Mozanbique; New Caledonia; Australia; New Guinea and Belau (Madang, Papua New Guinea).

Leslie Newman & Andrew Flowers
Terry Schuller

983. *Ophiarachna affinis* Lütken, 1869

Identification: The white marbling surrounded by thin black lines on the disk and the light and dark banded spines are distinctive. Clark and Rowe (1971) discuss the systematic confusion surrounding the various color forms of this probable species complex.

Natural History: This species is found in shallow water habitats in areas with algae and sea grasses.

Distribution: Depending on taxonomic considerations, Zanzibar; Solomon Islands; Fiji and Philippines (Batangas, Luzon, Philippines).

267

984. *Ophiarachna incrassata* (Lamarck, 1816)

Identification: This species is similar to the preceding one, but has a series of white spots surrounded by black rings. These spots form a series of triangular markings on the disk. Like *O. affinis*, the arm spines have light and dark bands.

Natural History: This species is commonly found among algae and seagrasses on shallow sand flats. It is more active at night.

Distribution: East Africa; Madagascar to Australia; New Guinea; Indonesia; New Caledonia; Philippines and Belau (Batangas, Luzon, Philippines).

David Reid

985. *Ophiarachnella gorgonia* (Müller & Troschel, 1842)

Identification: This species is brownish with distinct banded arms and a series of dark markings on the dorsal surface of the disk. A red spot may also be present in the center of the disk.

Natural History: This species is found under rocks on shallow reefs and areas of coral rubble.

Distribution: East Africa; Madagascar to Australia; New Caledonia; Philippines and Belau (Batangas, Luzon, Philippines).

T. M. Gosliner

986. *Ophiolepis superba* Clark, 1915

Identification: This species has extremely short arm spines and is tan with black markings on the disk and arms.

Natural History: *Ophiolepis superba* is commonly found under coral rubble on shallow reefs.

Distribution: Western Indian Ocean to Australia; New Caledonia; New Guinea, and Indonesia to Funafuti Atoll (Madang, Papua New Guinea).

T. M. Gosliner

T. M. Gosliner

987. *Ophioplocus imbricatus* (Müller & Troschel, 1842)

Identification: The disk is uniformly brownish. This color on the arms is interrupted by 8-10 black bands. The arm spines are short.

Natural History: This species is found in the open and under rocks in shallow rubble areas.

Distribution: East Africa; Madagascar to Philippines and Hawaiian Islands (Batangas, Luzon, Philippines).

988. *Amphiura* sp. 1

Identification: This species can be recognized by its extremely slender, lightly-banded arms and fine, elongate spines.
Natural History: This species is active at night and extends four of its arms erectly from the sand surface in order to filter feed. Depth: 10 m.
Distribution: Philippines (Batangas, Luzon, Philippines).

T. M. Gosliner

989. *Amphiura* sp. 2

Identification: This species has yellow, black and white transverse bands on its arms. It also has elongate spines on the arms.
Natural History: This species is also active at night and extends four arms from the sand in a sinusoidal posture for filter feeding. Depth: 15 m.
Distribution: Philippines (Batangas, Luzon, Philippines).

T. M. Gosliner

990. *Ophiactis* cf. *modesta* Brock, 1888

Identification: This species has green and white color on its elongate rays and small disk.
Natural History: Found on shallow reefs on sponges.
Distribution: Australia; Indonesia; Belau; Japan to Hawaiian Islands (Molokini Island, Hawaiian Islands).

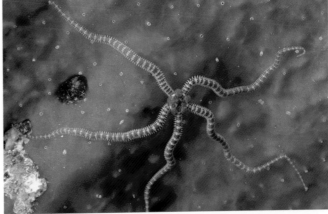
Mike Severns
G.C. Williams

991. *Ophiactis* cf. *savignyi* (Müller & Troschel, 1842)

Identification: This minute species is highly mottled and banded on the six arms. The disk has large polygonal plates over its surface. It may be *O. savignyi*, but positive identification requires microscopic examination of specimens.
Natural History: Found under coral slabs where it feeds upon colonial encrusting animals such as bryozoans and sponges. This species undergoes asexual reproduction by fission.
Distribution: East Africa; throughout the western Pacific to the Hawaiian Islands (Batangas, Luzon, Philippines).

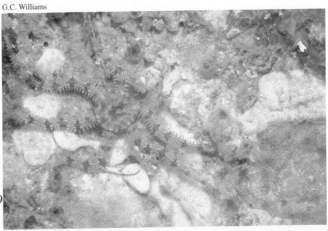

269

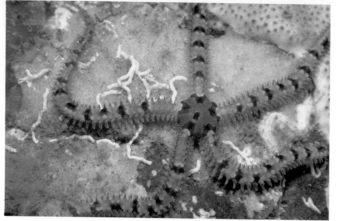

Carol Buchanan

992. *Ophiactis* sp.

Identification: This species is greenish with black and white bands on the disk and arms. It has short thick, conical arm spines.
Natural History: Found on moderately deep, rocky reefs at about 15 m depth.
Distribution: Australia (Solitary Island, New South Wales, Australia).

Echinoidea - Sea Urchins, Heart Urchins and Sand Dollars

993. *Eucidaris metularia* (Lamarck, 1816)

Identification: This species is a member of a group of sea urchins called the cidaroids, with club-shaped spines. This species has regular rows of low tubercles on each banded spine.
Natural History: Cidaroids are commonly found in crevices on living coral reefs, usually in shallow water, and *E. metularia* is no exception. It feeds on algae.
Distribution: East Africa; Red Sea to the western Pacific and Hawaiian and Society Islands (Bora Bora, Society Islands).

D.W. Behrens

994. *Phyllacanthus imperialis* (Lamarck, 1816)

Identification: *Phyllacanthus imperialis* is another common cidaroid with lightly and darkly banded spines. It differs from the preceding species in having longitudinal ridges on the spines rather than rows of tubercles.
Natural History: Commonly found on shallow water reefs in holes and crevices.
Distribution: East Africa; Red Sea to Australia; Solomon Islands; Indonesia; Philippines; southern Japan; Marshall Islands (Solomon Islands).

David Reid
Mike Severns

995. *Prionocidaris hawaiiensis* Agassiz & Clark, 1907

Identification: The spines of this cidaroid are covered with irregularly arranged, pointed spines. Frequently, as in this picture, the spines are covered with growth of algae or colonial animals. This may function as camouflage.
Natural History: Found on relatively deep reefs in crevices on living reefs.
Distribution: Apparently endemic to the Hawaiian Islands (Maui, Hawaiian Islands).

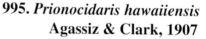

270

996. *Prionocidaris verticillata* (Lamarck, 1816)

Identification: This cidaroid has rows of small tubercles interrupted by crown-like swellings of larger spines on each club-shaped spine.

Natural History: Found on the undersurface of small coral heads and coral rubble on shallow water reefs.

Distribution: East Africa; western Pacific to Hawaiian Islands (Batangas, Luzon, Philippines).

T. M. Gosliner

997. *Asthenosoma varium* Grube, 1868
Fire Urchin

Identification: This urchin can be immediately recognized by its flexible rather than rigid test and the sharp spines with globular swellings below the tip. It has venom-tipped spines and can inflict a painful sting when handled. Intense pain lasts for 15 minutes to several hours after penetration of the spines into the skin.

Natural History: Found commonly in sandy or rubble habitats on open slopes or flats from shallow or deep water. Host to the commensal shrimp, *Allopontonia iaini* (#727) and *Periclimenes colemani* (#733) and the snail *Leutzenia asthenosomae* (#490).

Distribution: Red Sea to Australia; New Caledonia; Indonesia; Philippines and southern Japan (Batangas, Luzon, Philippines).

Nicholas Galluzzi

998. *Asthenosoma ijimai* Yoshiwara, 1897

Identification: This species has much more uniformly arranged spines that are more similar in size than those of *A. varium*. Also able to inflict painful stings.

Natural History: Whereas *A. varium* is found in sandy and rubble habitats, *A. ijimai* is found on shallow living reef and rock habitats. Depth: 1-20 m.

Distribution: Australia; Indonesia; Philippines and southern Japan (Batangas, Luzon, Philippines).

Mike Severns
T. M. Gosliner

999. *Astropyga radiata* (Leske, 1778)

Identification: This species is in the Diadematidae, a group of urchins with elongate, pointed spines. It has brownish banded spines with bright blue spots located in the grooves between the rows of spines. It has a translucent white anal sac visible on the top of the urchin.

Natural History: *Astropyga radiata* is found on shallow and deep water, sandy and rubble bottom habitats.

Distribution: Natal; South and East Africa to the western Pacific and Hawaiian Islands (Batangas, Luzon, Philippines).

T. M. Gosliner

1000. *Diadema savignyi* Michelin, 1845

Identification: This species has elongate black and white spines, often with transverse banding on the spines. The anal sac is dark with a light ring around the opening and bright blue pigment may also be found on the aboral surface.

Natural History: This species is commonly found in shallow water habitats, frequently abundant in areas that have been recently disturbed, such as wharves or habitats recovering from storm damage.

Distribution: Southern and eastern Africa to the entire western Pacific and Society Islands (Bohol, Philippines).

G.C. Williams

1001. *Diadema setosum* (Leske, 1778)

Identification: The spines of this species are generally more elongate than those of *D. savignyi*. Like other diadematids, it may inflict a painful injury to the hapless diver who is impaled by the spines. Most diagnostic is the bright orange or red ring surrounding the anal opening.

Natural History: This species is also abundant in shallow water in areas that have been recently disturbed, where it may form dense aggregations.

Distribution: East Africa; Red Sea to the entire western Pacific (Batangas, Luzon, Philippines).

1002. *Echinothrix calamaris* (Pallas, 1774)

Identification: *Echinothrix calamaris* has a dark and white spotted anal sac. The spines are tubular with open distal tips and may be banded frequently.

Natural History: Found under coral heads and coral rubble in shallow reef habitats. Also capable of inflicting painful puncture wounds.

Distribution: Natal; South Africa; Red Sea to the western Pacific and Hawai'i (Batangas, Luzon, Philippines).

T. M. Gosliner
G.C. Williams

1003. *Echinothrix diadema* (Linnaeus, 1758)

Identification: Like the preceding species, *E. diadmea* has a light and dark spotted anal sac. However, the spines are pointed and closed at their tips.

Natural History: Found in the open or under stones on shallow water reefs or rubble areas. Handling of this species should also be avoided.

Distribution: Natal; South Africa; Red Sea to the western Pacific and Hawai'i (Batangas, Luzon, Philippines).

1004. *Mespilia globulus* (Linnaeus, 1758)

Identification: The short, sharp spines of this species are separated by ten areas of blue where spines are absent.

Natural History: This species is found in shallow water coral rubble and patch reefs.

Distribution: Bay of Bengal; Australia; New Caledonia; Indonesia; Philippines; Marshall Islands (Batangas, Luzon, Philippines).

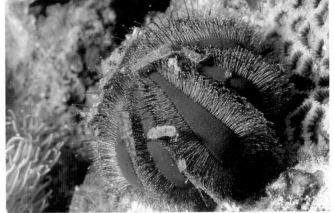

Terry Schuller

1005. *Microcyphus rousseaui* Agassiz in Agassiz & Desor, 1848

Identification: The five undulating areas of spines and bare areas are distinctive.

Natural History: Found in the open on shallow water reefs.

Distribution: East Africa; Red Sea and Arabian Sea (Red Sea).

Nicholas Galluzzi

1006. *Salmacis belli* Döderlein, 1902

Identification: This species has thin, sharp spines which are banded with a red base.

Natural History: Found in shallow water grass beds and rubble areas.

Distribution: Western Pacific: Australia; New Caledonia; Indonesia and Philippines (Batangas, Luzon, Philippines).

T. M. Gosliner
Mike Severns

1007. *Pseudoboletia indiana* (Michelin, 1862) Collector Urchin

Identification: This species has small whitish spines. Its common name originates from the fact that it collects rubble and debris on the aboral surface, probably as a means of camouflage.

Natural History: Occasionally found in shallow water rubble habitats.

Distribution: East Africa to Australia; Indonesia to Hawai'i (Maui, Hawaiian Islands).

273

Glenn Pollock

1008. *Pseudoboletia maculata* Troschel, 1869

Identification: Similar to the preceeding species, but with dark patches present on test.

Natural History: Found in rubble areas and grass beds in shallow waters.

Distribution: Sri Lanka to Australia; New Guinea; Indonesia; Philippines and Japan (Philippines).

T. M. Gosliner

1009. *Toxopneustes pileolus* (Lamarck, 1816)
Flower Urchin

Identification: This species is extremely venomous and has caused fatalities. The sting comes from the flower-like pedicillaria found all over the animal's surface. The spines are short and inconspicuous.

Natural History: *Toxopneustes pileolus* is found commonly in shallow sandy and rubble habitats where it is often partially buried in the substrate.

Distribution: South and East Africa to Australia; New Guinea; Indonesia; Philippines; southern Japan to the central Pacific (Batangas, Luzon, Philippines).

1010. *Tripneustes gratilla* (Linnaeus, 1758)

Identification: This is one of the most common and variable species of Indo-Pacific urchin. It has white and reddish spines, separated by areas of pedicillaria which are variable in width. The color of the animal may be black, white or greenish.

Natural History: Commonly found in shallow water lagoons and bays. This species, like *Pseudoboletia indiana*, frequently covers itself with debris and a wide variety of objects. One specimen, near a harbor entrance in Hawai'i, was observed covering itself with a pair of undershorts.

Distribution: South and East Africa; Red Sea to Hawaiian Islands (Batangas, Luzon, Philippines).

T. M. Gosliner
Leslie Newman & Andrew Flowers

1011. *Paraselinia gratiosa* Agassiz, 1863

Identification: This species can be distinguished by its ovoid shape and asymmetrically arranged plates on the aboral apex. A series of white beaded tubercles is present at the base of each spine.

Natural History: Found in shallow water reef habitats.

Distribution: Madagascar to Australia; New Guinea; Indonesia; Philippines and Tonga (Madang, Papua New Guinea).

T. M. Gosliner

1012. *Colobocentrotus atratus* (Linnaeus, 1758)

Identification: *Colobocentrotus atratus* is immediately recognized by its low, dome-shaped spines on the aboral surface and its uniform black color. A row of more elongate spines is found at the base of the urchin.

Natural History: This urchin is unique in that it inhabits intertidal rocky habitats in areas of considerable wave action. Its unusual shape is designed to reduce wave force, thus decreasing the chances that individuals will be pulled from rocks by the force of breaking waves.

Distribution: East Africa to Sri Lanka; India and Indonesia. It is apparently absent from most of the western Pacific but is found in the Hawaiian Islands (Kaena Point, Oahu, Hawaiian Islands).

G.C. Williams

1013. *Echinometra mathaei* (Blainville, 1825)

Identification: This species is related to *Colobocentrotus* and the following species, despite the fact that it has sharply pointed spines rather than club-shaped ones. The spines are reddish with a white ring around the base. The tissue surrounding the test is black.

Natural History: This species is found intertidally and in the shallow subtidal. It frequently erodes round holes into limestone reef, in which it seeks refuge from predators. These holes are formed by abrasion of the spines and jaws.

Distribution: Western Indian Ocean, western Pacific to Hawaiian Islands (Batangas, Luzon, Philippines).

1014. *Heterocentrotus mammillatus* (Linnaeus, 1758)

Slate Pencil Urchin

Identification: This species is extremely variable in color. The spines are reddish in color and may have transverse white bands. The secondary basal spines may be white or darkly pigmented. The related species, *H. trigonarius*, has more elongate, darker spines.

Natural History: Found in crevices and between coral heads on shallow water, living reefs.

Distribution: Southern and East Africa; Red Sea to western Pacific and Hawai'i (Australia).

Dan Gotshall
T. M. Gosliner

1015. *Echinoneus cyclostomus* Leske, 1778

Identification: Heart urchins and sand dollars are bilaterally rather than radially symmetrical. *Echinoneus cyclostomus* is a small bean-shaped species with uniformly short spines.

Natural History: Found buried under clean sand in shallow water sand flats.

Distribution: East Africa to the western Pacific and Hawaiian Islands (Msimbati, Tanzania).

D.W. Behrens

1016. *Clypeaster reticulatus* (Linnaeus, 1758)

Identification: This species of irregular urchin is flattened and has short spines.

Natural History: Found in silty sand in shallow water.

Distribution: East Africa; Red Sea throughout the western Pacific to Hawai'i (Batangas, Luzon, Philippines).

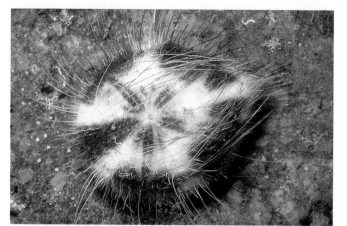

T. M. Gosliner

1017. *Maretia planulata* (Lamarck, 1816)

Identification: *Maretia planulata* may be entirely white or densely covered with dark pigment. It has a series of elongate spines which are directed to the posterior portion of the body.

Natural History: During the day this species is commonly buried under the sand. At night it emerges and may rapidly crawl over the surface of the sand. It inhabits silty or clean sandy areas and seagrass flats.

Distribution: Tanzania throughout the western Pacific to the Marshall Islands (Batangas, Luzon, Philippines).

Mike Severns
G.C. Williams

1018. *Brissus latercarinatus* (Leske, 1778)

Identification: This ovoid species has uniformly short spines and brownish color.

Natural History: Found in shallow water, where it is burried under the sand surface.

Distribution: East Africa; Red Sea throughout the western Pacific to Hawai'i (Maui, Hawaiian Islands).

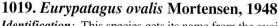

1019. *Eurypatagus ovalis* Mortensen, 1948

Identification: This species gets its name from the ovoid shape of the test, which bears uniformly distributed, elongate spines.

Natural History: Found buried in silty sand in shallow water.

Distribution: Known only from the type material from Luzon Island in 192 m depth and from the specimen depicted here. A specimen from Manado, Sulawesi photographed by Kathy deWet may also be the same species (Batangas, Luzon, Philippines).

1020. *Echinodiscus auritus* Leske, 1778

Identification: This true sand dollar is distinguished by its two notches which are open at the margins of the test.
Natural History: It inhabits shallow water in clean sand on the edge of slopes.
Distribution: Western Indian Ocean (Msimbati, Tanzania).

T. M. Gosliner

1021. *Echinodiscus tenuissimus* (Agassiz & Desor, 1847)

Identification: This species differs from the preceding one in that the two notches are not open to the margins of the test.
Natural History: Like *E. auritus*, *E. tenuissimus* inhabits clean sandy slopes in shallow water.
Distribution: New Caledonia; New Guinea; Indonesia; Cambodia; Malaysia; New Britain (Madang, Papua New Guinea).

T. M. Gosliner

Holothuroidea - Sea Cucumbers

1022. *Actinopyga lecanora* (Jaeger, 1833)

Identification: This yellow and brown mottled species is shaped like a football and bears thin elongate papillae over its dorsal surface.
Natural History: Found in shallow water coral rubble habitats.
Distribution: East Africa; Madagascar to Australia; New Caledonia; Indonesia; Philippines and southern Japan (Batangas, Luzon, Philippines).

David Reid
Terry Schuller

1023. *Actinopyga obesa* (Selenka, 1867)

Identification: This species can be recognized by its reddish brown or yellowish brown body color. Its dorsal surface is often partially covered by a dusting of sand particles.
Natural History: *Actinopyga obesa* is found in areas of rubble and sand interface in 15-30 m of water.
Distribution: Solomon Islands; Philippines; China and Hawaiian Islands (Batangas, Luzon, Philippines).

277

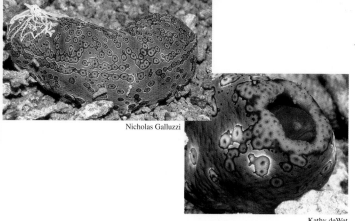

Nicholas Galluzzi

Kathy deWet

1024. *Bohadschia argus* (Jaeger, 1833)

Identification: This common species can be immediately recognized by its distinctive pattern of mottling.

Natural History: Like many holothurians, *B. argus* extrudes sticky defensive threads called Cuvierian tubules, seen here in the photo. On the right, the commensal fish, *Carapus* sp. lives in the anus of many sea cucumbers. This species inhabits shallow water reef and rubble habitats.

Distribution: Seychelles to Australia; New Caledonia; New Guinea; Indonesia; Malaysia; Philippines and southern Japan (Sipadan Island, Borneo; Manado, Indonesia).

1025. *Bohadschia graeffei* (Semper, 1868)

Identification: This species radically changes its appearance in its transition from a juvenile to an adult, shown in the three photos here. Adults are beige with black spots and short, white-tipped tubercles.

Natural History: The changes in appearance of this species are related to the mimicry of juveniles to several nudibranchs of the Phyllidiidae, including *Phyllidia coelestis* (#593) and *P. varicosa* (#596). Once the sea cucumber exceeds the maximum size of the nudibranch it begins altering its appearance, as the mimicry is no longer effective.

Distribution: Red Sea; Maldives to Australia; New Caledonia; New Guinea; Indonesia and Philippines (Batangas, Luzon, Philippines; Madang, Papua New Guinea; Madang, Papua New Guinea).

Adult

Terry Schuller

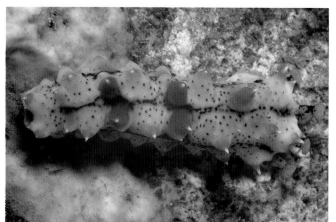

Subadult

G.C. Williams

Marc Chamberlain

T. M. Gosliner

Juvenile

1026. *Bohadschia marmorata* (Jaeger, 1833)

Identification: This species is yellowish with broad dark brown bands. Thin papillae are common all over the upper body.

Natural History: *Bohadschia marmorata* inhabits shallow water rubble and sand habitats.

Distribution: East Africa; Red Sea to Australia; Indonesia; Malaysia; Philippines and southern Japan (Sipadan Island, Borneo).

1027. *Bohadschia paradoxa* (Selenka, 1867)

Identification: This species is similar to the preceding one in its coloration and body form, but has darker circles and somewhat shorter papillae.

Natural History: Found in sandy habitats in relatively shallow water, generally less than 20 m depth.

Distribution: Australia; Philippines to Hawaiian Islands (Maui, Hawaiian Islands).

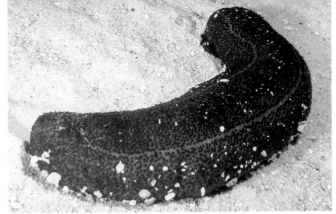
Mike Severns

1028. *Holothuria (Halodeima) atra* Jaeger, 1833

Identification: *Holothuria atra* can be recognized by its uniform black color and sausage shape. The body is frequently dusted with sand grains.

Natural History: This is one of the most common sea cucumbers in the Indo-Pacific tropics. Animals may form dense aggregations in shallow water sandy habitats, just below the low tide mark.

Distribution: East Africa; Red Sea; throughout the western Pacific to the Hawaiian Islands (Lembeh Strait, Sulawesi, Indonesia).

Kathy deWet

1029. *Holothuria (Halodeima) edulis* Lesson, 1830

Identification: As its name suggests, *Holothuria edulis* is one of the edible species of sea cucumbers served as "sea slugs," "trepang" or "beche de mer." It can be recognized by its reddish or beige undersurface and darker, often black, dorsal suface.

Natural History: A common inhabitant of shallow water rocky and sandy habitats.

Distribution: East Africa; Red Sea, throughout the western Pacific to the Hawaiian Islands and Tuamotus (Philippines).

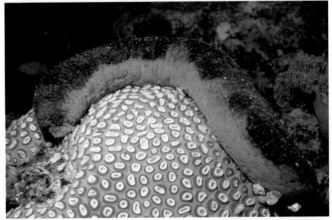
Bruce Watkins
Fred McConnaughey

1030. *Holothuria (Microthele) fuscogilva* Cherbonnier, 1980

Identification: This relatively recently described species is whitish with dark brown or black dorsal mottling.

Natural History: It inhabits shallow water, clean sandy areas.

Distribution: New Caledonia and New Guinea (Milne Bay, Papua New Guinea).

279

1031. *Holothuria (Microthele) fuscopunctata* Jaeger, 1833

Identification: This species is probably synonymous with *H. axiologa* which has been depicted in other field guides. Both can be recognized by the irregular transverse folds along the sides of the body.

Natural History: Found commonly in shallow water sandy habitats.

Distribution: Australia; New Caledonia; New Guinea; Belau and the Mariana Islands (Madang, Papua New Guinea).

Leslie Newman & Andrew Flowers

1032. *Holothuria (Thymiosycia) hilla* Lesson, 1830

Identification: The light brown color with scattered, white conical papillae distinguishes this species. The shape is narrow and elongate.

Natural History: *Holothuria hilla* inhabits shallow sandy rubble areas.

Distribution: Western Indian Ocean, including the Red Sea, throughout the western Pacific to the Hawaiian Islands (Batangas, Luzon, Philippines).

T. M. Gosliner

1033. *Holothuria (Thymiosycia) impatiens* (Forskål, 1775)

Identification: The rugose body with a uniform brownish color distinguishes this species. The body is covered with conical warts from which a filamentous appendage emerges.

Natural History: This species is commonly found in shallow water rubble habitats.

Distribution: East Africa; Red Sea to the Hawaiian Islands. This species is also known from the tropical Atlantic and the Mediterranean (Batangas, Luzon, Philippines).

T. M. Gosliner
Dave Zoutendyk

1034. *Holothuria (Mertensiothuria) leucospilota* (Brandt, 1835)

Identification: This elongate species is uniformly black in color with elongate tentacles surrounding the mouth.

Natural History: *Holothuria leucospilota* inhabits shallow patch reefs where the posterior end of the body is under rocks and the anterior portion is extended into sandy areas.

Distribution: Western Indian Ocean and Red Sea; western Pacific to the Hawaiian Islands (Cook Islands).

1035. *Stichopus chloronotus* Brandt, 1835

Identification: The body is uniformly dark green with numerous elongate papillae, each tipped with orange.

Natural History: Found in shallow water rocky and rubble habitats.

Distribution: East Africa; Madagascar to western Pacific and Hawaiian Islands (Batangas, Luzon, Philippines).

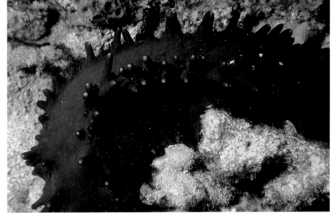

T. M. Gosliner

1033. *Stichopus horrens* Selenka, 1867

Identification: The body is tan with numerous greenish conical papillae. The papillae have a series of concentric green rings around their base and a filamentous apex.

Natural History: *Stichopus horrens* inhabits shallow water rubble and sand habitats.

Distribution: Maldives to Australia; New Caledonia; New Guinea; Indonesia; Malaysia; Philippines and southern Japan to the Hawaiian Islands (Sipadan Island, Borneo).

Marc Chamberlain

1037. *Stichopus noctivagus* Cherbonnier, 1980

Identification: This species is similar in appearance to the preceding one. It has a more transparent, off-white body with chocolate brown striations and orange markings on the base of the tubercles.

Natural History: *Stichopus noctivagus* is found in shallow water rubble areas where algae are present.

Distribution: New Caledonia; New Guinea; Philippines and Hawaiian Islands (Maui, Hawaiian Islands).

Mike Severns

Leslie Newman & Andrew Flowers

1038. *Stichopus variegatus* Semper, 1868

Identification: Unlike other species of *Stichopus, S. variegatus* lacks elongate tubercles. The body is burnt orange with darker low tubercles arranged in distinct rows.

Natural History: Found in shallow water rubble and sandy habitats.

Distribution: Western Indian Ocean; Red Sea to Australia; New Caledonia; Indonesia; Philippines; Japan and Belau (Heron Island, Great Barrier Reef, Australia).

281

Fred McConnaughey

1039. *Stichopus* sp.

Identification: This is an undescribed species of *Stichopus*. It is bright yellow with dark, brown stripes over the entire dorsal surface.

Natural History: Found on living coral reefs in relatively shallow water.

Distribution: New Guinea; Micronesia; Belau (Belau).

Ben Tetzner

1040. *Thelenota ananas* (Jaeger, 1833)

Identification: This is a large species, reaching half a meter in length. The body is black with orange conical papillae arranged in a stellate fashion.

Natural History: Found in 5-30 m depth, where it inhabits the interface between reefs and sand.

Distribution: Mauritius; Seychelles to Maldives; Australia; Fiji; New Caledonia; Indonesia; Japan and Guam (Fiji).

G.C. Williams
Bruce Watkins

1041. *Thelenota anax* Clark, 1921

Identification: This is the most massive holothurian in the Indo-Pacific, reaching one meter in length. The body is cream with scattered orange blotches. It has a warty texture.

Natural History: An inhabitant of sandy bottoms. Periodically more than half of the animal will rear up off the surface. In this spawning posture, it resembles an elephant's trunk.

Distribution: North of Madagascar to Australia; New Caledonia; Fiji; New Guinea; China; Guam and Marshall Islands (Madang, Papua New Guinea).

**1042. *Thelenota rubralineata*
Massin & Lane, 1991**

Identification: This large, recently described holothurian has a cream body color with numerous cherry red lines. It is one of the most beautiful holothurians in the Indo-Pacific.

Natural History: This species is unusual in that inhabits outer barrier reef slopes in 30 or more m of water.

Distribution: Indonesia; Solomon Islands and New Guinea (Solomon Islands).

1043. *Colochirus robustus* Östergren, 1898

Identification: This is a relatively small holothurian, rarely exceeding 60 mm in length. It is bright yellow with green pigment on the five highly branched tentacles.

Natural History: This species can form dense aggregations on reef faces and walls in areas with strong current.

Distribution: Indonesia; Belau and Philippines (Indonesia).

Bruce Watkins

1044. *Pseudocolochirus violaceus* (Théel, 1886)

Identification: This is one of the most brightly colored sea cucumbers. The body is dark inky blue. The tube foot rows are red and the highly branched tentacles are white.

Natural History: Found on relatively shallow reef flats.

Distribution: Known only from India, northern Australia and Philippines (Philippines).

Alex Kerstitch

1045. *Neothyonidium magnum* (Ludwig, 1882)

Identification: This species can be distinguished by its massive tentacular crown, often the only part of the animal that is visible. The body is creamy white with dark brown tentacles and tube feet.

Natural History: Found largely buried in clean sand, often with only the tentacles exposed.

Distribution: Western Pacific: New Caledonia; New Guinea; Indonesia and Philippines (Milne Bay, Papua New Guinea).

Ben Tetzner
G.C. Williams

1046. *Euapta godeffroyi* (Semper, 1868)

Identification: This species and the remaining holothurians are members of a group lacking tube feet. The body is translucent cream with brown bands in enlarged "beaded" areas of the skin. The pinnate tentacles are cream.

Natural History: Found in shallow water, sandy habitats and in grass beds. It is active at night and may extend its body length to 1.5 m.

Distribution: Western Indian Ocean to the Hawaiian Islands and eastern Pacific of Mexico and Panama (Batangas, Luzon, Philippines).

283

1047. *Synapta maculata* (Chamisso & Eysenhardt, 1821)

Identification: This species is similar in appearance to *Euapta godeffroyi* and may exceed 2 m in length. It is tan to brown with black markings. The tentacles are also tan with white lines along the margins of the pinnae.

Natural History: Found in shallow sandy habitats and grass beds.

Distribution: East Africa; Red Sea; throughout the western Pacific to the Society Islands (Sulawesi, Indonesia).

Kathy deWet

1048. *Synaptula media* Cherbonnier & Feral, 1985

Identification: This smaller cucumber is chocolate brown with white lines and dashes. The tentacles are cream colored with fine brown lines.

Natural History: Found in relatively shallow water, where it may form dense aggregations on a wide variety of massive sponges. It apparently gains at least part of its nutrition from the waste products of sponges.

Distribution: New Caledonia; New Guinea; Micronesia and Philippines (Batangas, Luzon, Philippines).

T. M. Gosliner

1049. *Synaptula lamperti* Heding, 1928

Identification: This species has an opaque white body with dark burgundy longitudinal stripes. The same colors are present on the tentacles.

Natural History: Like the preceding species, *Synaptula lamperti* is found in association with massive sponges in relatively shallow water.

Distribution: Known only from Australia and Indonesia (Indonesia).

Hemichordata - Acorn Worms

Acorn worms are primitive relatives of chordates. The body is divided into three distinct regions, the proboscis, the collar and the trunk. Most species are deposit feeders in sandy habitats.

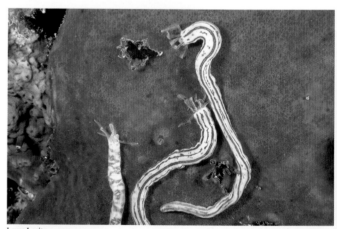
Laura Losito
Leslie Newman & Andrew Flowers

1050. *Ptychodera flava* Eschscholtz, 1835

Identification: Yellowish in color. The proboscis is short and rounded. The trunk is divided into distinct segments.

Natural History: Found under sand and under rocks in shallow habitats of 0-5 m depth. It is commonly found by fanning coarse, somewhat anaerobic sand.

Distribution: Tanzania and Mauritius, throughout the western Pacific to the central Pacific of the Hawaiian Islands and the Galápagos Islands (Lizard Island, Great Barrier Reef, Australia).

284

Chordata - Chordates

The chordates include the urochordates, the cephalochordates (lancelets), and the vertebrates (animals with backbones).

Urochordata - Tunicates

The urochordates include the benthic tunicates as well as the planktonic tunicates known as larvaceans (appendicularians) and the pelagic tunicates called thaliaceans (salps, doliolids, and pyrosomes). The larval stages of these three related groups are tadpole-like and exhibit the defining characteristics of all chordates at sometime during their development.

Ascidiacea - Ascidians, Benthic Tunicates or Sea Squirts

Ascidians are chordates, the animal group to which the vertebrates belong. Although ascidians lack a backbone, they do share three characteristics in their larval stages with all other chordates–a notochord or supporting rod, a dorsal tubular nerve cord, and pharyngeal gill slits.

The benthic tunicate fauna of the Indo-Pacific is far too large and incompletely known to include all species in this text. Most species cannot be identified by external characteristics alone, the diagnostic features being the internal anatomy of the zooid itself. An excellent source–Monniot, Monniot and Laboute (1991) covers species found in New Caledonia. Allen and Steene (1994) provide at least 37 species in color, and Colin and Arneson (1995) present a large collection of color photographs of tunicates, including approximately 96 species. See Kott (1985, 1990, and 1992) for an excellent technical account of the Australian ascidians.

In general, most forms are transparent, gelatinous, tubular in shape or encrusting. Sea squirts have an oral or incurrent siphon and a cloacal or excurrent siphon for water circulation and feeding. The oral siphon leads to an internal branchial sac, which is covered with ciliated slits called stigmata. The body of the adult sea squirt is covered with a tough tunic of polysaccharides and proteins. Many sea squirts are spectacularly colored. Most tunicates are hermaphroditic, producing tadpole-like larvae. Some species reproduce asexually by budding off new individuals.

Aplousobranchia - Aplousobranchs

The aplousobranchs are considered the most primitive sea squirts; their internal branchial sacs have the simplest structure of all ascidians. There are three large groups of aplousobranchs: the polyclinids, the polycitorids, and the didemnids.

1051. *Pseudodistoma* cf. *megalarva* Monniot & Monniot, 1996

Identification: This species forms soft gelatinous globular colonies up to 4 cm in diameter. The siphonal openings are relatively large. Color is usually uniformly orange.
Natural History: Encountered on shallow reef flats.
Distribution: Indonesia and Melanesia, including New Guinea (Indonesia).

1052. *Aplidium* cf. *tabascum* Kott, 1992

Identification: Colonies form mats or sheets or sometimes assume a globular growth habit. Color is mostly red-orange, often marbled with paler zooid-free areas.
Natural History: Infrequently encountered on shallow reef flats. Depth: 5-15 m.
Distribution: Australia and Hawaiian Islands (Hawai'i).

Bruce Watkins

Andy Sallmon

Leslie Newman & Andrew Flowers

1053. *Aplidium protectans* (Herdman, 1899)

Identification: This species forms firm, globular or spherical colonies, pale pinkish violet in color, with rings of whitish zooids; 3-6 cm in diameter.

Natural History: This ball-like ascidian has been recorded at depths ranging from 0-100 m.

Distribution: Eastern Australia and New Caledonia (Heron Island, Great Barrier Reef, Australia).

Roy Eisenhardt

1054. *Clavelina detorta* (Sluiter, 1904)

Identification: The tadpole-shaped zooids have a conspicuous gold or orange C-shaped or U-shaped intestine, easily seen through the clear tunic. The zooids have elongated stalks and are usually arranged in clusters. Some authors have placed this species in the genus *Pycnoclavella*.

Natural History: Infrequently encountered on reef flats and slopes.

Distribution: Red Sea to Australia; Indonesia; Philippines; Solomon Islands and New Caledonia (Solomon Islands).

1055. *Clavelina diminuta* Kott, 1957

Identification: Colonies form a mass of rounded heads, each on a narrow stalk up to 2 cm long. The heads are about 5 mm in diameter. Color usually golden-yellow with varying amounts of white pigment. Kott (1990) places this tunicate in the genus *Pycnoclavella*.

Natural History: This species is highly variable in color. Usually encountered in caves or under ledges. Depth: 5-20 m.

Distribution: Western Pacific: Australia; Indonesia; Philippines; and New Caledonia (Philippines; Indonesia).

Lovell & Libby Langstroth

Bruce Watkins

286

1056. *Clavelina* cf. *flava* Monniot, 1988

Identification: The zooids are mostly trapezoidal in shape with short stalks. They occur in clusters, and are highly variable in color, but usually yellow to orange, or sometimes white. Kott (1990) suggests that this species appears to have closer affinities to the genus *Euherdmania* than to *Clavelina*.

Natural History: Common but highly variable, on reef flats and slopes.

Distribution: Western Pacific: Indonesia and New Caledonia (Indonesia).

Nicholas Galluzzi

1057. *Clavelina moluccensis* (Sluiter, 1904)

Identification: Colonies are formed by numerous ovoid zooids on thick stalks. The wall of the zooids is transparent. Color is mostly pale blue or violet, sometimes darker blue. Darker pigment is usually present around the apertures.

Natural History: A striking but infrequently encountered tunicate on reef flats and slopes.

Distribution: Australia; Indonesia; Singapore; Philippines (Great Barrier Reef, Australia).

Leslie Newman & Andrew Flowers

1058. *Clavelina robusta* Kott, 1990

Identification: This species often occurs in dense clusters of cylindrical zooids. It is distinguished by the dark blue to smoke grey or black body with lemon yellow, green, or white rings around both siphons.

Natural History: This species is probably one of the most commonly encountered tunicates throughout its range.

Distribution: Western Pacific: Australia; Indonesia; New Guinea; Philippines; Japan; Belau; and Solomon Islands (Batangas, Luzon, Philippines).

G.C. Williams
Terry Schuller

1059. *Clavelina* cf. *viola*
Nishikawa & Tokioka, 1976

Identification: This form is distinguished by the gradually tapering body of the zooids and the numerous yellow speckles disposed throughout the length of the body of each individual. Bands are not present around the siphons as in *Clavelina robusta*.

Natural History: Infrequently encountered on reef flats and slopes.

Distribution: Western Pacific: southern Japan and Philippines (Batangas, Luzon, Philippines).

287

G.C. Williams

1060. *Clavelina* sp. 1

Identification: This form is similar in appearance to *Clavelina robusta*, but has two wing-shaped yellow blotches between the siphons, instead of having a continuous yellow ring surrounding the siphons. A few yellow speckles are also present below the yellow blotches.

Natural History: Infrequently found on reef flats and slopes.

Distribution: Philippines (Batangas, Luzon, Philippines).

G.C. Williams

1061. *Clavelina* sp. 2

Identification: Colonies are formed of congested bulbous zooids. The zooids are transparent. Color is a beautiful blue with lustrous white wing-like patches.

Natural History: Infrequently encountered on reef slopes and ledges. Fed upon by the nudibranchs, *Nembrotha* spp. (#601, 602). Depth: 15-25 m.

Distribution: Philippines (Batangas, Luzon, Philippines).

G.C. Williams
G.C. Williams

1062. *Clavelina* sp. 3

Identification: This species is similar in form to the previous one, but is mostly colorless instead of blue. The lustrous white patches often show a metallic iridescence.

Natural History: Infrequently encountered on reef slopes and reef margins. Depth: 10-20 m.

Distribution: Philippines (Batangas, Luzon, Philippines).

1063. *Clavelina* sp. 4

Identification: Barrel-shaped zooids 1-2 cm in height. Color is opaque white with chocolate brown rings around the siphons.

Natural History: Infrequently encountered on walls and slopes. Depth: 10-25 m.

Distribution: Philippines (Batangas, Luzon, Philippines).

1064. *Eudistoma* cf. *gilboviride* (Sluiter, 1909)

Identification: Firm, globular or ball-like colonies formed by wedge-shaped lobes. Color is bright green to very dark green, mottled with pale yellow blotches.

Natual History: Locally common, mostly on shallow reef flats.

Distribution: Australia; Melanesia and Indonesia (Solomon Islands).

Bruce Watkins

1065. *Eudistoma* sp.

Identification: Colonies form smooth globular masses. The star-shaped zooid systems are usually darker in color than the surrounding area. Color cream to light grey with smoke grey blotches.

Natual History: Frequently encountered covering large areas of rock surfaces, either on slopes or walls.

Distribution: Philippines (Verde Island, Philippines).

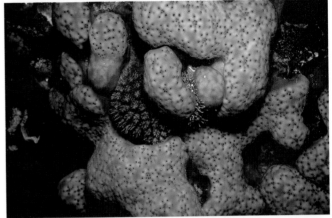
G.C. Williams

1066. *Sigillina signifera* (Sluiter, 1909)

Identification: Colonies grow as dark bluish-green mats, often with a lobular or hilly appearance. Very dark, almost black pigment is present around the siphons.

Natural History: On reef margins and slopes, often mixed with various seaweeds in channels where the current is often moderate to strong.

Distribution: Australia; Indonesia; Philippines and Micronesia (Batangas, Luzon, Philippines).

G.C. Williams
Mike Severns

1067. *Sycozoa* sp.

Identification: These are tall, club-shaped tunicates, up to 25 cm in height. The grape-like clusters of zooids are congested at the ends of slender stalks, forming elongated terminal heads. The zooids are transparent with striking lemon yellow markings.

Natural History: Locally abundant in deeper portions of muddy or sandy channels, often in areas with appreciable currents.

Distribution: Philippines (Batangas, Luzon, Philippines).

289

G.C. Williams

1068. *Oxycorynia fascicularis* Drasche, 1882

Identification: These tunicates appear as clusters of grapes at the end of long stalks. The zooids are tightly congested in the head. The colonies often form clusters and are mostly grey to dark green in color. Kott (1990) and some other authors place this species in the genus *Nephtheis*.

Natural History: Commonly encountered around overhangs, slopes, and walls. Fed upon by *Nembrotha lineolata* (#601) and *Nembrotha* sp. (#602).

Distribution: Australia; New Guinea; Indonesia; Philippines; New Caledonia and Micronesia (Batangas, Luzon, Philippines).

G.C. Williams

1069. *Didemnum molle* (Herdman, 1886)

Identification: Often urn-shaped and upright but sometimes globular or somewhat encrusting. The very large cloacal aperture is contrasted with the very numerous and minute oral siphons.

Natural History: This species is perhaps the most commonly encountered sea squirt in much of the Indo-Pacific. *Atriolum robustum* (#1075) is often confused with this species, but has much larger and fewer oral siphons. As in many tropical didemnid tunicates, the green coloration is produced by the symbiotic prokaryote *Prochloron*. See Kott (1980) for an account of symbiosis in Indo-Pacific didemnids.

Distribution: Zanzibar; Madagascar to Australia; New Guinea; Indonesia; Philippines; Belau and New Caledonia (Batangas, Luzon, Philippines).

1070. *Didemnum* cf. *moseleyi* (Herdman, 1886)
Strawberry Tunicate

Identification: This species forms squat, urn-shaped colonies that are uniformly orange to wine-red in color, usually less than 2 cm in diameter.

Natural History: Infrequently encountered on reef flats and slopes.

Distribution: Australia; New Guinea; Philippines; Japan; Micronesia; Belau and Kiribati (Milne Bay, Papua New Guinea).

Marc Chamberlain
Marc Chamberlain

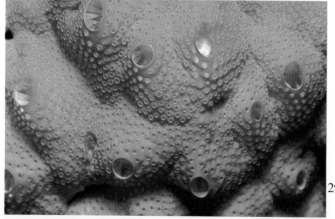

1071. *Didemnum* sp. 1

Identification: This encrusting tunicate has the cloacal apertures atop prominent domes or mounds. The numerous and smaller oral apertures cover the flanks of each mound. Color is orange to red.

Natural History: This species strongly resembles a form identified as *Didemnum rubium* by Colin and Arneson (1995). The shrimp *Periclimenes* often inhabit the cloacal siphons of tunicates similar to this species.

Distribution: Melanesia (Solomon Islands).

1072. *Didemnum* sp. 2

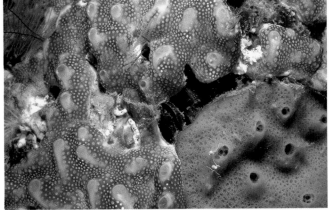

G.C. Williams

Identification: Colonies of these tunicates are membranous and spreading. They are most often deep magenta in color with white to pink areas around the cloacal siphons, resulting from varying concentrations of spicules in these areas. Figures 3K,L show typical didemnid spicules.

Natural History: Frequently observed on reef flats and slopes.

Distribution: Philippines and New Guinea (Madang, Papua New Guinea).

1073. *Didemnum* sp. 3

G.C. Williams

Identification: Membranous, sponge-like colonies with star-like patterns at each cloacal aperture. The smaller oral apertures are very numerous and cover the surface between the cloacal apertures. Color is vivid orange.

Natural History: Encrusting tunicates such as this species are superficially similar to poecilosclerid sponges (#20, 21) in that the light-colored, densely spiculated ridges radiating out from the cloacal apertures look very much like the large exhalant canals surrounding the excurrent apertures (oscula) of the sponges.

Distribution: Western Pacific: Melanesia (Solomon Islands).

1074. *Leptoclinides* cf. *reticulatus* (Sluiter, 1909)

G.C. Williams

Identification: A thin, membranous and spreading tunicate. Smoke-grey to almost black with bright white siphon interiors.

Natural History: A common form encountered on shallow-water reefs.

Distribution: Western Pacific: Philippines; Solomon Islands; New Caledonia and Micronesia (Solomon Islands).

G.C. Williams

1075. *Atriolum robustum* Kott, 1983

Identification: Urn-shaped colonies, usually less than 3 cm in height, with relatively large and few oral siphons. Color variable, often rust-orange or bright green.

Natural History: This species is easily and often confused with *Didemnum molle* (#1069), which has numerous and minute oral siphons. The sea star *Gomophia egyptiaca* (#952) and the snail *Gyrineum gyrinum* prey upon *Atriolum robustum*. The green color of this and other tropical didemnid species results from the symbiotic prokaryote *Prochloron*.

Distribution: South Africa to New Guinea; Philippines; New Caledonia (Madang, Papua New Guinea).

291

G.C. Williams

1076. *Lissoclinum patellum* (Gottschaldt, 1898)

Identification: These tunicate colonies are composed of gelatinous mats with transparent, milky white ridges and bright green valleys.

Natural History: This species produces a powerful anti-cancer agent known as ulithiacyclamide. The green color is produced by the symbiotic prokaryote *Prochloron*.

Distribution: Western Pacific: Australia; Indonesia; Malaysia; Philippines; Belau; New Guinea and New Caledonia (Madang, Papua New Guinea).

Leslie Newman & Andrew Flowers

1077. *Lissoclinum* cf. *vareau* Monniot & Monniot, 1987

Identification: Colonies are encrusting, soft, fleshy, and fragile. Both the oral and cloacal openings are numerous and distributed throughout the colony. Color is white with magenta siphon openings. Another species, *Leptoclinides dubius* (Sluiter, 1909), is superficially similar to this species, but forms globular cushions with only a few but relatively large cloacal openings.

Natural History: Pontoniine shrimp inhabit the cloacal cavities of *L. vareau*.

Distribution: Western Pacific: Australia; New Caledonia and Polynesia (Heron Island, Great Barrier Reef, Australia).

Spicule

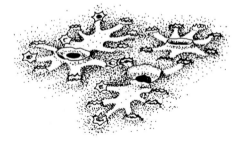

1078. Didemnid sp.

Identification: This strange didemnid tunicate is composed of an encrusting mass with stringy lobes. Color greyish-white.

Natural History: This species has been found on the ceilings of caves and overhangs.

Distribution: Philippines (Philippines).

Jerry Allen
Marc Chamberlain

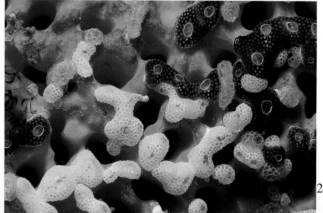

1079. Didemnid spp.

Identification: Three species of unidentified didemnid tunicates are shown in this photograph. They are membranous and somewhat globular. The yellow, pink, and bluish-black forms appear to be distinct.

Natural History: Encountered growing on sponges on reef flats.

Distribution: Malaysia (Sipadan, Borneo).

Phlebobranchia - Phlebobranchs

The phlebobranchs are intermediate in complexity between the primitive aplousobranchs and the more derived stolidobranchs. The stigmata or slits of the branchial sac may assume complex spiral shapes. Commonly encountered phlebobranchs include the cionids, the perophorids, and the ascidiids.

1080. *Rhopalaea crassa* (Herdman, 1880)

Identification: The cylindrical zooids are usually less than 5 cm in height, mostly transparent and colorless with thin sky-blue rings around the siphons.

Natural History: This species often has minute amphipods inhabiting the inside of the tunic or branchial chamber, several of which have proven to be new to science.

Distribution: Sri Lanka; Australia; New Guinea; New Caledonia; Indonesia; Philippines; Hong Kong; Japan (Madang, Papua New Guinea).

G.C. Williams

1081. *Rhopalaea* spp.

Identification: Several forms are similar in appearance to *Rhopalaea crassa* but are variably colored from colorless to transparent sky blue, or opaque blue to purple. In transparent forms, the sperm duct often appears as a white, brown, or blue longitudinal line running throughout most of the length of the tunicate body.

Natural History: The genus is very common throughout much of the Indo-Pacific on reef flats and slopes. One form is fed upon by nudibranchs of the genus *Nembrotha* (#601, 602).

Distribution: New Guinea; Philippines; Solomon Islands; Australia; New Caledonia (Batangas, Philippines; Sapawauhfik, Pohnpei; Indonesia; Batangas, Philippines).

Marc Chamberlain

T. M. Gosliner

Dave Zoutendyk
Bruce Watkins

293

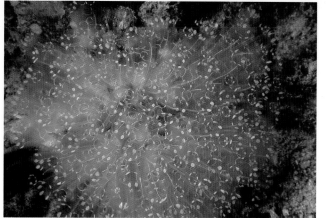

T. M. Gosliner

1082. *Diazona* sp.

Identification: This beautiful species presents delicate clusters of perfectly clear tubular tunicates, 1-2 cm in height, with six larger white spade-shaped dots around the lobes of the oral siphon and six smaller but similar dots around the cloacal siphon.

Natural History: This species occurs as colonies in quiet-water areas of reef flats and slopes.

Distribution: Western Pacific: Indonesia and Philippines (Batangas, Luzon, Philippines).

Terry Schuller

1083. *Perophora modificata* Kott, 1985

Identification: The zooids are disposed in dense clusters, each with a short stalk. They are usually transparent and lemon-yellow in color.

Natural History: Infrequently encountered in channels or lagoons. Depth: 17-27 m.

Distribution: Western Pacific: Australia; Indonesia; Philippines and New Caledonia (Manado, Indonesia).

Bruce Watkins
Lynn Funkhouser

1084. *Perophora namei* Hartmeyer & Michaelsen, 1928

Identification: A slender horn-like stalk less than 10 cm in length gives rise to numerous ovoid to somewhat triangular zooids at the end of slender stalks. The zooids are transparent, often bicolored blue and yellow, or uniformly pale yellow.

Natural History: Encountered at the base of slopes near sandy bottoms, often on vertical surfaces; around 20-26 m in depth.

Distribution: Coral Sea and Philippines (Philippines).

1085. *Ecteinascidia* sp.

Identification: Numerous gelatinous, cylindrical zooids are congested to form heads atop elongated stalks. The stalks are rough with epizoic growth. The zooids are mostly colorless with orange circles around the siphons; the stalks are yellowish. At least seven species of the genus have been recorded from the Indo-Pacific including *Ecteinascidia diaphanis* Sluiter, 1885 from Indonesia and *E. hataii* Tokioka, 1954 from Belau.

Natural History: Shallow subtidal on reef slopes and flats.

Distribution: Philippines (Philippines).

294

1086. *Phallusia julinea* Sluiter, 1915

Identification: Solitary, the body is transparent, 5 cm in length, with dense yellow patterns on and immediately below the siphons. Some individuals may be yellow throughout.

Natural History: These tunicates are often partially imbedded in sponges or brain corals with only the yellow siphons exposed; sometimes hidden underneath coral slabs or in crevices.

Distribution: Madagascar; Australia; Indonesia; Solomon Islands; New Caledonia and Micronesia (Kwajalein Lagoon, Marshall Islands; Heron Island, Great Barrier Reef, Australia).

Jack Randall

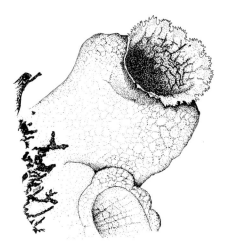

1087. Phlebobranch ascidians

Identification: These unidentified tunicates are globular, soft and gelatinous, with smooth surfaces. They probably represent the same species. Apricot orange or deep purple in color.

Natural History: Shallow subtidal on reef flats.

Distribution: Indian Ocean (Kuria Muria Islands, Oman).

Leslie Newman & Andrew Flowers

Stolidobranchia - Stolidobranchs

The stolidobranchs are the most structurally complex tunicates. The wall of the branchial sac is elaborated with numerous longitudinal vessels. Stolidobranchs comprise three common groups of sea squirts: the stylelids, pyurids, and molgulids.

1088. *Botryllus* cf. *tuberatus* Ritter & Forsyth, 1917

Identification: The zooids form circular or ovoid systems. Each zooid is 2-4 mm in length and has a tail-like appendage directed toward the inside of the circle. This species is extremely variable in color, ranging from pink or deep purple to pale green or cream, or markedly bicolored. This tunicate superficially resembles *Botryllus niger,* as well as *B. tuberatus.*

Natural History: A common and variable inhabitant of reef flats and slopes.

Distribution: Circumtropical: Australia; Indonesia; Philippines; China; Japan; New Caledonia and Micronesia; Atlantic Ocean (Batangas, Luzon, Philippines).

Jack Randall
Marc Chamberlain

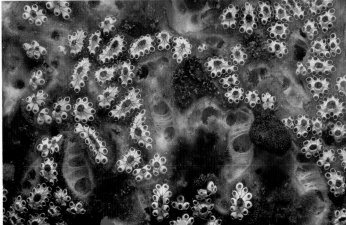

295

1089. *Botryllus* sp. 1

Identification: Several other botryllid tunicates have zooids that form circular systems similar to the previous species. *Botryllus tuberatus* and *B. niger* are examples, but unlike those species, each zooid in the species pictured here is globular in shape and does not have a tail-like appendage. In addition, the centers of the circles are mostly solid with numerous visible spicules, and do not form apertures. Color orange and dark grey. As is the case with other botryllids, these forms are highly variable in color.
Natural History: Encountered in a variety of reef habitats.
Distribution: Philippines (Batangas, Luzon, Philippines).

Marc Chamberlain

1090. *Botryllus* sp. 2

Identification: Colonies are puffy and encrusting, with circular oral siphons arranged around the prominent and larger cloacal siphons. Color is highly variable. This species superficially resembles *Botryllus leptus* (Herdman, 1899) from the western Pacific.
Natural History: Frequently encountered on reef flats.
Distribution: Philippines (Batangas, Luzon, Philippines).

G.C. Williams

1091. *Botryllus* spp.

Identification: Encrusting or thin and membranous to globular and lumpy colonies. The small zooids (1-4 mm in length) are crowded and arranged in chains or elliptical systems. The large common cloacal siphons are often conical and volcano-like or flush with the rest of the colony. These forms and other species of the genus *Botryllus* are highly variable in color.
Natural History: Common inhabitants on shallow reefs.
Distribution: The genus is very common throughout the Indo-Pacific (New Georgia Group, Solomon Islands; Philippines; Philippines).

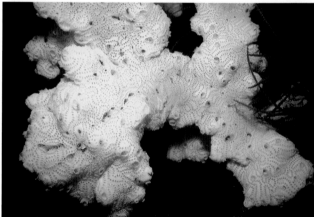

G.C. Williams
Mike Miller

D.W. Behrens

296

1092. *Botrylloides leachi* (Savigny, 1816)

Identification: The zooid systems are circular as well as elongate and chain-like. Color is highly variable as there seems to be no limit to possible color combinations. The genus *Botrylloides* differs from *Botryllus* by having elongate chain-like systems containing more than twenty zooids. The zooid systems of *Botryllus* include circular to ovoid or stellate systems usually with fewer than twenty zooids per system.

Natural History: Infrequently encountered on reef flats, ledges, and slopes.

Distribution: Red Sea and eastern Africa to Australia; New Zealand; Indonesia; Philippines and New Caledonia (Batangas, Luzon, Philippines).

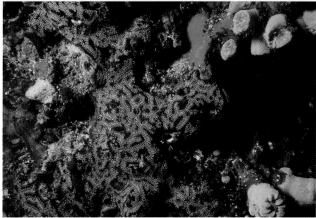
T. M. Gosliner

1093. *Eusynstyela latericius* (Sluiter, 1904)

Identification: Individual zooids are usually apricot-colored without a white mark between the siphons. The zooids are approximately 1 cm in length and are often highly congested.

Natural History: Frequently encountered covering a variety of objects such as coral slabs, rubble, and shells.

Distribution: Indo-West Pacific: including the Persian Gulf; Australia; Indonesia; Philippines; Vietnam and Solomon Islands (Solomon Islands).

G.C. Williams

1094. *Eusynstyela* cf. *misakiensis* (Watanabe & Tokioka, 1972)

Identification: A very similar form to the previous species. *Eusynstyela misakiensis* is orange-red to reddish purple with a conspicuous white mark between the siphons.

Natural History: Occasionally encountered covering objects such as shells or other tunicates.

Distribution: Indonesia; New Caledonia; Japan (Indonesia).

Bruce Watkins
Leslie Newman & Andrew Flowers

1095. *Eusynstyela* sp.

Identification: The zooids are often less congested than other forms of *Eusynstyela*. Color of the zooids is yellow with red or orange siphons.

Natural History: Numerous zooids often cover objects such as coral slabs.

Distribution: Western Pacific including parts of Melanesia (Papua New Guinea).

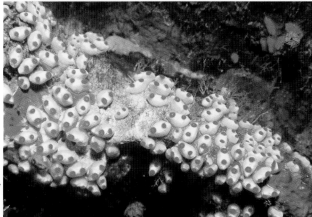

297

Bruce Watkins

1096. *Polycarpa aurata* (Quoy & Gaimard, 1834)

Identification: These are large bulbous solitary tunicates, commonly 4-10 cm in length, mottled with purple, white, and gold. The insides of the siphons are conspicuously yellow. The tunic is clean, without encrusting epizoic organisms.

Natural History: This species, along with *Didemnum molle*, are probably the two most common and conspicuous sea squirts encountered in the western Pacific. Depth: 3-20 m. The similar *Polycarpa aurita* (Sluiter, 1890) from the western Pacific and western Atlantic, often has the tunic covered with encrusting epizoic organisms.

Distribution: Widely distributed in the Indo-Pacific: Australia; New Guinea; Indonesia; Philippines; Solomon Islands and Micronesia (Solomon Islands).

1097. *Polycarpa clavata* Hartmeyer, 1919

Identification: This is one of the largest and most distinctive solitary reef tunicates. A stalked species up to 12 cm in length. The characteristic elongated sac-like body has a conspicuously bumpy surface and is attached by one end to the stalk. Color varies from lemon-yellow to rust-orange.

Natural History: Individuals often live in small groups on sandy or rocky bottoms. Depth: 5-40 m.

Distribution: Western Pacific: Australia; Coral Sea and New Caledonia (Great Barrier Reef, Australia).

Dan Gotshall

T. M. Gosliner

G.C. Williams

1098. *Polycarpa* cf. *contecta* (Sluiter, 1904)

Identification: A colonial species with short, rotund zooids having short siphons; tan to brownish orange in color. This species resembles *Polycarpa nigricans* Heller, 1878, from the Indo-West Pacific, which is smoke-grey to black in color.

Natural History: Occasionally encountered on reef slopes.

Distribution: Western Pacific: Australia; Indonesia; Philippines and New Caledonia (Batangas, Luzon, Philippines).

1099. *Polycarpa* cf. *papillata* (Sluiter, 1885)

Identification: A highly variable species in terms of shape, surface texture, and color patterns. The surface can be smooth or distinctly bumpy. Color white or grey and mottled to some degree with bright red.

Natural History: Often encountered in the midst of dense encrusting growths of a variety of benthic organisms.

Distribution: Madagascar; Sri Lanka; Australia; Indonesia; Philippines; Solomon Islands; New Caledonia and western Micronesia (Batangas, Luzon, Philippines; South Solitary Islands, New South Wales, Australia).

Stephen Smith

1100. *Polycarpa pigmentata* (Herdman, 1906)

Identification: Tunicates with a dark, thick and leathery tunic. The interior of the siphons are variously speckled with snow-white pigment.

Natural History: The tunic is cryptic when it is covered with thick epizoic growth.

Distribution: Indian Ocean to Australia; Indonesia; Philippines; Solomon Islands; New Caledonia and Micronesia; Japan (New Georgia Group, Solomon Islands).

G.C. Williams

1101. *Styela* sp.

Identification: Slender solitary tunicates with gradually tapering bodies, approximately 6 cm in height. The siphons have hood-like structures. Color often mottled black and white.

Natural History: Infrequently encountered on slopes and ledges.

Distribution: Philippines (Batangas, Luzon, Philippines).

T. M. Gosliner

1102. *Herdmania momus* (Savigny, 1816)

Identification: Globular, solitary tunicates with short trumpet-shaped siphons. The margin of the oral siphon is conspicuously flared. Color is tan, mottled with red and bluish-white. Some authors place this species in the genus *Pyura*.

Natural History: The external surface of this species often contain small epizoic encrustations. The shrimp *Pontonia sibogae* lives in the branchial sac of this tunicate. The species has been recorded to 100 m in depth.

Distribution: Western Atlantic and Indo-Pacific: Red Sea; eastern Africa to Sri Lanka; Australia; New Guinea; Indonesia; Philippines; Japan; Micronesia; New Caledonia; Fiji and Society Islands (Batangas, Luzon, Philippines).

Nicholas Galluzzi
Carol Buchanan

1103. *Pyura australis* (Quoy & Gaimard, 1834)
Sea Tulips

Identification: A remarkable species up to 30 cm in height with large globular heads at the end of slender stalks. Color is either yellow-orange or maroon-violet.

Natural History: Encountered in depths up to 20 m.

Distribution: Temperate Australia and southern Great Barrier Reef (Marsh Shoal, New South Wales, Australia).

ACKNOWLEDGMENTS

Compiling a field guide of 1100 species has been a task the three of us could not have done without help. Since the inception of the project, we have viewed it as a team effort. If we could, we would have included everyone's name on the cover with ours. Aside from the scientific and photographic contributors, we would like to especially acknowledge the support of our families. To our wives, Bonnie, Diana and Susie, and all our children, thank you.

We also acknowledge the contributors listed below, who have given generously of their time and expertise, whether photographic or scientific. They have provided us with enthusiasm to make this the scientific contribution it has become.

There have been a few who have gone beyond our requests for photographs and information. These folks have provided ongoing support, and given us leads to many of the resources we would never have tapped without them. We wish to recognize them separately: Carol Buchanan (New South Wales, Australia), Marc Chamberlain (San Diego, CA), Mike Miller (San Diego, CA), Leslie Newman and Andrew Flowers (Queensland, Australia), Terry Schuller (Oxnard, CA), and Mike Severns (Kihei, HA).

Many taxonomic experts were called upon to assist us in confirmation of species identification and manuscript review. Our sincere thanks to: Phil Alderslade, Museum and Art Gallery of the Northern Territories, Darwin (MNT) (*Subergorgia*), Charlie Arneson, Coral Reef Research Foundation (sponges), Ron Ates, Nationaal Natuurhistorisch Museum, Leiden (NNM) (*Verrillactis*), Fredrick Bayer, National Museum of Natural History (NMNH), Washinton D.C. (nephtheid soft coral), Maya Borel Best, NNM (faviid scleractinians), A.J. "Sandy" Bruce, previously at MNT (shrimp), Dale Calder, Royal Ontario Museum, Toronto (hydroids), Steven Cairns, NMNH (stylasterines), Roy Caldwell, Univ. California, Berkeley (stomatopods), Henry Chaney, Santa Barbara Museum of Natural History (SBMNH) (bryozoans), C. Allan Child, NMNH (pycnogonids), Dustin Chivers, California Academy of Sciences (CAS) (decapods), Daphne Fautin, Univ. of Kansas (*Phyllodiscus*),D. P. Gordon,

New Zealand Oceanographic Inst. (bryozoans), M. Grasshoff, Senckenburg Museum, Frankfurt (*Muricella*), Russell Hanley, MNT (polychaetes), J.C. den Hartog, NNM (*Actinostephanus* and *Pseudocorynactis*), Gordon Hendler, Los Angeles County Museum (brittle stars), Y. Hirano, Kominato Marine Laboratory (*Lipkea*), Eric Hochberg, SBMNH (cephalopods), Bert W. Hoeksema, NNM (fungiid scleractinians), L.B. Holthuis, NNM (*Scyllarus*), G. Jarms, Univ. of Hamburg (*Nausithoe*), Brian Kensley, NMNH (isopods), Alan Kabat, NMNH (naticids), C.C. Lu, Museum of Victoria (cephalopods), Christopher Mah, CAS (asteriods), Ray Manning, NMNH (stomatopods), Colin McLay, Canterbury Univ. (dromiids), George Metz, CAS (gastropods), Yasuhiko Miya, Nagasaki Univ. (alpheids), Claude Monniot, Museum National d'Histoire Naturelle Paris (tunicates), Rich Mooi, CAS (echinoids), Leslie Newman, Univ. of Queensland (flatworms), Peter K.L. Ng, National Univ. of Singapore (alpheids), Jon Norenburg, NMNH (nemerteans), Mark Norman, NMNH (cephalopods), Leen van Ofwegen, NNM (nephtheid soft corals), Junji Okuno, Natural History Museum, Chiba (*Rhynchocinetes*), David Pawson, NMNH (sea cucumbers), J. S. Ryland, Univ. of Wales, Swansea (zoanthids), Paul Scott, SBNHM (galeomatids), Robert Van Syoc, CAS (barnacles), Janet Voight, Field Museum (cephalopods), Tom Waller, NMNH (pectinids), Mary Wicksten, Texas A&M Univ. (decapods), Leigh Winsor, James Cook Univ. of North Queensland (aceol flatworms), Debra Zmarzly, Stephen Birch Aquarium, La Jolla (crinoids).

Field work was generously supported by the California Academy of Sciences, Christensen Research Institute, Katharine Stewart, the Smithsonian Institution and the South African Museum.

We also acknowledge the efforts of Diana Behrens, who assisted with the text processing and Hans Bertsch (National Univ., San Diego) and Katie Martin (CAS) who edited our text.

Lastly, we thank the following who, in addition to providing enthusiastic encouragement, also provided the photographic material that is the foundation on which this field guide stands. Their contributions, if not actually printed in this publication, were used in describing color variation and natural history observations:

Jerry Allen	Andrew Flowers	Bud Lee	Terry Schuller
Charlie Arneson	Lynn Funkhouser	Laura Losito	Stephen Smith
Glenn Barrall	Nicholas Galluzzi	Fred McConnaughey	Mike Severns
Jim Black	Dan Gotshall	Jonathan Mee	Pauline Fiene-Severns
Brian Boer	James Hargrove	Mike Miller	Dave Tarrant
Robert Bolland	Robert Herrick	David K. Mulliner	Ben Tetzner
Carol Buchanan	Roger Hess	Leslie Newman	Kathy Tubbenhauer
Neil Buchanan	Bert Hoeksema	Dot Norris	Tony Tubbenhauer
Roy Caldwell	Ken Howard	Greg Ochocki	Frank Viola
Marc Chamberlain	Mark Hughes	Junji Okuno	Bruce Watkins
Kathy deWet	Lauren Newberry Joy	Glenn Pollock	Robert Yin
Richard Dohner	Alex Kerstitch	John Randall	Debbie Zmarzly
Roy Eisenhardt	Libby Langstroth	David Reid	Dave Zoutendyk
	Lovell Langstroth	Andy Sallmon	

GLOSSARY

aboral plates: plates on side of body away from mouth.

acontia: thread-like structures that hang freely in the gut of an anemone which bear stinging cells and can serve in defense.

ahermatypic: corals that do not contribute directly to reef formation.

ambulacral groove: the space between rows of tube feet in echinoderms, used for feeding.

anchialine: a saltwater habitat that has no surface connection to the sea.

antero-lateral: on the side at the front or head end.

aposematic: conspicuous, warning color or display.

aposymbiotic: organisms that do not have symbiotic zooxanthellae in their tissues.

aragonite: a mineral of calcium carbonate crystallized in orthorhombic form, and having a density greater than calcite.

arborescent: bushy or tree-like in growth form.

asexual: reproduction by fragmentation or budding, not the combination of egg and sperm.

atrial siphon: incurrent siphon of a tunicate.

autotomy: the ability to dislodge or drop off a part of the body.

benthic: living in association with the bottom, or benthos.

bilateral symmetry: symmetry which divides a body along a single plane into identical halves.

biogenic: produced by living things.

calcite: a mineral of calcium carbonate crystallized in hexagonal form, with a density less than aragonite.

calcium carbonate: a chemical compound of which the skeletal material of many animals is based - abbreviated as $CaCO_3$. A single molecule is composed of one atom of calcium (Ca), one of carbon (C), and three of oxygen (O).

capitate: with a rounded tip, as in the tentacles of some hydroids.

capitulum: the broad disc-shaped to spherical distal portion of a soft coral that contains the polyps, usually arising from a basal stalk region.

carapace: chitinous shell enclosing the body of crustaceans.

caryophyllidia: specialized tubercles, surrounded by spicules, which have a sensory function.

caudal: referring to the tail of an animal.

cavity: chamber or internal space.

cerata: processes on the dorsal surface of aeolid nudibranchs which contain a branch of the liver diverticulum, and function as respiratory organs.

cheliped: the claw bearing appendages of decapod crustaceans.

cirri: the feeding legs of barnacles, or the projections on the bottom of a crinoid used in holding it attached to the substratum.

cloacal aperture (siphon): posterior end of intestinal tract; excurrent siphon in tunicates.

coelom: a fluid-filled cavity between the body wall and the gut which is completely enclosed by a thin cellular lining (the peritoneum).

concrement granules: deposits of inorganic materials in the living tissue.

cortex: the outer or surface tissue layers of animals or organs.

dactylozooids: the defensive polyp in hydrozoan colonies such as hydroids and hydrocorals.

deuterostome: a group of animals containing lophophorares, echinoderms and chordates, in which the mouth forms secondarily rather than from the larval mouth.

discoidal: appearing rounded or chopped off.

DNA: Deoxyribonucleic acid. The molecule that carries the genetic information contained within a gene.

ecosystem: a unit of biological organization composed of biotic (living) and abiotic (physical) components.

elytra: the flat plates on the dorsal side of a scale worm.

gastrozooids: the feeding polyp in hydrozoan colonies such as hydroids and hydrocorals.

hadal: depth zone including the deepest trenches and canyons of the sea.

hermatypic: corals that directly contribute to the building of a coral reef.

horn: a dark, tough, fibrous protein of high sulphur content that composes the axis of black corals and certain gorgonian corals.

karst: a landscape composed of rounded hills or towers of exposed reef formations, resulting from the solution of limestone rock by groundwater and terrestrial erosion.

korallion: the ancient Greek word from which the word "coral" is derived; refers to the Mediterranean precious red coral of commerce (*Corallium rubrum*).

lamina: a thin layer or plate.

lophophore: the circle or spiral of tentacles used for feeding in phoronids, brachiopods and bryozoans.

medusa: a free-swimming, bowl-shaped individual in the life history of many coelenterates.

merus (meral): middle segment of a crustacean walking leg.

mimic: a species, or individual, that closely imitates another.

model: the species, or individual, imitated by a mimic.

monophyletic: a natural group of organisms that includes all of the descendents of a common ancestor.

nematocyst: a minute stinging cell found in the epidermis of coelenterates, used for food capture and defense; composed of a spherical capsule with a coiled, hollow, thread-like tube and a reservoir of phenol or protein toxins.

notum: the back surface of an opisthobranch, the dorsum.

operculum: the door closing a snail shell or bryozoan zooid.

oral aperture (oral siphon): the incurrent siphon in tunicates.

osculum (oscula): the larger, excurrent opening of a sponge.

ossicle: spicules of calcium carbonate comprising the exoskeleton of sea cucumbers and other echinoderms (Fig. 3J).

ostia: the small, incurrent pores of a sponge.

ovicells: an expanded brood chamber on the zooids of bryozoans often containing fertilized eggs.

oviger: egg carrying leg in pycnogonids.

parapodia: flap-like lateral body extensions which arise ventrally.

pedicellaria: microscopic pincer-like structures on surface of some echinoderms (Fig. 3I).

periostracum: the brownish protein covering of the shell of many mollusks.

pinnate: with feather-like branches.

pinnae or pinnules: lateral branches of a pinnate structure.

planar: branching in one plane, flat or fan-like, not bushy.

polyp: any stationary or attached individual of a coelenterate colony, composed of a mouth surrounded by a ring of tentacles.

polytrophic: deriving nourishment in a variety of ways.

pores: surface openings for internal water circulation.

priority: the name that is correctly published before any other name.

proboscis: a long, flexible snout.

protostome: a group of animals in which the adult mouth arises from the embryonic mouth.

prokaryote: minute organisms such as bacteria and cyanobacteria that do not have nuclei or other organelles in their cells.

ptychocyst: specialized cells in tube anemones that function to construct the unique tube in which the animal resides.

radial cleavage: cleavage occurring along a central axis, one end to another; occurs in such animal groups as echinoderms, bryozoans and chordates.

radial symmetry: symmetry in which similar body regions are arranged around a central axis.

radula (radular teeth): a file-like extensible feeding structure found in most mollusks; it bears numerous chitinous teeth (Fig. 3G).

rank: a level of classification, such as order and phylum.

ray: an arm of a starfish.

RNA: Ribonucleic acid, a nucleic acid associated with the control of cellular chemical activities.

rostrum: anterior extension of the carapace in Crustacea.

sclerite: spicules of calcium carbonate that are imbedded in the tissue of most octocorals and represent the skeletal components of soft corals (Fig. 3E).

septa: walls or ridges of calcium carbonate that radiate outward from the center of a hard coral polyp.

seta: a bristle-like spine of a polychaete worm (Fig. 3F).

somite: a body segment of a segmented animal.

spiral cleavage: a complex form of early development occurring in animal groups such as flatworms, polychaetes, mollusks and crustaceans.

spicule: skeletal element, composed of calcium carbonate or silicon dioxide (Figs. 3A-C, E, K-L).

symbiotic: living in association with.

systematist: a scientist who studies the evolutionary relationships between organisms, and who describes new taxa as well.

taxon: a monophyletic group of organisms.

taxonomist: a person who studies the classification of organisms.

telson: tail portion of an arthropod.

test: the calcareous internal skeleton of a sea urchin.

thoracic: on the thorax of a crustacean.

thorax: middle portion of the body of an arthopod.

tunic: gelatinous or leathery skin covering the body of tunicates.

uropods: lateral segments of the telson, or tail, of crustaceans.

vicariance biogeography: the study of distribution and isolation of phylogenetically closely related species.

zooid: an individual in a colony, particularly in a coelenterate, tunicate or bryozoan (Fig. 3H).

zooxanthellae: unicellular symbiotic algae, referring mostly to the dinoflagellate *Symbiodinium microadriaticum*, often golden brown in color, which inhabits the internal tissues of many reef corals.

RECENT GEOGRAPHIC NAME CHANGES
(New Names in Bold)

Burma = **Myanmar**
Caroline Islands = **Federated States of Micronesia**
Ceylon = **Sri Lanka**
Ellis Islands = **Tuvalu**
Gilberts & Phoenix Group = **Kiribati**
Hawaii = **Hawai'i**
Laccadives = **Lakshadweep**
New Hebrides = **Vanuatu**
Palau = **Belau**
Ponape = **Pohnpei**
Truk = **Chuuk**

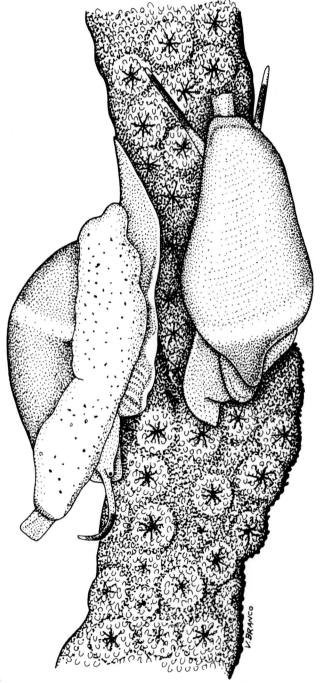

REFERENCES

Allen, G. R. & R. Steene. 1994. Indo-Pacific Coral Reef Field Guide. Tropical Reef Research: Singapore.

Alderslade, P. & Baxter. 1987. A new species of *Sinularia* (Coelenterata: Octocorallia) from western Australia. Records of the Western Australian Museum 13(2): 203-214.

Appleman, D.E. 1985. James Dwight Dana and Pacific geology. In: Viola, H.J. & Margolis, C. (eds.). Magnificent Voyagers - The U.S. Exploring Expedition, 1838-1842. Smithsonian Institution Press: Washington.

Auerbach, P.S. & E.C. Geehr. 1989. Management of wilderness and environmental emergencies. (second edition). The C.V. Mosby Company: St. Louis.

Bayer, F. M. 1949. The Alcyonaria of Bikini and other Atolls in the Marshall Group. Part I: The Gorgonacea. Pacific Science 3(3): 195-214.

Bayer, F.M. & Grasshoff, M. 1994. The genus group taxa of the family Ellisellidae, with clarification of the genera established by J.E. Gray (Cnidaria: Octocorallia). Senckenbergiana Biologica 74(1/2): 21-45.

Bergquist, P.R. 1965. The sponges of Micronesia, Part 1. The Palau Archipelago. Pacific Science 19: 123-204.

Bertsch, H. & S. Johnson. 1981. Hawaiian Nudibranchs. Oriental Publishing: Honolulu.

Borges, M.K. & P.R. Bergquist. 1988. Sponges from Motupore Island, Papua New Guinea. Indo-Malayan Zoology 5(2): 121-159.

Bruce, A.J. 1980. Pontoniine shrimps from the Great Astrolabe Reef, Fiji. Pacific Science 34(4): 389-400.

Bruce, A.J. 1989. A report on some coral reef shrimps from the Philippine Islands. Asian Marine Biology 6: 173-192.

Bruce, A.J. 1992. Two new species of *Periclimenes* (Crustacea: Decapoda: Palaemonidae) from Lizard Island, Queensland, with notes on some related taxa. Records of the Australian Museum 44: 45-84.

Bruce, A. J. 1993. Some coral reef pontoniine shrimps from Vietnam. Asian Marine Biology 10: 55-75.

Brunckhorst, D. 1993. The systematics and phylogeny of phyllidiid nudibranchs (Doridoidea). Records of the Australian Museum, supplement 16: 1-107.

Burgess, C.M. 1985. Cowries of the World. Seacomber Publications: Cape Town.

Carlgren, O. 1949. A survey of the Ptychodactiaria, Corallimorpharia and Actinaria. Kungl. Svenska Vetenskapsakademiens Handlingar 1(1): 1-121.

Cate, C. 1973. A systematic revision of the recent family Ovulidae (Mollusca: Gastropoda). Veliger 15 (suppl.): 1-116.

Cate, C. 1979. A review of the Triviidae (Mollusca: Gastropoda). Memoirs of the San Diego Society of Natural History. 10: 1-126.

Chase, F.A. & A.J. Bruce. 1993. The Caridean shrimps (Crustacea: Decapoda) of the Albatross Philippine Expedition 1907-1910, Part 6: Superfamily Palaemonoidea. Smithsonian Contributions to Zoology. 543: 1-152.

Clark, A.C. & F. Rowe. 1971. Monograph of shallow-water Indo-Pacific Echinoderms. British Museum (Natural History). Spec. Publ. 690: 1-238.

Clark, W.C. 1963. Australian Pycnogonida. Records of the Australian Museum 26(1): 1-81.

Coleman, N. 1989. Nudibranchs of the South Pacific. Neville Coleman's Sea Australia Resource Centre: Springwood, Australia.

Colin, P. & C. Arneson. 1995. Tropical Pacific Invertebrates. Coral Reef Press: Beverly Hills.

Couthouy, J.P. 1842. Remarks upon coral formations in the Pacific; with suggestions as to the causes of their absence in the same parallels of latitude on the coast of South America. Boston Journal of Natural History 4(1): 66-105, 137-162.

Cutler, E. 1994. The Sipuncula, their systematics, biology and evolution. Comstock Publications: Ithaca.

Daly, R.A. 1915. The glacial-control theory of coral reefs. Proceedings of the American Academy of Arts and Sciences 51: 155-251.

Dana, J.D. 1849. Geology. U.S. Exploring Expedition, Vol. 10. Philadelphia: C. Sherman; with Atlas. Geo. P. Putnam: New York.

Dana, J.D. 1853. On coral reefs and islands. New York.

Dana, J.D. 1872. Corals and coral islands. New York.

Darwin, C.R. 1984. The structure and distribution of coral reefs (foreword by M.T. Ghiselin). The University of Arizona Press: Tucson.

Davis, W.M. 1928. The coral reef problem. New York: American Geographical Society, Special Publ. No. 9.

Faulkner, D. & R. Chesher. 1979. Living Corals. Clarkson N. Potter, Inc.: New York.

Fautin, D.G. & G.R. Allen. 1992. Field guide to anemonefishes and their host sea anemones. Western Australian Museum: Perth.

Fromont, J. 1991. Descriptions of species of the Petrosida (Porifera: Demospongiae) occurring in the tropical waters of the Great Barrier Reef. The Beagle 8(1): 73-96.

George, J.D. & J.J. George. 1979. Marine life, an illustrated encyclopedia of invertebrates in the sea. J. Wiley & Sons: New York.

Gibson, R. 1979. Nemerteans of the Great Barrier Reef. 2. Anopla Heteronemertea (Baseodiscidae). Zoological of the Journal Linnean Society 60: 137-160.

Gosliner, T. 1987. Nudibranchs of Southern Africa. A Guide to the Opisthobranch Molluscs of Southern Africa. Sea Challengers and Jeff Hamann: Monterey.

Gosliner, T.M. 1989. Revision of the Gastropteridae (Opisthobranchia: Cephalaspidea) with descriptions of a new genus and six new species. Veliger 32: 333-381.

Gosliner, T.M. & R. Willan. 1990. Review of the Flabellinidae (Nudibranchia: Aeolidacea) from the tropical Indo-Pacific, with descriptions of five species. Veliger 34: 97-133.

Goy, J.W. & J.E. Randall. 1986. Redescription of *Stenopus devanyi* and *Stenopus earlei* from the Indo-West Pacific Region (Decapoda: Stenopodidae). Bishop Museum Occasional Papers 26: 81-101.

Hoeksema, B.W. 1989. Taxonomy, phylogeny and biogeography of mushroom corals (Scleractinia: Fungiidae). Zoologische Verhandelinger 254: 1-295.

Hooper, J.N.A. & F. Wiedenmayer. 1994. Zoological Catalogue of Australia, Vol. 12, Porifera. Australian Biological Resources Study. CSIRO: Melbourne, Australia.

Houart, R. 1992. The genus *Chicoreus* and related genera (Gastropoda: Muricidae) in the Indo-west Pacific. Memoires du Museum National d'Histoire Naturelle, ser. A. Zoologie 154: 1-188.

Hughes, R.N. 1983. Evolutionary ecology of colonial reef-organisms, with particular reference to corals. Biological Journal of the Linnean Society 20: 39-58.

Jangoux, M. 1973. Le genre *Neoferdina*. Revue Zoologique et Botanique de Afrique. 87: 775-794.

Kay, E.A. 1979. Hawaiian marine shells. Bernice P. Bishop Museum. Special Publication. 64(4): 1-653.

Kott, P. 1980. Algal-bearing didemnid ascidians in the Indo-west Pacific. Memoirs of the Queensland Museum 20(1): 1-48.

Kott, P. 1981. Didemnid-algal symbiosis: algal transfer to a new host generation. Proceedings of the Fourth International Coral Reef Symposium, vol. 2: 721-723.

Kott, P. 1983. Two new genera of didemnid ascidians from tropical Australian waters. The Beagle 1(2): 13-19.

Kott, P. 1985. The Australian Ascidiacea, Part 1. Phlebobranchia and Stolidobranchia. Memoirs of the Queensland Museum 23: 1-438.

Kott, P. 1990. The Australian Ascidiacea Part 2. Aplousobranchia (1). Memoirs of the Queensland Museum 29(1): 1-266.

Kott, P. 1992. The Australian Ascidiacea Part 3. Aplousobranchia (2). Memoirs of the Queensland Museum 32(2): 375-620.

Kuenen, P.H. 1950. Marine Geology. John Wiley & Sons, Inc.: New York.

Kükenthal, W. 1915. Pennatularia. Das Tierreich 43: 1-132.

Kükenthal, W. 1924. Gorgonaria. Das Tierreich 47: 1-478.

Larson, E.E. & P.W. Birkeland. 1982. Putnam's Geology (4th Ed.). Oxford University Press: New York.

Laubenfels, M.W. de. 1954. The sponges of the west-central Pacific. Oregon State Monographic Studies in Zoology. 7: 1-306.

Liltved, W. 1989. Cowries and their relatives of southern Africa. Gordon Verhoef, Seacomber Publications: Cape Town.

Lugo, A.E. 1990. Mangroves of the Pacific Islands: research opportunities. U.S. Dept. of Agriculture, Forest Service. Pacific southwest Research Station, General Technical Report, PSW 118: 1-13.

Manning, R.B. 1994. Stomatopod Crustacea of Vietnam: The legacy of Raoul Serene. Crustacean Research, Special No. 4. Carcinological Society of Japan.

Marshall, B. A. 1983. A revision of the recent Triphoridae of southern Australia. Records of the Australian Museum supplement 2: 1-119.

Mayer, A.G. 1910. Medusae of the world. Vol. III. The Scyphomedusae. Carnegie Institute of Washington.

McLaughlin, P.A. & R. Lemaitre. 1993. A review of the hermit crab genus *Paguritta* (Decapoda: Anomura: Paguridae) with descriptions of three new species. The Raffles Bulletin of Zoology 41(1): 1-29.

McLay, C.L. 1993. Crustacea decapods: the sponge crabs (Dromiidae) of New Caledonia and the Philippines with a review of the genera. In: A. Crosnier (ed), Resultats des Campagnes MUSORSTOM. Vol. 10 Memoirs Museum National d'Histoire Naturelle 156: 111-251.

Meñez, E.G., R.C. Phillips & H.P. Calumpong. 1983. Seagrasses of the Philippines. Smithsonian Contributions to the Marine Sciences 21: 1-40.

Ming, C.L. & P.M. Alino. 1992. An underwater guide to The South China Sea. Times Editions.

Monniot C., F. Monniot & P. Laboute. 1991. Coral reef ascidians of New Caledonia. Éditions de l'ORSTOM: Paris.

Newman, L. J. & L.R.G. Cannon. 1994a. Biodiversity of tropical polyclad flatworms from the Great Barrier Reef, Australia. Memoirs of the Queensland Museum 36(1): 159-163.

Newman, L. J. & L.R.G. Cannon. 1994b. *Pseudoceros* and *Pseudobiceros* (Platyhelminthes, Polycladida, Pseudocerotidae) from eastern Australia and Papua New Guinea. Memoirs of the Queensland Museum 37(1): 205-266.

Newman, L. J., L.R.G. Cannon & D.J. Brunckhorst. 1994. A new flatworm (Platyhelminthes: Polycladida) which mimics a phyllidiid nudibranch (Mollusca, Nudibranchia). Zoological Journal Linnean Society 110: 19-25.

Ofwegen, LP van. 1987. Melithaeidae (Coelenterata: Anthozoa) from the Indian Ocean and the Malay Archipelago. Zoologische Verhandelingen 239: 1-57.

Okuno, J. 1994a. *Rhynchocinetes concolor*, a new species (Caridea: Rhynchocinetidae) from the Indo-West Pacific. Proceedings Japanese Society Systematic Zoology 52: 65-74.

Okuno, J. 1994b. A new species of hinge-beak shrimp from the western Pacific (Crustacea, Decopoda, Rhynchocinetidae). The Beagle, Records of the Museum & Art Gallery of the Northern Territory 11: 29-37.

Okuno, J. & M. Takeda. 1992. Distinction between two hinge-beak shrimp, *Rhynchocinetes durbanensis* Gordon and *R. uritai* (Family Rhynchocinetidae). Revue française d' Aquariologie 19(3): 85-90.

Pearce, F. 1994. A paler shade of coral. New Scientist 142(1929): 19.

Phillips, R.C. & E.G. Meñez. 1988. Seagrasses. Smithsonian Contributions to the Marine Sciences 34: 1-104.

Rankin, J. J. 1955. The structure and biology of *Vallicula multiformis*, gen. et sp. nov., a platyctenid ctenophore. Journal Linnean Society Zoology 43: 55-71.

Roxas, H. A. 1933a. Philippine Alcyonaria. I. The families Cornulariidae and Xeniidae. Philippine Journal of Sciences 50(1): 49-110.

Roxas, H. A. 1933b. Philippine Alcyonaria. II. The families Alcyoniidae and Nephthyidae. Philippine Journal of

Sciences 50(4): 345-470.

Scheltema, A. and M. Jebb. 1994. Natural history of a solenogaster mollusc from Papua New Guinea, *Epimenia australis* (Thiele) (Aplacophora: Neomeniomorpha) Journal of Natural History 28: 1297-1318.

Schlichter, D. 1982. Nutritional strategies of cnidarians: the absorption, translocation and utilization of dissolved nutrients by *Heteroxenia fuscescens*. American Zoologist 22: 659-669.

Schumacher, H. & H. Zibrowius. 1985. What is hermatypic? A redefinition of ecological groups in corals and other organisms. Coral Reefs (1985) 4: 1-9.

Stephen A. & S. Edmonds. 1972. The phyla Sipuncula and Echiura. British Museum: London.

Trench, R.K. 1986. Dinoflagellates in non-parasitic symbioses. In: The biology of dinoflagellates. F.J.R. Taylor, Ed. Blackwell: Oxford. p. 513-570.

Trench, R.K. & H. Winsor. 1987. Symbiosis with dinoflagellates in two pelagic flatworms, *Amphiscolops* sp. and *Haplodiscus* sp. Symbiosis 3: 1-22.

Tudge, C.C. 1995. Hermit crabs of the Great Barrier Reef and coastal Queensland. University of Queensland: Brisbane.

Veron, J.E.N. 1986. Corals of Australia and the Indo-Pacific. Angus & Robertson Publishers.

Verseveldt, J. 1977. Octocorallia from various localities in the Pacific Ocean. Zoologische Verhandelingen 150: 1-42.

Verseveldt, J. 1980. A revision of the genus *Sinularia* May (Octocorallia, Alcyonacea). Zoologische Verhandelingen 179: 1-128.

Verseveldt, J. 1982. A revision of the genus *Sarcophyton* Lesson (Octocorallia, Alcyonacea). Zoologische Verhandelingen 192: 1-91.

Verseveldt, J. 1983. A revision of the genus *Lobophytum* von Marenzeller (Octocorallia, Alcyonacea). Zoologische Verhandelingen 200: 1-103.

Verseveldt, J. & Bayer, F. 1988. Revision of the genera *Bellonella*, *Eleutherobia*, *Nidalia* and *Nidaliopsis* (Octocorallia: Alcyoniidae and Nidalliidae), with descriptions of two new genera. Zoologische Verhandelingen 245: 1-131.

Wells, S.M. (ed.) 1988. Coral reefs of the world. Volume 1, Atlantic & eastern Pacific; Volume 2, Indian Ocean, Red Sea & Gulf; Volume 3, central and western Pacific. United Nations Environment Programme, International Union for Conservation of Nature and Natural Resources.

Wells, F. & C. Bryce. 1993. Sea slugs and their relatives of western Australia. Western Australian Museum: Perth.

Willan, R. C. and N. Coleman. 1984. Nudibranchs of Australasia. Australian Marine Photographic Index: Sydney

Williams, G.C. 1986. What are corals? Sagittarius - Magazine of the South African Museum, Cape Town 1(2): 11-15.

Williams, G.C. 1989. The pennatulacean genus *Cavernularia* Valenciennes (Octocorallia: Veretillidae). Zoological Journal of the Linnean Society 95: 285-310.

Williams, G.C. 1992. Revision of the soft coral genus *Minabea* (Octocorallia: Alcyoniidae) with new taxa from the Indo-West Pacific. Proceedings of the California Academy of Sciences 48(1): 1-26.

Williams, G.C. 1993. Coral reef octocorals - an illustrated guide to the soft corals, sea fans and sea pens inhabiting the coral reefs of northern Natal. Durban Natural Science Museum.

Williams, G.C. 1995. Living genera of sea pens (Coelenterata: Octocorallia: Pennatulacae): illustrated key and synopsis. Zoological Journal of the Linnean Society 113: 93-140.

Winsor, L. 1990. Marine Turbellaria (Acoela) from North Queensland. Memoirs of the Queensland Museum 28(2): 785-800.

Wood, E.M. 1983. Corals of the World. T.F.H. Publications.

Zmarzly, D.L. 1984. Distribution and ecology of shallow-water crinoids at Enewetak Atoll, Marshall Islands, with an annotated checklist of their symbionts. Pacific Science 38(2): 105-122.

Zmarzly, D.L. 1985. The shallow water crinoid fauna of Kwajalein Atoll, Marshall Islands: Ecological observations, interatoll comparisons, and zoogeographical affinities. Pacific Science 39(4): 304-358.

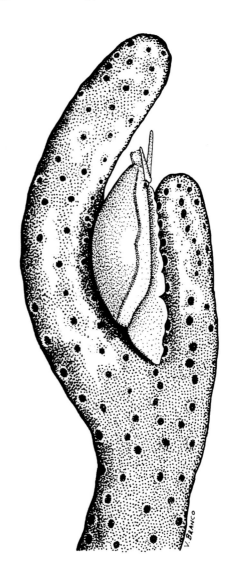

Systematic List of Genera

Porifera

Calcarea
CLATHRINIIDAE
  Clathrina
LEUCETTIDAE
  Leucetta
  Pericharax
LEUCOSOLENIIDAE
  Leuconia
  Leucosolenia
Demospongiae
PLAKINIDAE
  Plakortis
TETILLIDAE
  Cinachyra
ANCORIDAE
  Stelletinopsis
LATRUNCULIIDAE
  Diacarnus
POLYMASTIIDAE
  Atergia
THEONELLIDAE
  Theonella
COELOSPHAERIDAE
  Coelocarteria
CRAMBIDAE
  Monanchora
DESMACELLIDAE
  Desmacella
  Liosina
MICROCIONIDAE
  Clathria
  Echinochalina
AXINELLIDAE
  Auletta
HALOCHONDRIIDAE
  Axinyssa
CALLYSPONGIIDAE
  Callyspongia
CHALINIDAE
  Haliclona
  Kallypilidion
  Nara
NIPHATIDAE
  Amphimedon
  Cribrochalina
  Niphates
PETROSIDAE
  Xestospongia
OCEANAPIIDAE
  Oceanapia
SPONGIIDAE
  Carteriospongia
  Phyllospongia
IRCINIIDAE
  Ircinia
DYSIDEIDAE
  Dysidea
IANTHELLIDAE
  Ianthella

Coelenterata

Hydrozoa
Hydroida
TUBULARIIDAE
  Ralpharia
  Tubularia
HALOCORDYLIDAE
  Pennaria
SOLANDERIIDAE
  Solanderia
ZANCLEIDAE
  Zanclea
EUDENDRIIDAE
  Eudendrium
HYDRACTINIIDAE
  Hydractinia
LOFOEIDAE
  Zygophylax
PLUMULARIIDAE
  Aglaophenia
  Gymnangium
  Macrorhynchia
Milleporina
MILLEPORIDAE
  Millepora
Stylasterina
STYLASTERIDAE
  Distichopora
  Stylaster
Anthozoa
Octocorallia
Helioporacea
HELIOPORIDAE
  Heliopora
Alcyonacea
CLAVULARIIDAE
  Carijoa
  Clavularia
TUBIPORIDAE
  Pachyclavularia
  Tubipora
ALCYONIIDAE
  Cladiella
  Eleutherobia
  Minabea
  Lobophytum
  Sarcophyton
  Sinularia
NEPHTHEIDAE
  Capnella
  Dendronephthya
  Lemnalia
  Litophyton
  Nephthea
  Paralemnalia
  Scleronephthya
  Stereonephthya
NIDALIIDAE
  Chironephthya
  Nidalia
  Siphonogorgia
XENIIDAE
  Anthelia
  Cespitularia
  Heteroxenia
  Sympodium
  Xenia
BRIAREIDAE
  Briareum
ANTHOTHELIDAE

  Semperina
  Subergorgia
MELITHAEIDAE
  Acabaria
  Melithaea
ACANTHOGORGIIDAE
  Acalycigorgia
  Muricella
PLEXAURIDAE
  Echinogorgia
GORGONIIDAE
  Rumphella
ELLISELLIDAE
  Ctenocella
  Junceella
Pennatulacea
VERETILLIDAE
  Cavernularia
  Veretillum
KOPHOBELEMNIDAE
  Sclerobelemnon
VIRGULARIIDAE
  Scytalium
  Virgularia
PTEROEIDIDAE
  Pteroeides
Hexacorallia
Actiniaria
BOLOCEROIDIDAE
  Boloceroides
  Bunodeopsis
ALICIIDAE
  Alicia
  Triactis
  Phyllodiscus
ACTINIIDAE
  Condylactis
  Entacmaea
  Macrodactyla
ACTINODENDRONIDAE
  Actinodendron
  Actinostephanus
THALASSIANTHIDAE
  Actineria
  Cryptodendrum
STICHODACTYLIDAE
  Heteractis
  Stichodactyla
HORMATHIIDAE
  Calliactis
  Nemanthus
  Verrillactis
Zoanthidea
PARAZOANTHIDAE
  Parazoanthus
ZOANTHIDAE
  Palythoa
  Protopalythoa
Corallimorpharia
CORALLIMORPHIDAE
  Pseudocorynactis
ACTINODISCIDAE
  Amplexidiscus
  Discosoma
Scleractinia

POCILLOPORIDAE
  Pocillopora
  Seriatopora
  Stylophora
ACROPORIDAE
  Acropora
  Anacropora
  Astreopora
  Montipora
PORITIDAE
  Alveopora
  Goniopora
  Porites
AGARICIIDAE
  Pachyseris
  Pavona
FUNGIIDAE
  Ctenactis
  Fungia
  Halomitra
  Heliofungia
  Herpolitha
  Polyphyllia
OCULINIDAE
  Galaxea
PECTINIDAE
  Oxypora
  Mycedium
  Pectinia
MUSSIDAE
  Cynarina
  Lobophyllia
  Symphyllia
FAVIIDAE
  Diploastrea
  Echinopora
  Favia
  Favites
  Goniastrea
  Leptoria
  Moseleya
  Platygyra
CARYOPHYLLIIDAE
  Euphyllia
  Cataphyllia
  Physogyra
  Plerogyra
FLABELLIDAE
  Flabellum
DENDROPHYLLIIDAE
  Balanophyllia
  Dendrophyllia
  Tubastraea
  Turbinaria
Ceriantipatharia
Antipatharia
ANTIPATHIDAE
  Cirripathes
  Stichopathes
  Antipathes
Ceriantharia
CERIANTHIDAE &
ARACHNANTHIDAE
  Cerianthids
Scyphozoa

Rhizostomeae
CASSIOPEIDAE
  Cassiopeia
MASTIGIIDAE
  Mastigias
Coronatae
EPHYROPSIDAE
  Nausithoe
Stauromedusae
ELEUTHEROCARPIDAE
  Lipkea

Ctenophora

Platyctenea
COELOPLANIDAE
  Coeloplana
  Lyrocteis

Platyhelminthes

Acoela
CONVOLUTIDAE
  Amphiscolops
  Convoluta
  Waminoa
Polycladida
CALLIOPLANIDAE
  Acanthozoon
  Bulaceros
  Callioplana
EURYLEPTIDAE
  Eurylepta
PSEUDOCEROTIDAE
  Maiazoon
  Paraplanaria
  Phrikoceros
  Prosthiosomum
  Pseudobiceros
  Pseudoceros
  Thysanozoon

Nemertea

BASEODISCIDAE
  Baseodiscus
LINEIDAE
  Lineus
  Notospermus
EMPLECTONEMATIDAE
  Emplectonema

Annelida

Polychaeta
APHRODITIDAE
  Laetmonice
POLYNOIDAE
  Asterophila
  Gastrolepidia
  Iphione
  Lepidonotus
  Paralepidonotus
TEREBELLIDAE
  Amphitrite
  Loimia
CHAETOPTERIDAE
  Chaetopterus
AMPHINOMIDAE
  Chloeia
  Pherecardia
EUNICIDAE

Eunice
Palola
HESIONIDAE
Hesione
SYLLIDAE
Odontosyllis
SPIONIDAE
Polydorella
SERPULIDAE
Filograna
Filogranella
Pomatostegus
Protula
Spirobranchus
SABELLIDAE
Bispira
Myxicola
Sabellastarte

Sipuncula

PHASCALOSOMATIDAE
Phascolosoma
THEMISTIDAE
Themiste

Echiura

BONELLIIDAE
Bonellia
Archibonellia

Mollusca

Aplacophora
EPIMENIIDAE
Epimenia
Polyplacophora
CHITONIDAE
Acanthopleura
Chiton
Tonicia
CRYPTOPLACIDAE
Cryptoplax
Gastropoda
Patellogastropoda
PATELLIDAE
Cellana
Vetigastropoda
FISSURELLIDAE
Emarginula
Scutus
HALIOTIDAE
Haliotis
TROCHIDAE
Tectus
TURBINIDAE
Ethalia
Phasianella
Turbo
Neritacea
NERITIDAE
Nerita
Titiscania
Caenogastropoda
VERMITIDAE
Dendropoma
Serpulorbis
HIPPONICIDAE
Sabia
CASSIDAE
Cassis

Semicassis
NATICIDAE
Naticarius
Neverita
Polinices
Tanea
STROMBIDAE
Lambis
Strombus
Terebellum
CYPRAEIDAE
Cypraea
OVULIDAE
Aclyvolva
Calpurnus
Crenavolva
Cymbovula
Delonovolva
Dentiovula
Diminovula
Hiatovula
Ovula
Phenacovolva
Primovula
Prosimnia
Volva
TRIVIIDAE
Trivia
LAMELLARIIDAE
Coriocella
Lamellaria
EULIMIDAE
Luetzenia
Thyca
TRIPHORIDAE
Euthymella
EPITONIIDAE
Epitonium
CYMATIIDAE
Charonia
Cymatium
VOLUTIDAE
Cymbiola
HARPIDAE
Harpa
NASSARIIDAE
Nassarius
TONNIDAE
Malea
Tonna
MURICIDAE
Chicoreus
Coralliophila
Murex
OLIVIDAE
Oliva
MARGINELLIDAE
Granulina
Marginella
MITRIDAE
Mitra
TURRIDAE
Lienardia
Turris
CONIDAE
Conus
Heterobranchia

ARCHITECTONICIDAE
Architectonica
Heliacus
Opisthobranchia
HYDATINIDAE
Hydatina
Micromelo
Cephalaspidea
BULLIDAE
Bulla
HAMINOEIDAE
Haminoea
AGLAJIDAE
Chelidonura
Philinopsis
GASTROPTERIDAE
Sagaminopteron
Siphopteron
Anaspidea
APLYSIIDAE
Aplysia
Dolabella
Paraplysia
Stylocheilus
Sacoglossa
CYLINDROBULLIDAE
Volvatella
JULIIDAE
Berthelinia
OXYNOIDAE
Lobiger
Oxynoe
CALIPHYLLIDAE
Cyerce
ELYSIIDAE
Elysia
Thuridilla
Notaspidea
UMBRACULIDAE
Umbraculum
PLEUROBRANCHIDAE
Berthella
Berthellina
Euselenops
Pleurobranchus
Nudibranchia
DISCODORIDIDAE
Discodoris
ASTERONOTIDAE
Asteronotus
Halgerda
KENTRODORIDIDAE
Jorunna
PLATYDORIDIDAE
Platydoris
HEXABRANCHIDAE
Hexabranchus
CHROMODORIDIDAE
Ardeadoris
Ceratosoma
Chromodoris
Glossodoris
Hypselodoris
Miamira
Risbecia
Thorunna
PHYLLIDIIDAE

Ceratophyllidia
Reticulidia
Phyllidia
Phyllidiella
Phyllidiopsis
DENDRODORIDIIDAE
Dendrodoris
POLYCERATIDAE
Nembrotha
Roboastra
GONIODORIDIDAE
Okenia
GYMNODORIDIDAE
Gymnodoris
NOTODORIDIDAE
Notodoris
BORNELLIDAE
Bornella
TETHYIDAE
Melibe
DOTOIDAE
Doto
TRITONIIDAE
Marionia
Tritonia
ARMINIDAE
Dermatobranchus
ZEPHYRINIDAE
Janolus
FLABELLINIDAE
Flabellina
TERGIPEDIDAE
Cuthona
Phestilla
FACELINIDAE
Caloria
Phyllodesmium
PTERAEOLIDIIDAE
Pteraeolidia
AEOLIDIIDAE
Berghia
Cerberilla
Scaphopoda
DENTALIIDAE
Fustiaria
Bivalvia
MYTILIDAE
Modiolius
PTERIIDAE
Pteria
Pinctada
PINNIDAE
Atrina
Pedum
PECTINIDAE
Decatopecten
Mirapecten
SPONDYLIDAE
Spondylus
LIMIDAE
Limaria
CARDIIDAE
Corculum
Ostreacea
OSTREIDAE
Alectryonella
Hyotissa

Lopha
GALEOMATIDAE
Amphilepida
Scintilla
TRIDACNIDAE
Hippopus
Tridacna
Cephalopoda
NAUTILIDAE
Nautilus
SEPIOLIDAE
Eupyrmna
LOLIGINIDAE
Sepioteuthis
SEPIIDAE
Metasepia
Sepia
OCTOPODIDAE
Hapalochlaena
Octopus

Arthropoda

Pycnogonida
PHOXICHILIDIIDAE
Anoplodactylus
ENDEIDAE
Endeis
NYMPHONIDAE
Nymphon
AMMOTHEIDAE
Nymphopsis
Crustacea
Cirripedia
BALANIDAE
Balanus
ARCHAEOBALANIDAE
Conopea
LEPADIDAE
Lepas
TETRACLITIDAE
Tetraclita
Stomatopoda
GONODACTYLIDAE
Gonodactylaceus
Gonodactylellus
Gonodactylus
LYSIOSQUILLIDAE
Lysiosquilla
ODONTODACTYLIDAE
Odontodactylus
PROTOSQUILLIDAE
Echinosquilla
PSEUDOSQUILLIDAE
Pseudosquilla
SQUILLIDAE
Oratosquilla
TAKUIDAE
Mesacturoides
Mysidacea
Isopoda
IDOTEIDAE
Idotea
Santia
EPICARIDAE
CYMATHOIDAE
Amphipoda
Maera
GAMMARIDAE

CAPRELLIDAE
Decapoda
PENAEIDAE
  Metapenaeopsis
  Heteropenaeus
  Penaeus
PALAEMONIDAE
  Allopontonia
  Dasycaris
  Hamodactylus
  Leander
  Periclimenes
  Pliopontonia
  Pontonides
  Stegopontonia
  Urocaridella
  Vir
ALPHEIDAE
  Alpheopsis
  Alpheus
  Synalpheus
STENOPODIDAE
  Microprosthema
  Stenopus
HIPPOLYTIDAE
  Hippolyte
  Koror
  Lysmata
  Lysmatella
  Parhippolyte
  Saron
  Thor
  Tozeuma
GNATHOPHYLLIDAE
  Gnathophylloides
  Gnathophyllum
  Hymenocera
RHYNCHOCINETIDAE
  Rhynchocinetes
PROCESSIDAE
  Processa
ENOPLOMETOPIDAE
  Enoplometopus
  Hoplometopus
PALINURIDAE
  Justitia
  Palinurella
  Panulirus
SCYLLARIDAE
  Arctides
  Ibacus
  Parribacus
  Scyllarides
  Scyllarus
Anomura
COENOBITIDAE
  Coenobita
DIOGENIDAE
  Aniculus
  Calcinus
  Dardanus
  Trizopagurus
PAGURIDAE
  Paguritta
GALATHEIDAE
  Allogalathea
  Galathea

Lauriea
PORCELLANIDAE
  Neopetrolisthes
  Petrolisthes
  Porcellanella
HIPPIIDAE
  Emerita
Brachyura
CALAPPIDAE
  Calappa
RANINIDAE
  Ranina
DORIPPIDAE
  Dorippe
  Ethusa
DROMIIDAE
  Cryptodromia
  Dromia
  Lauridromia
HYMANOSOMATIDAE
  Trigonoplax
MAJIDAE
  Achaeus
  Camposcia
  Cyclocoeloma
  Hoplophrys
  Huenia
  Hyastenus
  Micippa
  Naxioides
  Schizophrys
  Xenocarcinus
LATRELLIDAE
  Latreilla
  Eplumra
PORTUNIDAE
  Charybdis
  Lissocarcinus
  Podophthalmus
  Portunus
  Thalamita
XANTHIDAE
  Atergatis
  Carpilius
  Etisus
  Lybia
  Lophozozymus
  Neoliomera
  Polydectus
  Trapezia
  Xanthias
  Zosymus
EUMEDONIDAE
  Zebrida
PARTHENOPIDAE
  Harrovia
  Lambrus
GRAPSIDAE
  Plagusia
  Percnon

Phoronida
PHORONIDAE
  Phoronis

Brachiopoda
DALLINIDAE
  Frenulina

Bryozoa
Ctenostomata
ALCYONIDIIDAE
  Alcyonidium
MIMOSELLIDAE
  Bantariella
Cyclostomata
LICHENOPORIDAE
  Lichenopora
TUBULIPORIDAE
  Idmidronea
Cheilostomata
PHIDOLIPORIDAE
  Iodictyum
  Reteporella
  Reteporellina
  Rhynchozoon
  Triphyllozoon
BUGULIDAE
  Bugula
SCRUPOCELLARIIDAE
  Canda
  Scrupocellaria
  Tricellaria
  Tropidozoum
CELLOPORARIIDAE
  Celloporaria
SCHIZOPORELLIDAE
  Calyptotheca

Echinodermata
Crinoidea
COMASTERIDAE
  Comantheria
  Comanthina
  Comaster
  Comissa
MARIAMETRIDAE
  Lamprometra
  Liparometra
  Oxycomanthus
  Oxymetra
  Stephanometra
COLOBOMETRIDAE
  Petasometra
Asteroidea
LUIDIIDAE
  Luidia
ASTROPECTINIDAE
  Astropecten
GONIASTERIDAE
  Bothriaster
  Iconaster
  Pentagonaster
  Stellaster
  Tosia
OREASTERIDAE
  Choriaster
  Culcita
  Halityle
  Monachaster
  Pentaceraster
  Pentaster
  Protoreaster
OPHIDIASTERIDAE
  Ferdina
  Fromia
  Gomophia

Leiaster
Linckia
Nardoa
Neoferdina
Ophidiaster
ASTERINIDAE
  Asterina
ACANTHASTERIDAE
  Acanthaster
MITHRODIIDAE
  Mithrodia
  Thromidia
ECHINASTERIDAE
  Echinaster
Ophiuroidea
GORGONOCEPHALIDAE
  Astroboa
  Conocladus
OPHIOMYXIDAE
  Ophiomyxa
OPHIOTRICHIDAE
  Ophiomaza
  Ophiopteron
  Ophiothrix
OPHIOCOMIDAE
  Ophiomastix
OPHIODERMATIDAE
  Ophiarachna
  Ophiarachnella
OPHIURIDAE
  Ophiolepis
  Ophioplocus
AMPHIURIDAE
  Amphiura
OPHIACTIDAE
  Ophiactis
Echinoidea
CIDARIDAE
  Eucidaris
  Phyllacanthus
  Prionocidaris
ECHINOTHURIIDAE
  Asthenosoma
DIADEMATIDAE
  Astropyga
  Diadema
  Echinothrix
TEMNOPLEURIDAE
  Mespilia
  Microcyphus
  Salmacis
TOXOPNEUSTIDAE
  Pseudoboletia
  Toxopneustes
  Tripneustes
PARASALENIIDAE
  Paraselinia
ECHINOMETRIDAE
  Colobocentrotus
  Echinometra
  Heterocentrotus
ECHINONEIDAE
  Echinoneus
CLYPEASTERIDAE
  Clypeaster
SPATANGIDAE
  Maretia

BRISSIDAE
  Brissus
  Eurypatagus
SCUTELLIDAE
  Echinodiscus
Holothuroidea
HOLOTHURIIDAE
  Actinopyga
  Bohadschia
  Holothuria
STICHOPODIDAE
  Stichopus
  Thelenota
CUCUMARIIDAE
  Colochirus
  Psuedocolochirus
PHYLLOPHURIIDAE
  Neothyonidium
SYNAPTIDAE
  Euapta
  Synapta
  Synaptula

Hemichordata
PTYCHODERIDAE
  Ptychodera

Urochordata
Ascidiacea
Aplousobranchia
POLYCLINIDAE
  Aplidium
  Pseudodistoma
POLYCITORIDAE
  Clavelina
  Eudistoma
  Oxycorynia
  Sigillina
  Sycozoa
DIDEMNIDAE
  Atriolum
  Didemnum
  Leptoclinides
  Lissoclinum
Phlebobranchia
CIONIDAE
  Diazona
  Rhopalaea
PEROPHORIDAE
  Ecteinascidia
  Perophora
ASCIDIIDAE
  Phallusia
Stolidobranchia
STYELIDAE
  Botryllus
  Botrylloides
  Eusynstyela
  Polycarpa
  Styela
PYURIDAE
  Herdmania
  Pyura

310

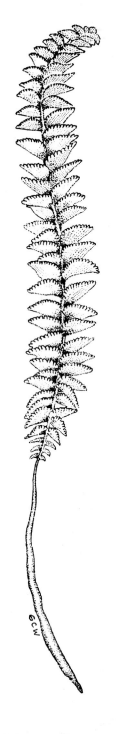